T0225275

Urban Remote Sensing

Second Edition

Remote Sensing Applications

Series Editor

Qihao Weng

Indiana State University
Terre Haute, Indiana, U.S.A.

Urban Remote Sensing

Second Edition

Edited by
Qihao Weng, PhD
Dale Quattrochi, PhD
Paolo Gamba, PhD

CRC Press
Taylor & Francis Group
Boca Raton London New York

CRC Press is an imprint of the
Taylor & Francis Group, an **informa** business

CRC Press
Taylor & Francis Group
6000 Broken Sound Parkway NW, Suite 300
Boca Raton, FL 33487-2742

First issued in paperback 2020

© 2018 by Taylor & Francis Group, LLC
CRC Press is an imprint of Taylor & Francis Group, an Informa business

No claim to original U.S. Government works

ISBN-13: 978-0-367-57204-4 (pbk)
ISBN-13: 978-1-138-05460-8 (hbk)

This book contains information obtained from authentic and highly regarded sources. Reasonable efforts have been made to publish reliable data and information, but the author and publisher cannot assume responsibility for the validity of all materials or the consequences of their use. The authors and publishers have attempted to trace the copyright holders of all material reproduced in this publication and apologize to copyright holders if permission to publish in this form has not been obtained. If any copyright material has not been acknowledged, please write and let us know so we may rectify in any future reprint.

Except as permitted under U.S. Copyright Law, no part of this book may be reprinted, reproduced, transmitted, or utilized in any form by any electronic, mechanical, or other means, now known or hereafter invented, including photocopying, microfilming, and recording, or in any information storage or retrieval system, without written permission from the publishers.

For permission to photocopy or use material electronically from this work, please access www.copyright .com (http://www.copyright.com/) or contact the Copyright Clearance Center, Inc. (CCC), 222 Rosewood Drive, Danvers, MA 01923, 978-750-8400. CCC is a not-for-profit organization that provides licenses and registration for a variety of users. For organizations that have been granted a photocopy license by the CCC, a separate system of payment has been arranged.

Trademark Notice: Product or corporate names may be trademarks or registered trademarks, and are used only for identification and explanation without intent to infringe.

Library of Congress Cataloging-in-Publication Data

Names: Weng, Qihao, editor. | Quattrochi, Dale A., editor. | Gamba, Paolo, editor.
Title: Urban remote sensing / [edited by] Qihao Weng, Dale Quattrochi, and Paolo Gamba.
Description: Second edition. | Boca Raton, FL : CRC Press, 2018. | Series: Remote sensing applications series | Includes index.
Identifiers: LCCN 2017040164 | ISBN 9781138054608 (hardback : alk. paper)
Subjects: LCSH: City planning--Remote sensing. | Land use, Urban--Remote sensing. | Urban geography--Remote sensing.
Classification: LCC HT166 .U74523 2018 | DDC 307.1/216--dc23
LC record available at https://lccn.loc.gov/2017040164

Visit the Taylor & Francis Web site at
http://www.taylorandfrancis.com

and the CRC Press Web site at
http://www.crcpress.com

Contents

SECTION III   Monitoring, Analyzing, and Modeling Urban Growth

SECTION IV   Urban Planning and Socioeconomic Applications

Preface

EARTH OBSERVATION FOR A SUSTAINABLE EARTH

Earth observation (EO) technology, in conjunction with in situ data collection, has been used to observe, monitor, measure, and model many of the components that comprise natural and human ecosystem cycles (Weng 2012a). Driven by societal needs and improvement in sensor technology and image processing techniques, we have witnessed a great increase in research and development, technology transfer, and engineering activities worldwide since the turn into the twenty-first century. Commercial satellites acquire imagery at spatial resolutions previously only possible to aerial platforms, but these satellites have advantages over aerial imageries including their capacity for synoptic coverage, shorter revisit time, and capability to produce stereo image pairs conveniently for high-accuracy 3D mapping thanks to their flexible pointing mechanism (Weng 2012b). Hyperspectral imaging affords the potential for detailed identification of materials and better estimates of their abundance in the Earth's surface, enabling the use of remote sensing data collection to replace data collection that was formerly limited to laboratory testing or expensive field surveys (Weng 2012b). While LiDAR technology provides high-accuracy height and other geometric information for urban structures and vegetation, radar technology has been re-invented since the 1990s due greatly to the increase of space-borne radar programs (Weng 2012b). These technologies are not isolated at all. In fact, their integrated uses with more established aerial photography and multispectral remote sensing techniques have been the main stream of current remote sensing research and applications (Weng 2012b). With these recent advances, techniques of and data sets from remote sensing and EO have become an essential tool for understanding the Earth, monitoring of the world's natural resources and environments, managing exposures to natural and man-made risks and disasters, and helping the sustainability and productivity of natural and human ecosystems (Weng 2012b).

The 2002 World Summit on Sustainable Development in Johannesburg highlighted the urgent need for coordinated observation relating to the state of the Earth. The First Earth Observation Summit in Washington, D.C., in 2003 adopted a declaration to establish the ad hoc intergovernmental Group on Earth Observations (ad hoc GEO) to draft a 10-Year Implementation Plan. Since 2003, GEO has been working to strengthen the cooperation and coordination among global observing systems and research programs for integrated global observations. The GEO process has outlined a framework document calling for Global Earth Observation System of Systems (GEOSS) and defined nine areas of societal benefits (http://www.earthobservations.org/about_geo.shtml). On September 25, 2015, the United Nations adopted a set of sustainable development goals (SDGs), each of which has specific targets to be achieved over the next 15 years (United Nations Development Programme 2015). These goals represent the UN's response to numerous societal challenges and efforts to build a sustainable Earth. Through large-scale, repetitive acquisition of the Earth surface image data, remote sensing can provide essential information and knowledge to supplement statistical analyses in the assessment of indicators toward the attainment of the SDGs. Because EO offers an indispensable tool to measure and monitor progresses toward SDGs, in the recently developed "GEO Strategic Plan 2016–2025: Implementing GEOSS," GEO has determined to develop a concerted direction with the SDGs (Group on Earth Observations 2015). To address the current state of remote sensing knowledge for sustainable development and management, Weng (2016) published an edited volume in this same book series entitled "Remote Sensing for Sustainability," which intended to contribute to the GEO's Strategic Plan by addressing and exemplifying a number of societal benefit areas using remote sensing data sets, methods, and techniques for sustainable development.

ADDRESSING THE NEEDS IN DEVELOPING COUNTRIES

GEO includes in its work plan a "Global Urban Observation and Information" (GUOI) Initiative since 2012. The leaders of this initiative set the following goals for the 2012–2015 period: (1) improving the coordination of urban observations, monitoring, forecasting, and assessment initiatives worldwide; (2) supporting the development of a global urban observation and analysis system; (3) producing up-to-date information on the status and development of the urban system—from a local to a global scale; (4) filling existing gaps in the integration of global urban observation with data that characterize urban ecosystems, environment, air quality and carbon emission, indicators of population density, environmental quality, quality of life, and the patterns of human environmental and infectious diseases; and (5) developing innovative techniques in support of effective and sustainable urban development. These goals have been extended and expanded for the GEO Work Programme of 2017–2019 (Weng et al. 2014). The goals support GEO's objectives on Sustainable Urban Development well, which advocate the value of EOs, engage communities, and deliver data and information by assisting in the development of resilient cities and assessment of urban footprints. By accomplishing these objectives, GEO hopes to make cities and human settlements inclusive, safe, resilient, and sustainable through identifying economic externalities; managing environmental, climate, and disaster risks; and building capacity to participate, plan, and manage based on objective information regarding urban development. The GUOI Initiative, in particular, supports the development of urban resilience (including coastal resilience) by supplying objective data and information on the footprints of global urbanization and cities, developing indicators for sustainable cities (supporting UN's SDGs), and developing innovative methods and techniques in support of effective management of urban environment, ecosystems, natural resources, and other assets, and the mitigation of adverse impacts caused by urbanization.

A major strategic shift of the GUOI Initiative for the 2017–2019 period is to extend urban mapping methods and EO data sets and technologies to developing countries. There are several activities in connection with this strategic focus. First, the GUOI team initiated a joint project of "Impervious Surface Mapping in Tropical and Subtropical Cities (ISMiTSC)," aiming at providing EO technologies and data sets to Asia, Africa, and South America. A preliminary research has been conducted in selected cities in the three continents with data support from German Aerospace Center (DLR) and research collaboration between Chinese University of Hong Kong, Indiana State University, and DLR, with a grant support from Hong Kong Research Grants Council (2016–2017). Preliminary result was published via a book entitled *Remote Sensing of Impervious Surfaces in Tropical and Subtropical Areas* (Zhang et al. 2015). Since most developing countries are located in tropical and subtropical regions, continuing urbanization in these regions has important implications in biodiversity, rainforest ecosystem, and global climate change (Weng 2015). Optical remote sensing faces more environmental challenges than it does in a temperate zone, due to frequent cloudy and rainy days and complex hydrological systems in association with strong seasonal change in water surface area, vegetation phenology, and morphological and species complexity (Weng 2015). To fully utilize long-term archives of medium-resolution satellite imagery, researchers have developed new algorithms and methods to overcome these limitations (Fu and Weng 2016; Zhang and Weng 2016).

Second, to facilitate online and off-line learning and knowledge sharing, a website has been created for sharing computer codes, algorithms, systems, products, and publications that support remote sensing observations and applications, digital image processing, and the extraction of geophysical and biophysical information. It is our hope that through co-learning, sharing, and collaborating, an e-community can be built among researchers, practitioners, teachers, and students worldwide, which is named Remote Sensing E-community for Digital imaGe procESsing (RS-EDGES, http://rs-edges.net/).

Finally, we strive to train and to educate students and young researchers worldwide to become tomorrow's leaders in EO technologies by disseminating GUOI ideas and goals and sharing outcomes of various activities through annual symposium, summer school, joint field works, and publications. Since 2012, the GUOI team has held annual workshops/symposia in conjunction with various international conferences. The GUOI symposia were held in conjunction with the conference series IEEE-sponsored EORSA in Shanghai, Changsha, and Guangzhou, China, respectively, and with IEEE/ISPRS jointly sponsored JURSE conference series in Sao Paulo, Brazil, and Lausanne, Switzerland. In 2014, the International Workshop on Global Urban Observation and Monitoring from Space was held in Athens, Greece, sponsored by the European Space Agency. Furthermore, during annual conferences of the American Association of Geographers, there was an annual Global Urban Observation Symposium with multiple sessions (Chicago, Illinois, 2015; San Francisco, California, 2016; and Boston, Massachusetts, 2017). Additional GUOI sessions were also organized in 2016 both in the IEEE IGARSS conference in Beijing, China, and in the ISPRS Congress in Prague, Czech Republic. A joint field work on current land use and land cover (LULC) and urban morphology was conducted in the Pearl River Delta (PRD), China, from January 4 to 8, 2016. The 5-day field campaign was made possible by grants from the Research Grants Council of Hong Kong and Natural Science Foundation of China. The field work was directed by Prof. Qihao Weng, Indiana State University, joined by researchers and graduate students from Chinese University of Hong Kong, Wuhan University, and South China Normal University. The joint field campaign aimed to obtain up-to-date "ground truth" data on all LULC types in PRD and to verify the accuracy of LULC maps that were derived from satellite imagery, especially on urban impervious surfaces in the delta region (see Zhang et al. 2017 for details).

SYNOPSIS OF THE BOOK

To meet the growing interests in applications of remote sensing technology to urban and suburban areas, Drs. Weng and Quattrochi assembled a team of experts to edit a book on Urban Remote Sensing in 2006. That book, for the first time, systematically examined various aspects of the field. Since its inception, the book had been used as a textbook in many universities and also served as a reference book for researchers in academia, governmental, and commercial sectors. When the acquisition editor at CRC Press expressed a strong interest for us to publish a second edition, we thought it would almost be impossible to update any chapter. Remote sensing technology in general has changed significantly since then, so has urban remote sensing. Thus, we decide to edit a new volume, instead of updating the 2006 volume. In addition to Drs. Weng and Quattrochi, Dr. Paolo Gamba (University of Pavia, Italy) was invited to be a co-editor.

The second edition reflects new developments in satellite sensors, image processing methods and techniques, and wider applications of urban remote sensing in order to meet societal and economic challenges. This book is divided into four sections. Section I focuses on data, sensors, and systems considerations and algorithms for urban feature extraction; Section II illustrates applications in assessing and modeling urban landscape compositions, patterns, and structures; Section III presents methods for monitoring, analyzing, and modeling urban growth; and Section IV demonstrates urban planning and socioeconomic applications. For each section, we are particularly interested in addressing the following issues:

- Methods for upscaling urban feature extraction to the global scale (Section I)
- New methods in mapping and detecting urban landscape features and structures (Section II)
- Mapping and monitoring urbanization in developing countries (Section III)
- Urban sustainability and environmental issues (Section IV)

Section I includes three chapters concerning methods and algorithms for extracting urban extents that may be applied globally. Chapter 1 introduces the generation of global urban footprint (GUF) based on data from TerraSAR-X and TanDEM-X at a spatial resolution of 12 m. GUF aims at deriving a GUF map with automation. Chapter 2 presents an experimental development of an on-demand system for human settlement mapping using the Landsat archive. The system was implemented with automated algorithms of satellite data selection for user's preference and human settlement mapping using a machine learning–based method called Learning with Local and Global Consistency. In Chapter 3, a novel morphological building index (MBI) and its improved versions are introduced. MBI utilized the spectral–spatial properties of buildings (e.g., contrast, size, and directionality) by a set of morphological operators (e.g., top-hat by reconstruction, granulometry, and directionality) and can be applied in a large area of complex building patterns.

Section II introduces novel methods for mapping and detecting urban features and structures. Very high resolution (VHR) satellite images provide an ideal data for building mapping. However, these images lack the height information and are usually acquired off-nadir. These limitations pose challenges for mapping buildings in off-nadir VHR satellite images. Chapter 4 identifies the challenges associated with building detection from off-nadir VHR images based on stereo 3D information and presents a few solutions through a case study. In Chapter 5, two recent projects—the World Urban Database and Portal Tools (WUDAPT) and the Global Human Settlement Layer (GHSL) project—are compared to find their agreement on mapping built-up and built density. WUDAPT uses the Local Climate Zone (LCZ) scheme, a generic typology of urban structures, and supervised classification, while GHSL-LABEL is derived from physical characteristics of settlements such as vegetation cover and building height. The result of cross-comparison proved useful to identify both doubtful LCZ maps and areas of low confidence within the maps. Chapter 6 explores the use of off-nadir satellite images for urban change detection. Close-to-nadir satellite images are commonly used in previous studies to avoid the mis-registration caused by image relief displacements. To use off-nadir images, a change detection procedure is presented in this chapter that uses the Patch-Wise Co-Registration method to overcome the mis-registration problem and integrates other methods to enable accurate change detection using images taken from different sensors and platforms.

Over the past decade, remote sensing technology has been increasingly employed for developing countries for mapping and monitoring urbanization and associated environmental changes. Section III illustrates a few cases of this direction of application. Chapter 7 analyzes urban growth in four megacities in India to understand land use changes over the past three decades and associated environmental deterioration. Further, this chapter models future land uses to examine different scenarios of urban growth and their implications for sustainable development. Chapter 8 continues to examine urbanization in Asia, but focuses on Vietnam. Urbanization is a major trend in the Asia-Pacific region where many cities are threatened by natural hazards such as urban flooding, typhoon, tsunami, and sea-level rise. In Chapter 8, an annual impervious surface map was generated for the greater Hanoi area by using time series Landsat imagery from 1988 to 2015. The rapid but uneven increase in impervious surfaces over time shows agreement with major events of economic transitions in Vietnam. In Chapter 9, the derivation of impervious surfaces in a desert environment is explored. The authors applied spectral mixture analysis and a machine learning method to map the subpixel distribution of urban impervious surfaces in Dubai, United Arab Emirates. The main sources of errors in the estimations were found to relate to the spectral confusion of impervious surfaces with dark sand and certain types of desert plants. Chapter 10 intends to extract urban densities and to model urban sprawl in several cities in Argentina, South America. The methodology consisted of extracting urban density classes using spectral mixture signatures and applying a predictive simulation (LanduseSIM) for a 30-year period to reveal the process of sprawl in the nation.

The last part of the book, Section IV, examines case studies of urban planning and socioeconomic applications of remote sensing technology. Chapter 11 reports on the evolution of different methodologies used to develop HEAT Scores as urban energy consumption metrics. HEAT Scores are defined for each house using high-resolution thermal infrared imagery obtained from the Thermal Airborne Broadband Imager (TABI-1800) and are developed as a part of the HEAT (Heat Energy Assessment Technologies) research project, initially developed as a public GeoWeb service designed to help residents improve their home energy efficiency and reduce greenhouse gas emissions. This project was conducted on 9000+ houses in 12 communities in the southwest region of Calgary, Alberta, Canada. In Chapter 12, a study was conducted to study the interplay between air pollution (as estimated from remotely sensed data) and clinical records, and to find out the relationship among black particulate concentration, micro- and macrovascular disease onsets, and hospitalization tracks. Experimental results show that effective connections between the estimated air quality and the clinical data behavior can be accurately derived by the methods for data mining over large-scale heterogeneous records. In the last chapter of this book, Chapter 13, EO satellite data and spatial analysis coupled with qualitative survey-based assessments are used to tackle the challenges of urban green planning and monitoring in the city of Salzburg, Austria, in the frame of a longer-term monitoring endeavor using VHR satellite data in 5-year intervals.

ACKNOWLEDGMENTS

We thank all the contributors for making this endeavor possible. Furthermore, we offer our deepest appreciation to all the reviewers who have taken precious time from their busy schedules to review the chapters submitted to this book. Finally, we are indebted to our families for their love and support. It is our hope that the publication of this book will provide fresh stimulation to students, researchers, and practitioners to conduct more in-depth studies on urban remote sensing, and will open up new opportunities for EO technology transfer and data services to developing countries. The realization of the societal and economic benefits of EO technology and sustaining of the Earth requires cooperation between developed and developing countries through such a framework as GEO.

The reviewers of the chapters for this book are as follows (in alphabetical order): Raid Al-Tahir, Christoph Aubrecht, Gang Chen, Xuefei Hu, Xiuping Jia, Alexander Keul, Kourosh Khoshelham, Wenzhi Liao, Linlin Lu, Hiroyuki Miyazaki, María Teresa Camacho Olmedo, Tonny J. Oyana, Dale A. Quattrochi, Stevan Savic, Yang Shao, Limin Yang, Nithiyanandam Yogeswaran, and Hongsheng Zhang.

Qihao Weng
Dale A. Quattrochi
Paolo Gamba

REFERENCES

Fu, P. and Q. Weng. 2016. A time series analysis of urbanization induced land use and land cover change and its impact on land surface temperature with Landsat imagery. *Remote Sensing of Environment*, 175(4), 205–214.

Group on Earth Observations. 2015. The GEO 2016–2025 Strategic Plan: Implementing GEOSS, https://www.earthobservations.org/geoss_wp.php, last accessed March 5, 2016.

United Nations Development Programme. 2015. Sustainable Development Goals (SDGs), http://www.undp.org/content/undp/en/home/sdgoverview/post-2015-development-agenda.html, last accessed March 5, 2016.

Weng, Q. 2012a. Remote sensing of impervious surfaces in the urban areas: Requirements, methods, and trends. *Remote Sensing of Environment*, 117(2), 34–49.

Weng, Q. 2012b. *An Introduction to Contemporary Remote Sensing.* New York: McGraw-Hill Professional, 320 pp.

Weng, Q. 2015. Remote sensing for urbanization in tropical and subtropical regions—Why and what matters? In Zhang, H., Lin, H. Zhang, Y., and Q. Weng. *Remote Sensing of Impervious Surfaces in Tropical and Subtropical Areas.* Boca Raton, FL: CRC Press/Taylor & Francis, pp. xvii–xxi.

Weng, Q. 2016. *Remote Sensing for Sustainability.* Boca Raton, FL: CRC Press/Taylor & Francis, 366 pp.

Weng, Q., Esch, T., Gamba, P., Quattrochi, D. A. and G. Xian. 2014. Global urban observation and information: GEO's effort to address the impacts of human settlements. In Weng, Q. editor. *Global Urban Monitoring and Assessment through Earth Observation.* Boca Raton, FL: CRC Press/Taylor & Francis, Chapter 2, pp. 15–34.

Zhang, H. Lin, H. Zhang, Y. and Q. Weng. 2015. *Remote Sensing of Impervious Surfaces in Tropical and Subtropical Areas.* Boca Raton, FL: CRC Press/Taylor & Francis, 174 pp.

Zhang, L. and Q. Weng. 2016. Annual dynamics of impervious surface in the Pearl River Delta, China, from 1988 to 2013, using time series Landsat data. *ISPRS Journal of Photogrammetry and Remote Sensing*, 113(3), 86–96.

Zhang, L., Weng, Q. and Z. F. Shao. 2017. An evaluation of monthly impervious surface dynamics by fusing Landsat and MODIS time series in the Pearl River Delta, China, from 2000 to 2015. *Remote Sensing of Environment*, 201(11), 99–114.

Editors

 **Qihao Weng, PhD,** is the director of the Center for Urban and Environmental Change and a professor of Remote Sensing and GIS at Indiana State University, and worked as a Senior Fellow at the National Aeronautics and Space Administration from December 2008 to December 2009. He earned his PhD in geography from the University of Georgia in 1999. Dr. Weng is currently the Lead of Group on Earth Observation (GEO) Global Urban Observation and Information Initiative and serves as editor-in-chief of *ISPRS Journal of Photogrammetry and Remote Sensing* and the series editor of Taylor & Francis Series in Remote Sensing Applications. He has been the organizer and program committee chair of the biennial IEEE/ISPRS/GEO-sponsored International Workshop on Earth Observation and Remote Sensing Applications conference series since 2008; he was a national director of the American Society for Photogrammetry and Remote Sensing from 2007 to 2010 and a panelist of U.S. DOE's Cool Roofs Roadmap and Strategy in 2010.

In 2008, Dr. Weng received a prestigious NASA senior fellowship. He received the Outstanding Contributions Award in Remote Sensing in 2011 from the American Association of Geographers in 2011 as well as the Willard and Ruby S. Miller Award in 2015 for his outstanding contributions to geography. In 2005 at Indiana State University, he was selected as a Lilly Foundation Faculty Fellow, and in the following year, he also received the Theodore Dreiser Distinguished Research Award. In addition, he was the recipient of the 2010 Erdas Award for Best Scientific Paper in Remote Sensing (first place) and the 1999 Robert E. Altenhofen Memorial Scholarship Award, which were both awarded by the American Society for Photogrammetry and Remote Sensing. He was also awarded the Best Student-Authored Paper Award by the International Geographic Information Foundation in 1998. Dr. Weng has been invited to give more than 90 talks by organizations and conferences held in the United States, Canada, China, Brazil, Greece, UAE, and Hong Kong, and is honored with distinguished/chair/guest professorship at 11 top universities in China, which includes Peking University.

Dr. Weng's research focuses on remote sensing applications to urban environmental and ecological systems, land use and land cover changes, urbanization impacts, environmental modeling, and human–environment interactions. Through a serial invention of innovative algorithms, techniques, methods, and theories for urban remote sensing, he focuses research efforts on fostering the understanding of remote sensing in geographical applications and narrowing down the gap between geography and landscape ecology. Dr. Weng is the author of 206 articles (journal articles, chapters, and others) and 10 books. According to Google Scholar, as of July 2017, his SCI citation reached 11,568 (H-index of 49), and 28 of his publications had more than 100 citations each. Dr. Weng's research has been supported by funding agencies that include NSF, NASA, USGS, USAID, NOAA, National Geographic Society, European Space Agency, and Indiana Department of Natural Resources.

Dale Quattrochi, PhD, is a senior research scientist with the NASA Marshall Space Flight Center in Huntsville, Alabama, and has more than 26 years of experience in the field of Earth science remote sensing research and applications. Dr. Quattrochi's research interests focus on the application of thermal remote sensing data for analysis of heating and cooling patterns across the diverse urban landscape as they affect the overall local and regional environment. He is also conducting research on the applications of geospatial statistical techniques, such as fractal analysis, to multiscale remote sensing data.

Dr. Quattrochi is the recipient of numerous awards including the NASA Exceptional Scientific Achievement Medal, NASA's highest science award, which he received for his research on urban heat islands and remote sensing. He is also a recipient of the Ohio University College of Arts and Science, Distinguished Alumni Award. Dr. Quattrochi is the co-editor of two books: *Scale in Remote Sensing and GIS* (with Michael Goodchild) published in 1997 by CRC/Lewis Publishers and *Thermal Remote Sensing in Land Surface Processes* (with Jeffrey Luvall) published in 2004 by CRC Press. He earned his PhD degree from the University of Utah, his MS degree from the University of Tennessee, and his BS degree from Ohio University, all in geography.

Paolo Gamba, PhD, (SM'00, F'13) is professor of telecommunications at the University of Pavia, Italy, where he leads the Telecommunications and Remote Sensing Laboratory and serves as deputy coordinator of the PhD School in Electronics and Computer Science. He earned his Laurea degree in electronic engineering (cum laude) from the University of Pavia, Italy, in 1989, and his PhD in electronic engineering from the same university in 1993.

He served as editor-in-chief of the *IEEE Geoscience and Remote Sensing Letters* from 2009 to 2013, and as chair of the Data Fusion Committee of the IEEE Geoscience and Remote Sensing Society from October 2005 to May 2009. Currently, he serves as GRSS executive vice president.

Dr. Gamba has been the organizer and technical chair of the biennial GRSS/ISPRS Joint Workshops on "Remote Sensing and Data Fusion over Urban Areas" since 2001. He also served as technical co-chair of the 2010 and 2015 IGARSS conferences in Honolulu (Hawaii) and Milan (Italy), respectively.

Dr. Gamba has been the guest editor of special issues of *IEEE Transactions on Geoscience and Remote Sensing, IEEE Journal of Selected Topics in Remote Sensing Applications, ISPRS Journal of Photogrammetry and Remote Sensing,* and *International Journal of Information Fusion and Pattern Recognition Letters* on the following topics: urban remote sensing, remote sensing for disaster management, and pattern recognition in remote sensing applications.

He has been invited to give keynote lectures and tutorials on several occasions about urban remote sensing, data fusion, and EO data for exposure and risks. He published more than 130 papers in international peer-review journals and presented more than 250 research works in workshops and conferences.

Contributors

Bharath Haridas Aithal
Energy & Wetlands Research Group
Centre for Ecological Sciences
Indian Institute of Science
Bangalore, India
and
RCGSIDM
IIT Kharagpur
West Bengal, India

Benjamin Bechtel
Universität Hamburg
Center for Earth System Research
   and Sustainability
Hamburg, Germany

Thomas Blaschke
Department of Geoinformatics—Z_GIS
University of Salzburg
Salzburg, Austria

Mysore Chandrashekar Chandan
RCGSIDM
IIT Kharagpur
West Bengal, India

Isabelle Couloigner
Foothills Facility for Remote Sensing
   and GIScience
Department of Geography
University of Calgary
Calgary, Alberta, Canada

Chengbin Deng
Department of Geography
State University of New York at
   Binghamton
Binghamton, New York

Thomas Esch
German Aerospace Center (DLR)
   German Remote Sensing Data
   Center (DFD)
Oberpfaffenhofen, D-82234
Weßling, Germany

Aneta J. Florczyk
Space, Security and Migration
Disaster Risk Management Unit
European Commission
Joint Research Centre
Ispra, Italy

Tak S. Fung
Information Technologies
University of Calgary
Calgary, Alberta, Canada

Paolo Gamba
Telecommunications and Remote
   Sensing Lab
Department of Electrical, Computer and
   Biomedical Engineering
University of Pavia
Pavia, Italy

Hung Q. Ha
Center for Urban and Environmental
   Change
Department of Earth & Environmental
   Systems
Indiana State University
Terre Haute, Indiana

Geoffrey J. Hay
Foothills Facility for Remote Sensing
   and GIScience
Department of Geography
University of Calgary
Calgary, Alberta, Canada

Wieke Heldens
German Aerospace Center (DLR)
  German Remote Sensing Data
  Center (DFD)
Oberpfaffenhofen, D-82234
Weßling, Germany

Bharanidharan Hemachandran
Foothills Facility for Remote Sensing
  and GIScience
Department of Geography
University of Calgary
Calgary, Alberta, Canada

Andreas Hirner
German Aerospace Center (DLR)
  German Remote Sensing Data
  Center (DFD)
Oberpfaffenhofen, D-82234
Weßling, Germany

Daniel Hölbling
Department of Geoinformatics—Z_GIS
University of Salzburg
Salzburg, Austria

Xin Huang
School of Remote Sensing and
  Information Engineering and
State Key Laboratory of Information
  Engineering in Surveying, Mapping
  and Remote Sensing
Wuhan University
Wuhan, China

Shabnam Jabari
Department of Geodesy and Geomatics
  Engineering
Faculty of Engineering, University of
  New Brunswick
Fredericton, New Brunswick, Canada

Bilal Karim
Foothills Facility for Remote Sensing
  and GIScience
Department of Geography
University of Calgary
Calgary, Alberta, Canada

Gyula Kothencz
Department of Geoinformatics—Z_GIS
University of Salzburg
Salzburg, Austria

Christopher D. Kyle
Foothills Facility for Remote Sensing
  and GIScience
Department of Geography
University of Calgary
Calgary, Alberta, Canada

Stefan Lang
Department of Geoinformatics—Z_GIS
University of Salzburg
Salzburg, Austria

Weiying Lin
Department of Geography
State University of New York at
  Binghamton
Binghamton, New York

Santiago Linares
Instituto de Geografía, Historia y
  Ciencias Sociales, CONICET/
  UNCPBA
Centro de Investigaciones Geográficas
  and
Departamento de Geografía
Facultad de Ciencias Humanas,
  UNCPBA
Tandil, Buenos Aires, Argentina

Andrea Marinoni
Telecommunications and Remote
    Sensing Lab
Department of Electrical, Computer
    and Biomedical Engineering
University of Pavia
Pavia, Italy

Gerald Mills
University College Dublin
Dublin, Ireland

Hiroyuki Miyazaki
Center for Spatial Information Science
The University of Tokyo
Tokyo, Japan

Martino Pesaresi
Space, Security and Migration
Disaster Risk Management Unit
European Commission
Joint Research Centre
Ispra, Italy

Natasha Picone
Instituto de Geografía, Historia y
    Ciencias Sociales, CONICET/
    UNCPBA
Centro de Investigaciones Geográficas
    and
Departamento de Geografía
Facultad de Ciencias Humanas,
    UNCPBA
Tandil, Buenos Aires, Argentina

Mir Mustafiz Rahman
Foothills Facility for Remote Sensing
    and GIScience
Department of Geography
University of Calgary
Calgary, Alberta, Canada

T.V. Ramachandra
Energy & Wetlands Research Group
Centre for Ecological Sciences
Indian Institute of Science
Bangalore, India

Ryosuke Shibasaki
Center for Spatial Information Science
The University of Tokyo
Tokyo, Japan

Alaeldin Suliman
Department of Geodesy and Geomatics
    Engineering
University of New Brunswick
Fredericton, New Brunswick, Canada

Shivamurthy Vinay
Energy & Wetlands Research Group
Centre for Ecological Sciences
Indian Institute of Science
Bangalore, India

Qihao Weng
Center for Urban and Environmental
    Change
Department of Earth & Environmental
    Systems
Indiana State University
Terre Haute, Indiana

Tao Zhang
State Key Laboratory of Information
    Engineering in Surveying, Mapping
    and Remote Sensing
Wuhan University
Wuhan, China

Yilong Zhang
Foothills Facility for Remote Sensing
    and GIScience
Department of Geography
University of Calgary
Calgary, Alberta, Canada

Yun Zhang
Department of Geodesy and Geomatics
    Engineering
Faculty of Engineering
University of New Brunswick
Fredericton, New Brunswick, Canada

Section I

Data, Sensors, and Systems Considerations and Algorithms for Urban Feature Extraction

1 The Global Urban Footprint

Thomas Esch, Wieke Heldens, and Andreas Hirner

CONTENTS

1.1 INTRODUCTION

One of the most urgent present and future challenges is global urbanization. The real dimension of this phenomenon is still not completely understood. Particularly, precise worldwide information on the location and distribution of human settlements in urban and in rural areas is lacking. This chapter presents the Global Urban Footprint (GUF), which aims to close this information gap. The GUF is an inventory of human presence on Earth in the form of a raster map (Esch et al. 2017) that reflects the human settlements pattern in a thus far unique spatial resolution of 12 m.

In 1950, the urban population was only half the size of the rural population. In 2008, for the first time, the urban population became larger than the rural population. This observation, commonly known as the global urban transition (UN 2014), indicates that the majority of people on Earth inhabit some kind of urban environment. Various publications and reports on the vast urbanization (Birch and Wachter 2011; UN 2001, 2004, 2006) have made it clear that cities play a primary role as drivers of all social, economic, and environmental systems. Currently, there is considerable evidence that global urbanization affects the entire spectrum of human and natural systems, in particular with respect to energy, water, food, biodiversity, climate, or human health (Grimm et al. 2008; Kaufmann et al. 2007; Moore et al. 2003; Tilman et al. 2011; Zhou et al. 2004). However, many fundamental questions about the spatial dimension of urbanization could still not be answered to satisfaction: Which proportion of the global land surface is covered with built-up area? What is the ratio between urban and rural settlements area? How many cities are on Earth? Many studies agree that estimations of the effects of human presence on Earth are strongly biased (Potere and Schneider 2007; Potere et al. 2009) and that the dynamics of urban growth and its economic and social effects are poorly understood (Batty 2008). The lack of a shared definition of urban as opposed to rural areas as well as a missing spatially detailed and up-to-date inventory of the entirety of urban and rural settlements on Earth further complicates such analysis. This data gap is now closed by the GUF, which is described in this chapter.

1.2 GENERATING THE GUF

1.2.1 GLOBAL URBAN MAPPING USING EARTH OBSERVATION

Earth observation (EO) imagery certainly represents an effective approach to overcome the lack of objective spatial information on the structure and spatiotemporal development of human settlements on Earth (Esch et al. 2010). The global classification of human settlements is a very specific topic in urban remote sensing because of the necessary trade-off between spatial resolution of the available EO data and ability to collect a global coverage within a reasonable period of time. A comprehensive overview of the available EO-based and EO-supported global geo-information layers on human settlements is provided by Potere et al. (2009), Gamba and Herold (2009), and Ban et al. (2015a). As they report, the majority of these data sets are generated from medium-resolution optical EO data, as, for instance, the largely established MODIS 500 (Schneider et al. 2010) and GlobCover 2009 (Bontemps et al. 2011) land cover maps with a spatial resolution of 500 and 300 m, respectively. However, their capabilities to accurately detect and delineate small and scattered villages and towns are quite limited.

More recent initiatives aim to provide spatially more accurate human settlement layers based on high-resolution EO data. NASA, for example, released in 2013 a new global nighttime light product derived from imagery of the Visible Infrared Imaging Radiometer Suite (VIIRS) on board the Suomi NPP satellite (NASA 2012). The European Joint Research Center (JRC) with the Global Human Settlements Layer (GHSL) presented a procedure for an automatic extraction of built-up areas by analyzing global Landsat coverage for several time steps (Pesaresi et al. 2016). Wieland and Pittore (2016) also proposed a method based on object-based analysis and SVM to classify urban large areas from Landsat 8. Miyazaki et al. (2013) propose a method based on the integrated analysis of ASTER satellite images and geographic information system (GIS) data to produce a new global high-resolution settlement mask. By means of Envisat-ASAR radar imagery, Gamba and Lisini (2012) and Ban et al. (2015b) derived a built-up area layer that aims at improving the GlobCover 2009 urban class.

1.2.2 FROM TanDEM-X IMAGERY TO GUF

In 2007 and 2010, the German Aerospace Center (DLR) launched the EO radar satellites TerraSAR-X and TanDEM-X, respectively (Krieger et al. 2007; Werninghaus

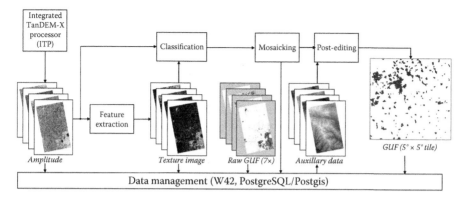

FIGURE 1.1 Urban footprint processor (UFP) processing environment to generate the GUF.

and Buckreuss 2010). Because the two missions collect a global coverage of very high resolution SAR imagery within a comparably short period, they are optimal to support global environmental monitoring activities. Encouraged by the promising outcomes of previous studies (Esch et al. 2010, 2011), DLR's German Remote Sensing Data Center (DFD) started the internal GUF initiative. The goal of this activity was the development of a fully automated processing framework to produce a worldwide map of human settlements in a thus far unique spatial detail by analyzing a global coverage of TerraSAR-X and TanDEM-X images collected in the context of the TanDEM-X mission (Esch et al. 2012).

The production of the GUF layer is based on the Urban Footprint Processor (UFP). This is a fully automatic, generic, and autonomous processing environment, orchestrating an extensive suite of processing and analysis modules. This system ensures an effective processing of the 182,249 TerraSAR-X and TanDEM-X single look complex (SSC) image products mostly collected in 2011 and 2012 (93%) in Stripmap mode with 3 m ground resolution. To fill data gaps, single scenes collected in 2013 and 2014 were included as well. Figure 1.1 provides a schematic overview on the UFP processing environment. The UFP consists of five main technical modules covering functionalities for data management, feature extraction, unsupervised classification, mosaicking, and automatic post-editing.

1.2.2.1 Feature Extraction

Esch et al. (2011) demonstrated that built-up areas show a characteristic small-scale heterogeneity of local backscatter in TSX Stripmap images that can efficiently be used to delineate built-up areas. This effect is related to the specific properties of SAR data for built environments that exhibit strong scattering due to double bounce effects and direct backscattering from the vertical structures (e.g., buildings, bridges, traffic signs). At the same time, shadow effects occur on those sides of vertical structures that are facing away from the incoming radar beams. To define this local image heterogeneity, or texture, the UFP calculates the speckle divergence feature defined as the ratio between the local standard deviation and local mean of the backscatter computed in a defined local neighborhood. A detailed description of the feature extraction algorithm is provided in Esch et al. (2013b).

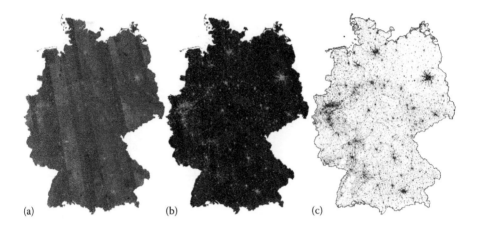

(a)   (b)   (c)

FIGURE 1.2   Backscattering amplitude (a), speckle divergence or texture image (b), and GUF (c).

Figure 1.2 shows TSX Stripmap amplitude data, the speckle divergence texture image derived from the amplitude data and the resulting GUF layer for the area of Germany next to each other. In the amplitude data (Figure 1.2a), urban areas show high values represented by bright gray tones, but other land cover types do so as well. The high-texture regions appearing as bright spots in the speckle divergence image (Figure 1.2b) show the urban areas in Germany very distinctively. Since the high local image heterogeneity originates from intense backscatter plus shadow effects around vertical structures, the texture directly relates to the presence of buildings or any structure with a distinct vertical component. This characteristic is used to derive the GUF (Figure 1.2c).

1.2.2.2   Unsupervised Classification

The classification step combines the analysis of the amplitude and the speckle divergence images. For that purpose, an unsupervised classification method based on advanced Support Vector Data Description (SVDD) one-class classification was implemented as described in detail in Esch et al. (2013b). The result of the classification procedure is a binary raster layer assigning the class built-up area and, for all regions not assigned to this class, the category non–built-up. The implemented classification method proved to be very robust, although over- or underestimation of a built-up area might still occur when scenes show very specific land cover distributions—for instance, in the case of scenes covering a coastline and mostly showing water with only few areas representing land surface. Such specific land cover configurations lead to extreme distributions of the amplitude and texture statistics of the corresponding images, which finally hinders the proper definition of accurate classification settings. To compensate for the resulting deviations between the GUF masks, a total of six additional GUF raw versions with systematically altered classification settings—that can be considered as confidence levels—are generated. Three versions are based on constantly stricter thresholds compared to the automatically defined version (leading to the assignment of less built-up area) and three versions with more relaxed settings compared to the original definition (thus showing an increased amount of potential built-up regions).

1.2.2.3 Mosaicking

The individual GUF raw masks in their original image geometry were merged to tiles of 5° × 5° geographical latitude and longitude for each of the seven GUF confidence levels, to provide more manageable working units for the post-editing procedure. As a result, 1,284,743 GUF masks (182,249 masks for each of the seven confidence levels) were reduced to 8309 GUF tiles (1187 masks for each of the seven confidence levels). The chosen tile size presents a reasonable trade-off between file size, number of tiles, and practicability in terms of data and file handling. The conformity of the merge is also retained as a quality measure. During the mosaicking, all multiple GUF accounts in the overlapping areas of neighboring scenes were aggregated by means of a majority vote for each single pixel and each separate GUF versions 1–7. As a result, each tile comprises seven GUF bands in the geometric resolution of 0.4 arc seconds (arcsec) (~12 m), or approaching the poles, in correspondingly lower longitudinal resolutions (e.g., north of 50°N up to 60°N in 0.6 arcsec).

1.2.2.4 Automated Postprocessing

The final stage of the GUF production is postprocessing. This highly automated procedure consists of two steps. First, image segmentation is conducted, which transfers all clusters of connected pixels classified as built-up in each of the seven GUF raw raster layers (confidence levels) into individual image objects. Second, the appropriate local GUF confidence level is selected and all GUF segments from the resulting collection that most likely represent false alarms are removed. For this second, rule-based step, auxiliary data, such as relief and water masks, are used. The postprocessing and the auxiliary data are described in detail in Esch et al. (2017).

1.3 THE GUF DATA SET

Figure 1.3 shows four urban footprints of Tokyo, Cairo, Minneapolis, and Ho Chi Minh City. Tokyo is a very densely built megacity with about 36 million people. Also, the center of the mega region of Cairo (estimated 80 million people, with estimated 24 million in Cairo City) is very densely built. The urban footprint of the whole Egyptian mega region follows the Nile and its delta. The urbanization of the Tokyo area is shaped by the surrounding mountains. Compared to these two megacities, Minneapolis (USA) is quite small. Together with its twin city Saint Paul, it has a dense urban core and extensive residential areas with many parks. The surroundings of Minneapolis are characterized by medium-sized cities along the main streets and single farms in between.

Ho Chi Minh City in Vietnam is a megacity with around 7 million people. In the north east, there is a mangrove area that is almost uninhabited. The people live, apart from in the city core, mainly along the many rivers and canals in the region, resulting in a characteristic settlement pattern.

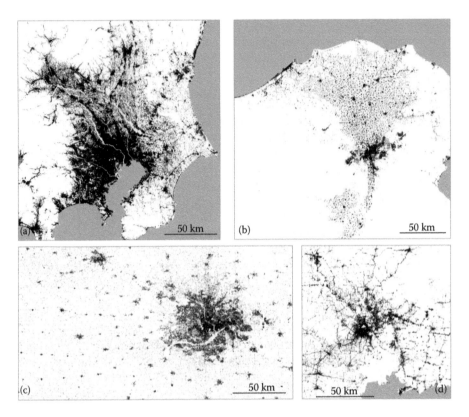

FIGURE 1.3   Urban footprints of (a) Tokyo (Japan), (b) Cairo (Egypt), (c) Minneapolis (USA), and (d) Ho Chi Minh City (Vietnam).

Because of the high spatial detail, not only (mega) cities but also rural areas with far fewer inhabitants can be studied with the GUF. Figure 1.4 shows four "rural" regions around the world. The name "rural" is actually not really appropriate, because built-up areas cover still a large part of the area in some of these regions. Each region has its own pattern of villages and/or single houses. The analysis of settlement patterns and structures provides unique and exciting insights regarding the roots of historic and cultural origin of settlements and man-made landscapes.

The GUF layer is provided as binary raster data sets in 8-bit, LZW-compressed GeoTiff format with a value of 255 indicating a built-up area, a value of 0 representing all non–built-up areas, and no data assigned by the value 128. The projection is geographic coordinates (Lat, Lon). The geometric resolution of the original GUF data is 0.4 arcsec, which corresponds to 12 m near the equator, while the GUF layer for any nonscientific/noncommercial use shows a reduced resolution of 2.8 arcsec (84 m near the equator). Toward the poles, the spatial resolution decreases to 0.6 arcsec between 50° and 60°N/S, 0.8 arcsec from 60° to 70°N/S, and 1.2 arcsec in areas >80°N/S. The generalized GUF version in 2.8 arcsec is directly derived from the 0.4-arcsec version by assigning a value of 255 (=built-up) to all pixels whose coverage contains a proportion of >25% GUF area as defined by the original 0.4-arcsec data.

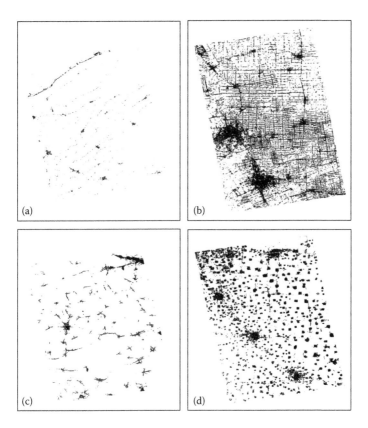

FIGURE 1.4 Footprints of the built-up areas in "rural" regions: (a) Saint-Henri/Saint-Gervais, Quebec, Canada; (b) Hai'an, Jiangsu, China; (c) Diakowar, Osijek-Baranja, Croatia; (d) Zhaoxian, Hebei, China.

On a regional scale, the thematic accuracy and the strengths and weaknesses of the GUF layer have already been documented in detail by comparisons to ground truth data in several studies and projects such as those presented by Esch et al. (2013a), Felbier et al. (2014), Gessner et al. (2015), and Klotz et al. (2016), who conducted a detailed multi-scale cross-comparison between the GUF layer and existing low-resolution (MOD500, GlobCover) and high-resolution (GHSL-SPOT2.5m) human settlement data derived from EO imagery. A statistical comparison of MOD500, GlobCover, GHSL, and GUF at the local and global scale is discussed in Esch et al. (2017).

Figure 1.5 shows the GUF of three cities in comparison to MOD500, GlobCover, and the FTS European Soil Sealing Layer (ESS). Figure 1.5a and b shows that the GUF displays a high level of detail, even with 2.8″ spatial resolution. Especially in the surroundings of New Delhi and Montreal, the GUF could identify much more settlements in comparison to MOD500 and GlobCover, respectively. This can be mainly attributed to the higher spatial resolution of the GUF. Figure 1.5c shows the GUF in comparison to the high-resolution (20 m) ESS. The similarity is very high. However, the ESS only covers the EU member states. Other regions such as Africa or Asia

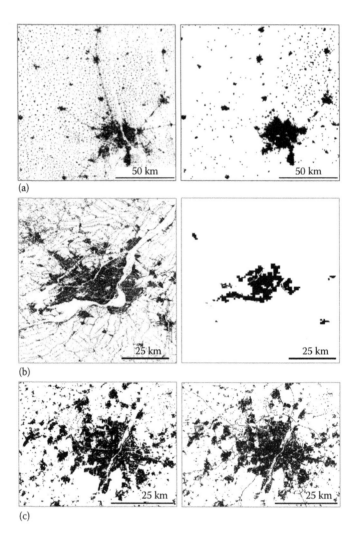

(a)

(b)

(c)

FIGURE 1.5 Comparison of the GUF (2.8″) on the left to other urban layers: (a) GUF New Delhi (India) compared to MODIS 500; (b) GUF of Montreal (Canada) compared to GlobCover2009; (c) GUF of Munich (Germany) compared to FTS European Soil Sealing.

often lack such detailed large area data sets. The ESS also shows roads, which are, by definition, not included in the GUF because roads do not have vertical structures.

1.4   ANALYZING URBAN FOOTPRINTS

With the GUF data set, for the first time, urban footprints from all over the world can be analyzed in comparison. The different cities in Figure 1.3 allow questions to the total area of the city, or the urbanization of the surroundings, to be answered. And the urban footprints in Figure 1.4 allow a comparison of the urbanization of

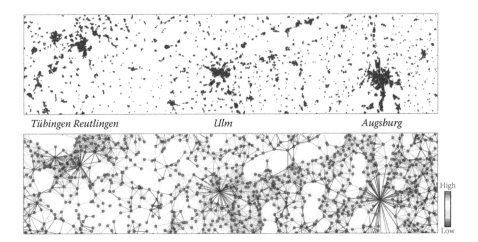

FIGURE 1.6 Settlement analysis of Tübingen, Reutlingen, Ulm, and Augsburg in southern Germany. Top: GUF settlement mask as input. Bottom: local significance measure.

rural areas. Even more interesting analysis can be carried out if additional data are used. For example, in combination with population data, population density can be calculated. Additional historical data sets show urban growth.

The binary GUF map can also be used to derive various spatial metrics, for example, on the compactness or dispersion of the settlements. Figure 1.6 shows the result of a network analysis based on the GUF. The urban areas were segmented, nodes were assigned, and the edges between the nodes were calculated. Subsequent derived statistics allow the visualization of various network characteristics, such as the connectivity and centrality of urban settlements within a region. In Figure 1.6, the local connectivity is shown. This measure characterizing the connection between the nodes combines the size of the two connected nodes with the distance between them (Esch et al. 2014). Thus, the connection (arcs) between two large urban areas (nodes) that are close together receives the highest local significance (colored red). When visualizing this network measure, the urban centers (here Tübingen, Reutlingen, Ulm, and Augsburg) are highlighted in red, whereas the areas with small dispersed settlements become blue.

1.5 SUMMARY AND OUTLOOK

With 12-m pixel spacing, the GUF data represent a precise inventory of both large urban agglomerations and the dispersed small-scale built-up areas in rural regions. It allows detailed quantitative and qualitative analyses and comparisons of settlement properties and patterns from the local municipal level up to the global scale. Urban areas over the whole globe are mapped with the same, comparable method. In this context, the GUF layer also benefits from the fact that the global input data could be collected within just 2 years. The GUF data set can help to acquire a better understanding of the urbanization phenomenon and to respond appropriately to future challenges related to sprawling cities, population explosion, poverty reduction, economic growth, climate change and carbon emissions, and the loss to biodiversity.

Especially in remote and underdeveloped regions of the Earth, where suitable geographical data are frequently scarce, the GUF adds valuable information.

DLR has released the global GUF data set to be used open and free of charge at full spatial resolution for any scientific use and at a generalized resolution of about 84 m for any nonprofit applications. Currently available by e-mail request from DLR (guf@dlr.de), the data sets are also accessible on the European Space Agency's Urban Thematic Exploitation Platform (via: https://urban-tep.eo.esa.int/) and at DLR-EOC's geoservice (via: https://geoservice.dlr.de/web/maps/eoc:guf:4326).

DLR plans to release regular updates of the GUF layer based on Sentinel-1/-2 and Landsat data. At the same time, there will be an extended suite of GUF+ products covering further thematic detail compared to the current binary GUF mask. This includes a GUF-NetS layer generated on the basis of a spatial network analysis (Esch et al. 2014) and describing the global settlements pattern and properties. A GUF-DenS layer will further specify the built-up density and urban green within those areas assigned as settlements by the conventional GUF based on a combination of the GUF mask and imperviousness/greenness information derived from TimeScan data described by Esch et al. (2016). Another product of the GUF+ suite is the GUF-TrendS product that aims at the provision of information of the spatio-temporal development of the human settlements based on the analysis of Landsat and Envisat-ASAR archive data. Finally, the generation of a GUF-VolumeS layer is planned, which estimated the average volume of the built-up areas from TanDEM-X DEM data based on a procedure introduced by Marconcini et al. (2014).

ACKNOWLEDGMENTS

We would like to thank the TerraSAR-X and TanDEM-X Science Teams for providing the global SAR data used to derive the GUF layer. We also thank the World Bank and the Swiss State Secretariat for Economic Affairs for their financial support in the context of the GUF quality assessment and enhancement activities (Contract 7174285).

REFERENCES

Ban, Y., Gong, P., Giri, C., 2015a. Global land cover mapping using Earth observation satellite data: Recent progresses and challenges. *ISPRS Journal of Photogrammetry and Remote Sensing* 103, 1–6. 10.1016/j.isprsjprs.2015.01.001.

Ban, Y., Jacob, A., Gamba, P., 2015b. Spaceborne SAR data for global urban mapping at 30 m resolution using a robust urban extractor. *ISPRS Journal of Photogrammetry and Remote Sensing* 103, 28–37. 10.1016/j.isprsjprs.2014.08.004.

Batty, M., 2008. The size, scale, and shape of cities. *Science* 319 (5864), 769–771. 10.1126/science.1151419.

Birch, E.L., Wachter, S.M., 2011. *Global Urbanization*. Pennsylvania Press.

Bontemps, S., Defourny, P., Bogaert, E.V., Arino, O., Kalogirou, V., Perez, J.R., 2011. GLOBCOVER 2009—Products description and validation report.

Esch, T., Asamer, H., Boettcher, M., Brito, F., Hirner, A., Marconcini, M., Mathot, E., Metz, A., Permana, H., Soukup, T., others, 2016. Earth Observation-Supported Service Platform for the Development and Provision of Thematic Information on the Built Environment—The TEP-Urban Project.

Esch, T., Felbier, A., Heldens, W., Marconcini, M., Roth, A., Taubenbock, H., 2013a. Spatially detailed mapping of settlement patterns using SAR data of the TanDEM-X mission, in: *Urban Remote Sensing Event (JURSE)*, 2013 Joint, pp. 041–044.

Esch, T., Heldens, W., Hirner, A., Keil, M., Marconcini, M., Roth, A., Zeidler, J., Dech, S., Strano, E., 2017. Breaking new ground in mapping human settlements from space. The Global Urban Footprint. *ISPRS Journal of Photogrammetry and Remote Sensing* 134, 30–42. 10.1016/j.isprsjprs.2017.10.012.

Esch, T., Marconcini, M., Felbier, A., Roth, A., Heldens, W., Huber, M., Schwinger, M., Taubenböck, H., Müller, A., Dech, S. 2013b. Urban footprint processor; fully automated processing chain generating settlement masks from global data of the TanDEM-X mission. *IEEE Geoscience and Remote Sensing Letters* 10 (6), 1617–1621. 10.1109/LGRS.2013.2272953.

Esch, T., Marconcini, M., Marmanis, D., Zeidler, J., Elsayed, S., Metz, A., Müller, A., Dech, S., 2014. Dimensioning urbanization—An advanced procedure for characterizing human settlement properties and patterns using spatial network analysis. *Applied Geography* 55, 212–228. 10.1016/j.apgeog.2014.09.009.

Esch, T., Schenk, A., Ullmann, T., Thiel, M., Roth, A., Dech, S., 2011. Characterization of land cover types in TerraSAR-X images by combined analysis of speckle statistics and intensity information. *IEEE Transactions on Geoscience and Remote Sensing* 49, 1911–1925. 10.1109/TGRS.2010.2091644.

Esch, T., Taubenböck, H., Roth, A., Heldens, W., Felbier, A., Thiel, M., Schmidt, M., Müller, A., Dech, S., 2012. TanDEM-X mission—New perspectives for the inventory and monitoring of global settlement patterns. *Journal of Applied Remote Sensing* 6 (1), 061702-1-061702-21. 10.1117/1.JRS.6.061702.

Esch, T., Thiel, M., Schenk, A., Roth, A., Muller, A., Dech, S., 2010. Delineation of urban footprints from TerraSAR-X data by analyzing speckle characteristics and intensity information. *IEEE Transactions on Geoscience and Remote Sensing* 48 (2), 905–916. 10.1109/TGRS.2009.2037144.

Felbier, A., T. Esch, W. Heldens, M. Marconcini, J. Zeidler, A. Roth, M. Klotz, M. Wurm, H. Taubenböck, 2014. The global urban footprint; processing status and cross comparison to existing human settlement products, in: *2014 IEEE Geoscience and Remote Sensing Symposium*, pp. 4816–4819.

Gamba, P., Herold, M. (Eds.), 2009. *Global Mapping of Human Settlement: Experiences, Datasets, and Prospects*. CRC Press, Boca Raton, FL, USA.

Gamba, P., Lisini, G., 2012. A robust approach to global urban area extent extraction using ASAR Wide Swath Mode data, 1–5. 10.1109/TyWRRS.2012.6381093.

Gessner, U., Esch, T., Tillack, A., Naeimi, V., Kuenzer, C., Dech, S., 2015. Multi-sensor mapping of West African land cover using MODIS, ASAR and TanDEM-X/TerraSAR-X data. *Remote Sensing of Environment* 164, 282–297. 10.1016/j.rse.2015.03.029.

Grimm, N.B., Faeth, S.H., Golubiewski, N.E., Redman, C.L., Wu, J., Bai, X., Briggs, J.M., 2008. Global change and the ecology of cities. *Science* 319 (5864), 756–760. 10.1126/science.1150195.

Kaufmann, R.K., Seto, K.C., Schneider, A., Liu, Z., Zhou, L., Wang, W., 2007. Climate response to rapid urban growth: Evidence of a human-induced precipitation deficit. *Journal of Climate* 20 (10), 2299–2306. 10.1175/JCLI4109.1.

Klotz, M., Kemper, T., Geiß, C., Esch, T., Taubenböck, H., 2016. How good is the map? A multi-scale cross-comparison framework for global settlement layers: Evidence from Central Europe. *Remote Sensing of Environment* 178, 191–212. 10.1016/j.rse.2016.03.001.

Krieger, G., Moreira, A., Fiedler, H., Hajnsek, I., Werner, M., Younis, M., Zink, M., 2007. TanDEM-X: A satellite formation for high-resolution SAR interferometry. *IEEE Transactions on Geoscience and Remote Sensing* 45 (11), 3317–3341. 10.1109/TGRS.2007.900693.

Marconcini, M., Marmanis, D., Esch, T., Felbier, A., 2014. A novel method for building height estimation using TanDEM-X data, in: *2014 IEEE Geoscience and Remote Sensing Symposium*, pp. 4804–4807.

Miyazaki, H., Shao, X., Iwao, K., Shibasaki, R., 2013. An automated method for global urban area mapping by integrating ASTER satellite images and GIS data. *IEEE Journal of Selected Topics in Applied Earth Observations and Remote Sensing* 6 (2), 1004–1019. 10.1109/JSTARS.2012.2226563.

Moore, M., Gould, P., Keary, B.S., 2003. Global urbanization and impact on health. *International Journal of Hygiene and Environmental Health* 206 (4–5), 269–278. 10.1078 /1438-4639-00223.

NASA, 2012. National Aeronautics and Space Administration: Night Lights 2012.

Pesaresi, M., Ehrlich, D., Ferri, S., Florczyk, A.J., Freire, S., Halkia, M., Julea, A., Kemper, T., Soille, P., Syrris, V., 2016. Operating procedure for the production of the Global Human Settlement Layer from Landsat data of the epochs 1975, 1990, 2000, and 2014. European Joint Research Center.

Potere, D., Schneider, A., 2007. A critical look at representations of urban areas in global maps. *GeoJournal* 69 (1), 55–80. 10.1007/s10708-007-9102-z.

Potere, D., Schneider, A., Angel, S., Civco, D.L., 2009. Mapping urban areas on a global scale: Which of the eight maps now available is more accurate? *International Journal of Remote Sensing* 30 (24), 6531–6558. 10.1080/01431160903121134.

Schneider, A., Friedl, M., David Potere, 2010. Mapping global urban areas using MODIS 500-m data: New methods and datasets based on 'urban ecoregions'. *Remote Sensing of Environment* 114 (8), 1733–1746. 10.1016/j.rse.2010.03.003.

Tilman, D., Balzer, C., Hill, J., Befort, B.L., 2011. Global food demand and the sustainable intensification of agriculture. *Proc. Natl. Acad. Sci. USA* 108 (50), 20260–20264.

UN, 2001. *Cities in a Globalizing World: Global Report on Human Settlements 2001.* Earthscan.

UN, 2004. *The State of the World Cities 2004/5—Globalization and Urban Culture.* Routledge.

UN, 2006. *The State of the World Cities 2006/7—The Millennium Development Goals and Urban Sustainability.* Routledge.

UN, 2014. *World Urbanization Prospects: The 2014 Revision.*

Werninghaus, R., Buckreuss, S., 2010. The TerraSAR-X mission and system design. *IEEE Transactions on Geoscience and Remote Sensing* 48 (2), 606–614. 10.1109/TGRS.2009 .2031062.

Wieland, M., Pittore, M., 2016. Large-area settlement pattern recognition from Landsat-8 data. *ISPRS Journal of Photogrammetry and Remote Sensing* 119, 294–308. 10.1016 /j.isprsjprs.2016.06.010.

Zhou, L., Dickinson, R.E., Tian, Y., Fang, J., Li, Q., Kaufmann, R.K., Tucker, C.J., Myneni, R.B., 2004. Evidence for a significant urbanization effect on climate in China. *Proceedings of the National Academy Sciences USA* 101 (26), 9540–9544.

2 Development of On-Demand Human Settlement Mapping System Using Historical Satellite Archives

Hiroyuki Miyazaki and Ryosuke Shibasaki

CONTENTS

2.1 INTRODUCTION

Urban sprawl is an important global issue of sustainable development (Angel et al. 2005, Foley et al. 2005). Monitoring urban formation is required for control of urban development and related issues, such as disaster risk management (Dasgupta et al. 2009; Doocy et al. 2007), public health (Brooker et al. 2006; Omumbo et al. 2005), transportation (Schneider et al. 2003), and food security (Balk et al. 2005). Because geographic data of urban development, such as topographic maps and infrastructure maps, are rarely affordable in lower and lowest income countries, innovative methods to develop such data are urgently needed.

Satellite earth observation has an important role in human settlement mapping, which is a typical classification of built-up and non–built-up areas, and has contributed to better urban monitoring and planning (Schneider and Woodcock 2008; Schneider et al. 2003), especially in less documented regions.

While coarse-resolution satellite data, such as DMSP-OLS and MODIS with a spatial resolution of 500–1000 m (Center for International Earth Science Information Network et al. 2004; Schneider et al. 2010), have a role of urban monitoring on a global scale, development of global data with a spatial resolution of 10–100 m was recently achieved using ASTER with 15-m resolution (Miyazaki et al. 2014), Landsat with 30-m resolution (European Commission Joint Research Centre 2014), and TerraSAR-X with 12-m resolution (Esch et al. 2013). The data provided detailed insights of urban structures, which are closely connected with the socio-economy of urban growth that was not identified by the coarse-resolution satellite data (Angel et al. 2005; Nelson and Robertson 2007).

Time-series data sets of urban extent are required to quantify urban growth (Taubenböck et al. 2014; Yang et al. 2017), which is closely connected with the Sustainable Development Goal 11 "Make cities inclusive, safe, resilient and sustainable safeguard in urban development" (United Nations 2015). Use of Global Land Survey, a collection of time-series Landsat data acquired in 1970, 1990, 2000, 2005, and 2010 with good qualities for all the tiles (Global Land Cover Facility 2014), helps development of the time-series human settlement data, such as the Global Human Settlement Layer (GHSL; Pesaresi et al. 2015). The GHSL is an important data set of baselines for measuring urban growth in the world. However, even use of the GHSL may not satisfy users' requirements, especially for studies of rapidly growing cities, which have considerable impacts to realizing the Sustainable Development Goals. There is a need for flexibility in acquisition of time-series data sets, in addition to fixed time-series data sets. An on-demand system with automated selection of time-series data for users' preferences is a solution to meet the needs.

This chapter presents experimental development of a system for time-series human settlement mapping using historical archives with on-demand data processing. The subsequent sections comprise the following:

   i. A methodology for automated development of time-series human settlement maps using historical satellite data archives
   ii. An implementation of the methodology using open-source software
   iii. Experimental results of the time-series human settlement mapping
   iv. An implementation of an on-demand mapping system using web-based technology

2.2 METHODOLOGY OF AUTOMATED TIME-SERIES HUMAN SETTLEMENT MAPPING

2.2.1 INTRODUCTION

Development of human settlement maps was initiated using coarse-resolution satellite data, such as DMSP-OLS and MODIS. Although the map data contributed to the socio-economic analysis of urban development with the consistency of world-wide data (Montgomery 2008), the need for more spatially detailed data emerged after applying the data to the analysis of urban formation, such as sprawl and compactness of cities (Angel et al. 2005; Schneider and Woodcock 2008). Satellite data of Landsat and ASTER are useful data resources to represent forms and networks of human settlement, including buildings, paved areas, and other man-made structures (Esch et al. 2014; Small 2005). To use higher-resolution satellite data, such as Landsat and ASTER, for human settlement mapping in traditional methods, the development of training data, the most labor-intensive process in remote sensing projects, is required for accurate classification; however, this process is not feasible to extend the data development on a global scale. Some research initiatives developed automated algorithms using machine-learning techniques to use existing coarse-resolution human settlement maps as training data (Duan et al. 2015, European Commission Joint Research Centre 2014; Miyazaki et al. 2014). The automated algorithms enabled global coverage of higher-resolution human settlement maps without extensive human labor.

While classification algorithms were successfully automated, there have been problems caused by uncertainties of satellite data quality, such as cloud contamination, which are considerable constraints for applications of Landsat data (Ju and Roy 2008). The limited availability of satellite data is a considerable constraint to ensure the regularity of a time-series data set. For example, cloud-free satellite data acquired in 2000, 2005, and 2010 are available for some regions of interest while not available for other regions. Cloud-free satellite data sets have to be chosen from other years instead of target years (e.g., 2001, 2004, and 2011), although these choices do not completely satisfy users' requirements. Better usability is expected if target years are flexibly tuned along with users' preferences, such as equal intervals and specific dates.

To overcome this kind of availability constraint, we propose a method to combine data sets of multiple dates into a single target date. For example, to develop a data set for 2005, supplemental data for August 2004 and September 2006 are used if very little data for 2005 are available. The following sections describe details of the automated method of human settlement mapping, combining the outputs from supplemental data of multiple dates into a single data set. Overview of the data processing for a target year is shown in Figure 2.1.

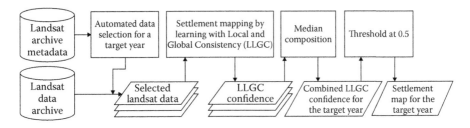

FIGURE 2.1   Overview of the data processing for a target year.

2.2.2 LEARNING WITH LOCAL AND GLOBAL CONSISTENCY

Classification of satellite image pixels as settlements or non-settlements requires two basic steps: clustering and labeling. For automated clustering, an unsupervised clustering method, such as ISODATA, has been employed in the past for land cover classification (Angel et al. 2005; Koeln et al. 2000). To label clusters correctly, the classifier requires external training data. In the conventional method, training data for labeling clusters have been acquired by human visual interpretations. However, it is not feasible to conduct human visual interpretations if a study needs data sets for all major cities of the world, which is possible in assessing indicators of Sustainable Development Goals.

To automate cluster labeling, we employed existing global human settlement maps as training data. Well-classified human settlement maps should be good training data for clustered satellite images of high resolution because of their overall classification accuracy rates of 0.83 to 0.98 (Potere et al. 2009). However, most global maps have been developed at coarse resolution, such as 300–1000 m (Bartholome and Belward 2005; Bicheron et al. 2008; Center for International Earth Science Information Network et al. 2004; Schneider et al. 2009), and the gap in spatial resolution causes inconsistencies in reflectance values and labels. For example, if a cluster that is likely to be settlements and a neighboring cluster that is likely to be non-settlements are covered within a settlement pixel at coarse resolution, each cluster will include training data classified as settlements even though they should be separated into settlements and non-settlements at a fine resolution.

For the roughly labeled training data, iterative machine-learning algorithms should work effectively. One of the classic machine-learning methods employs the perceptron learning rule, where iterative error correction converges into a discriminant classification of linearly separable patterns (Haykin 1998).

A popular method with such an approach is AdaBoost, which iterates data classification and weights training data so that the effect of overfitting is reduced (Friedman et al. 2000). With such an error correction mechanism, the roughly labeled training data set can be refined into a newly labeled data set based on the pixel values of the high-resolution satellite images.

When applying an iterative machine-learning algorithm, the primary parameter is the number of iterations. In general, an increase in the number of iterations improves the classification accuracy. However, the number cannot be increased infinitely because of limited computer resources. In addition, a greater number of iterations does not assure higher classification accuracy (Mease and Wyner 2008). To achieve the best accuracy within the conditional constraints, many trials are required with adjusted parameters. For global human settlement mapping, such trials had to be avoided because the variety of environmental conditions of the cities around the world suggested that an uncountable number of combinations would be needed to determine the parameters for the greatest accuracy.

To save costs for determining the number of iterations, we introduced the Learning with Local and Global Consistency (LLGC; Zhou et al. 2003), which determined classifiers with an infinite number of iterations of spectral clustering. Zhou et al. presented an analytic solution to the problem of incorporating infinite iterations into the operations of linear algebra, thus reducing computation costs through a few operations of linear algebra.

LLGC is principally an iterative spectral clustering method, in which data are classified based on the graph structure defined by the data set's affinity matrix. In spectral clustering, the Laplacian matrix plays an important role for graph-based clustering. It is defined by

$$L = D - W \qquad (2.1)$$

$$W_{ij} = \exp\left(-x_i - x_j^2 / 2\sigma^2\right) \text{ if } i \neq j \text{ and } W_{ij} = 0, \qquad (2.2)$$

where W is an affinity matrix, an $n \times n$ matrix; σ^2 is the distance weight of the feature vector among the data (smaller values reduce the distance effect); and D is a diagonal matrix, called the degree matrix, with its (i, i) element equal to the sum of the ith row of W (von Luxburg 2007). The normalized Laplacian is defined as

$$L_n = I - D^{-1/2}WD^{-1/2}. \qquad (2.3)$$

For spectral clustering, Equation 2.3 could be replaced with (Ng et al. 2002)

$$L_n' = D^{-1/2}WD^{-1/2}. \qquad (2.4)$$

The normalized Laplacian matrix represents the approximate probabilities of transition from one node to other nodes. By multiplying with a matrix L_n' that represents an initial classification, including distorted classifications, denoted as Y, class information is transferred based on affinity among the nodes. As a result, distorted classifications are corrected to a neighbor's class.

LLGC is a method of iterative multiplication of the normalized Laplacian with a constraint from the initial classification. Here, we present the algorithm of labeling with LLGC. Given a feature vector set $X = \{x_1,..., x_n\}$ and a label set $L = \{1, 2\}$ indicating non-settlements and settlements, x_i are labeled $y_i \in L$. An $n \times 2$ matrix F, with which x_i is labeled $y = \text{argmax } F_{ij}$, is defined. An $n \times 2$ matrix Y_{ij} is also defined, in which, at the initial step, if x_i is labeled $y_i = j$, $Y_{ij} = 1$; otherwise, $Y_{ij} = 0$.

- Form an $n \times n$ matrix as W shown in Equation 2.2.
- Construct a normalized Laplacian L_n' defined as in Equation 2.4.
- Iterate $F(t+1) = \alpha L_n' F(t) + (1-\alpha)Y$ until convergence, where t is the number of iterations and α indicates the learning rate, a parameter that ranges from 0 to 1 specifying the relative amount of the information from the affinity matrix.
- Let F^* denote the limit of the sequence $\{F(t)\}$. Label each point x_i as $y_i = 1$ if $F_{i1}^* > F_{i2}^*$; otherwise, $y_i = 2$.

$F(t)$ converges to $F^* = \left(I - \alpha L_n'\right)^{-1} Y$, where $n \times n$ is a unit matrix (Zhou et al. 2003). For x_i to be assigned as settlements, F_{i2}^* values must be greater than F_{i1}^*, and vice versa. In the case of satellite images in which the number of pixels is usually more than a million, we could not apply LLGC to the image data because of computer resource constraints. To avoid such limitations, we applied the algorithm to clustered satellite images. We assigned the mean surface reflectance of the clusters to feature vectors x_i, where i is the index of a cluster, W and Y weighted with the numbers of pixels in the clusters. Thus, L'_n and $F*$ were replaced with the following:

$$L'_n = D^{-1/2} N^{-1/2} W N^{-1/2} D^{-1/2} \tag{2.5}$$

$$F* = \left(I - \alpha L'_n\right)^{-1} NY, \tag{2.6}$$

where N is a diagonal matrix, in which element (i, i) indicates the number of pixels in cluster i. The proposed algorithm was implemented by the following steps:

- Perform ISODATA clustering on image data of surface reflectance or digital numbers, and calculate mean values for each band by cluster. The feature vectors (mean values) were normalized by the following:

$$x_{ij} = (\rho_{ij} - \mu_j)/\sigma_j, \tag{2.7}$$

  where ρ_{ij} is the mean surface reflectance of band j for cluster i, μ_j is the mean value of band j among clusters, and σ_j is the standard deviation of band j among clusters.
- Partition the clusters with the boundary of settlement and non-settlement pixels in a coarse-resolution settlement map, called initial reference map (IRM).
- Form W, D, and L'_n defined by Equations 2.2 and 2.5 for the cluster layer. i and j in Equation 2.2 indicate the index of a cluster of the partitioned cluster layer, and x_i is the feature vector, or mean values, for cluster i.
- Calculate $F*$ of Equation 2.6, where the ith row of $F*$ and Y is for the ith cluster of the partitioned cluster layer.
- Classify clusters into settlement or non-settlement based on $F*$. For the ith cluster to be assigned as settlement, F^*_{i2} must be greater than F^*_{i1}, and vice versa.

Basically, the clusters were classified as settlement or non-settlement by comparing F_{i2} and F_{i1}; however, that classification discarded information on the compositions of settlement and non-settlement classifications in the clusters. Retaining that information was useful in identifying settlement areas in regions where relatively little built-up clusters are dominant. We proposed a confidence value that ranged from 0 to 1.

$$\text{LLGC Confidence} = F_{i2} / (F_{i2} + F_{i1}) \tag{2.8}$$

The confidence value was calculated for each cluster to introduce it into the map composition discussed below; we call it the LLGC confidence map.

The algorithm was applied to satellite images for extents around settlement clusters in the IRM. A buffer around each settlement cluster with an area equal to the settlement cluster was used for the algorithm because the confidence values were biased by proportions of settlement and non-settlement in the IRM.

2.2.3 COMPOSITION OF LLGC RESULTS

Because of uncertainty in satellite data, such as cloud contamination and hazes, the results of the classification are not always stable in automated algorithms. To reduce such uncertainties, we proposed to compose several results of LLGC confidence for a certain period around the target date into a single output. The LLGC confidence results were combined by calculation median values by pixel. This operation was expected to strengthen robustness of the result against obstacles caused by uncertainties such as cloud contamination. An example is shown in Figure 2.2. The example chose satellite data acquired in August 2004, July 2005, and September 2006 to develop a human settlement map for 2005. The results for August 2004 and September 2006 captured the built-up areas, while the results for July 2005 could not identify the settlements due to some obstacles, such as cloud contamination. If only the data for July 2005 were used for the human settlement map for 2005, the built-up areas could not be identified. The median values calculated with supplement data for August 2004 and September 2006 captured the built-up areas.

Selection of supplemental data should be automated because the system is to be fully automated for better usability of on-demand data processing. For the automation,

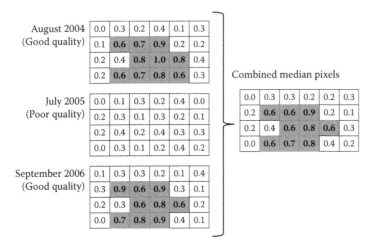

FIGURE 2.2 An example of combining supplemental data by median values. (From Miyazaki, H., M. Nagai and R. Shibasaki. 2016. An automated method for time-series human settlement mapping using Landsat data and existing land cover maps. In *2016 IEEE International Geoscience and Remote Sensing Symposium (IGARSS)*, Beijing, China.)

we proposed to calculate a selection score for each satellite data. If a metadata database of satellite data is available, a selection score of satellite data for an extent, such as path-row tile, can be calculated by

$$\text{score} = \sum (x - \mu_x) / \sigma_x + (YD - \mu_{YD}) / \sigma_{YD}, \tag{2.9}$$

where x is a data quality indicator where less value is preferable, such as cloud coverage, μ_x is the mean of x among the metadata, σ_x is the standard deviation of x among the metadata, YD is the difference of dates between the target date and the observation date, μ_{YD} is the mean of the YD of the metadata, and σ_{YD} is the standard deviation of the YD of the metadata. The score is simply a sum of scores normalized to an average of 0 and a standard deviation of 1. In this experiment, we used cloud coverage percentile for the quality indicator.

2.3 EXPERIMENTAL IMPLEMENTATIONS OF TIME-SERIES MAPPING SYSTEM

2.3.1 DATA AND PROCESSING

We conducted an experiment of the proposed method using Landsat scenes covering cities with populations more than 1 million covered by 838 WRS2 path-row tiles for 1990, 2000, 2005, and 2010. The Landsat data were retrieved from the US Geological Survey's Landsat data archive (US Geological Survey 2015b). We used the built-up layer extracted from MCD12Q1 (Land Processes Distributed Active Archive Center 2014), a MODIS-based global land cover map, for the IRM of settlement areas.

Because most settlement areas would be irreversible to natural land covers once developed, we designed the algorithm to classify the pixels of settlement areas for an earlier date to be settlement areas in the following dates. For example, in the case of time slices for 1990, 2000, 2005, and 2010, which are presented below, if a pixel in the result for 1990 is classified as a settlement area, the pixels at the location for the following dates (2000, 2005, and 2010) will be classified as settlement areas regardless of the results of the LLGC and median composition. This step could reduce inconsistencies of land cover changes among time-series human settlement maps, especially omission errors in the results for more recent dates.

2.3.2 SYSTEM IMPLEMENTATION

We implemented the algorithms on a high-performance computing system in the University of Tokyo called Data Integration and Analysis System (DIAS; Koike et al. 2015). The DIAS provided parallel computing with a few tens of processors. For better flexibility and configurability, we implemented the algorithms and constructed the programs using open-source software only. As the DIAS is a Linux-based system, which is originated by open-source development, construction of the system was open-source friendly so that the system could be portable to other Linux-based computing systems. Table 2.1 lists the open-source software used for the system.

TABLE 2.1

List of Software Used for the Human Settlement Mapping System

Software	Functions	Functions of the System	Website
GRASS	Geospatial data analysis suite	General operations of raster data	https://grass.osgeo.org/
PostgreSQL	Relational database management system	Analysis of Landsat metadata for choosing scenes	http://www.postgresql.org/
PostGIS	Geospatial data extension of PostgreSQL		http://postgis.net/
SQLite	File-based database management system	General operations of vector data	http://www.sqlite.org/
SpatiaLite	Geospatial data extension of SQLite		http://www.gaia-gis.it /gaia-sins/
GDAL	Library for geospatial data abstraction	Conversion of data formats and projections	http://gdal.org/
GNU Octave	Mathematic operations	Implementation of LLGC	https://gnu.org/software /octave/
R	Statistical analysis software	Composition of LLGC results	https://www.r-project.org
LEDAPS	Production of surface reflectance data from Landsat data.	Acquiring surface reflectance data from Landsat archive data	https://github.com/usgs-eros /espa-surface-reflectance

2.3.3 Experiment Results and Discussions

Some examples of the results are presented in Figure 2.3. The results well represented urban expansion of the cities in the last decades. However, some results for 1990 showed major areas of human settlement extent although the urban development in the cities rapidly progressed later than 2000. This might be due to overestimation of settlement areas for 1990, where the results for 2000, 2005, and 2010 were classified as non-settlement. This sort of inconsistency can be corrected by comparison of the results among the target years by majority decision or Bayesian inference. For example, if a pixel is classified as a settlement area for 1990 and as a non-settlement area for 2000, 2005, and 2010, the pixel should be classified as a non-settlement area also for 1990.

While the target years after 2000 are split by 5 years, the first interval of the time-series data is 10 years. This might have emphasized the change of human settlement more in the period of 1990–2000 than in other periods. It indicates that the time slices would need to be flexible to monitor the growth of human settlement extents by an equal interval.

While such overestimations were observed for some cities, the results for some other cities had underestimations. This was due to underestimations of settlement

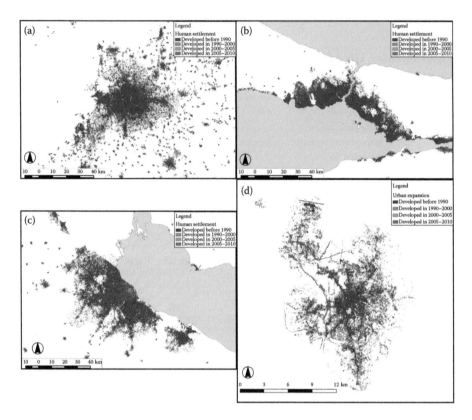

FIGURE 2.3 Examples of the results. (a) Beijing, China; (b) Istanbul, Turkey; (c) Buenos Aires, Argentina; (d) Antananarivo, Madagascar. (Panels a to c from Miyazaki, H., M. Nagai and R. Shibasaki. 2016. An automated method for time-series human settlement mapping using Landsat data and existing land cover maps. In *2016 IEEE International Geoscience and Remote Sensing Symposium (IGARSS)*, Beijing, China.)

areas in MCD12Q1 for such cities because the algorithm depended on the IRM of human settlement (the built-up areas in MCD12Q1 for this case).

We added supplemental data to the IRM of human settlement, which were quickly prepared in 10 min by visual interpretation of the Landsat false color composite for respective target years (Figure 2.4). The result was improved by the addition of the supplemental data regardless of the very rough visual interpretation in a coarse scale for Landsat's spatial resolution. The results indicated that the system should have a function of collecting visual interpretation data that supplements omitted built-up areas in the MCD12Q1.

Figure 2.5 shows an example result indicating urban development in Metro Manila, the Philippines, between 1990 and 2010. The results detected coast reclamation constructed in the target period. While a major portion of the reclamation was completed before 2000, additional reclamation was newly constructed between 2005 and 2010 (Figure 2.5b). The result also depicted process of urban expansion between 1990 and 2010 for a city scale. For example, the result for Taguig City,

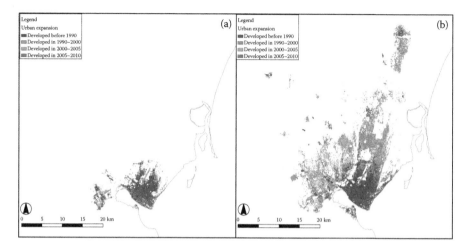

FIGURE 2.4 Improvement by supplemental data of initial built-up areas for Maputo, Mozambique. (a) Result without supplemental data; (b) result with supplemental data.

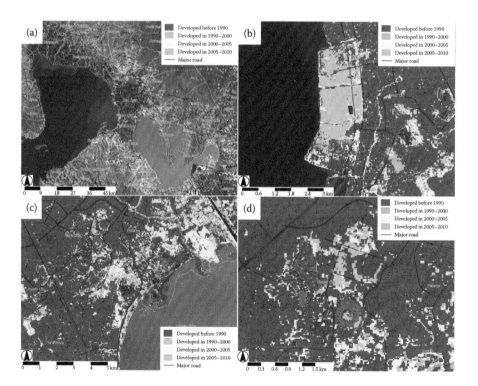

FIGURE 2.5 Examples of the results for Metro Manila, the Philippines. (a) Metro Manila; (b) land reclamation around Manila Bay; (c) urban developments around Taguig City; (d) parks and golf fields in Taguig City. The background images are from World Imagery in ArcGIS Online.

which has been rapidly developed since around 2003, indicated the progress of urban expansion around 2000 (Figure 2.5c). The result data also depicted urban structure of built-up and non–built-up areas owing to the 30-m resolution, which enabled to identify green areas in the city, such as parks and golf fields (Figure 2.5d).

We experimentally conducted an accuracy assessment of a result for Delhi, India, based on an error matrix (Foody 2002). Because of a limitation in preparation of reference data, the accuracy assessment was carried out only for 2000 and 2010. For 2000, overall accuracy was 98%, producer's accuracy was 15%, user's accuracy was 81%, and kappa coefficient was 0.25. For 2010, overall accuracy was 97%, producer's accuracy was 27%, user's accuracy was 61%, and kappa coefficient was 0.36. The results indicate that built-up areas are rather underestimated for 2000 and 2010 while the result for 1990 was visually assessed as underestimation of built-up areas as well as other cities presented above. For improvements of the method, further accuracy assessment is needed with soundly sampled cities to validate quality for the global coverage.

2.4 DEVELOPMENT OF ON-DEMAND SYSTEM

2.4.1 WEB-BASED MAPPING AND DATA PROCESSING STANDARDS

Web-based systems have major roles in systems that involve parameters from users interactively. Web Mapping Service (WMS; Open Geospatial Consortium 2006) and Web Feature Service (WFS; Open Geospatial Consortium 2010) are core technologies of web-based GIS, where users can interactively set geographical extent to browse map data products. In addition to WMS and WFS, Web Processing Service (WPS; Open Geospatial Consortium 2015) provides data publishers with frameworks to manage users' parameters, not limited to geographical extent, to process data products. A combination of these frameworks enables the systems to flexibly handle users' parameters of target date as well as geographical extent.

We designed an on-demand settlement mapping system using the frameworks of WPS shown as Figure 2.6. In the system, users can set parameters of geographical extent and target date through WPS. The parameters were passed to a metadata database of data products. If existing data products satisfy user requests with the parameters, WPS returns data products for the requested extent and year. If any existing data product does not satisfy the user's request, the parameters are passed to the settlement mapping algorithms to create the requested map products. The settlement

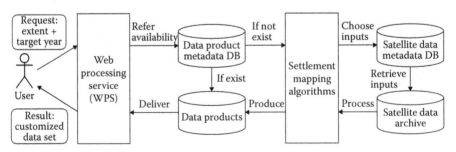

FIGURE 2.6 Overview of the on-demand settlement mapping system.

mapping algorithm refers to a metadata database of satellite data and retrieves the input data from satellite data archives. Once the data processing of settlement mapping is completed, the outputs are stored as a data product and registered in the metadata database of data products. While the data processing is undertaken, WPS watches the progress of data processing so that data products are delivered as soon as the data processing is completed. WPS helps automate these processes so that the system operator does not have to check the status of data availability and processing.

2.4.2   IMPLEMENTATION OF ON-DEMAND DATA PROCESSING SERVICE

The system was implemented using PyWPS (http://pywps.org/), a server software of WPS implemented by Python, with the database management systems based on PostgreSQL and PostGIS. The settlement mapping algorithms were implemented using open-source software as presented in Section 2.3.2. The metadata database of input satellite data was acquired from Landsat Bulk Metadata Service (US Geological Survey 2015a).

All the systems were constructed on a cloud service accessible web-based interface via web browsers so that the users are not required to install stand-alone software on their computers. The implemented system successfully delivered the data products based on availability of data products by users' parameters from GET parameters that are embedded in the URL of the WPS. The system can improve efficiency of data product management in disseminating customized data products because the system can save computing resources by avoiding data processing for unnecessary regions and target years as the system processes the data only for target years and regions of users' preferences.

By accumulating the users' requests with the parameters, the system can recognize popular target regions and time. The information will be useful to plan the development of ready-to-use data products by users' preferences. Such a system would facilitate interactions more efficiently and effectively between data providers and users on thematic data platforms such as TEP-Urban (Esch et al. 2016).

2.5   CONCLUSIONS

This chapter presented an experimental development of an on-demand time-series human settlement mapping system with a series of automated algorithms, which chose Landsat satellite data for users' preferences and identified human settlement from the chosen Landsat data. The data selection was structured by a scoring algorithm, which used quality indicators, especially cloud contamination in this experimental development, and observation dates to compare with users' preferences of target year. The selection algorithm was implemented with the USGS' Landsat metadata database. The chosen data were sent to the automated algorithm of human settlement mapping using the LLGC modified for remote sensing image data. The algorithms were applied to 838 WRS2 tiles of Landsat data for 1990, 2000, 2005, and 2010. The results effectively showed maps of human settlements because of the robustness strengthened by combining satellite data of multiple dates. The quality needs to be validated in statistical ways, such as accuracy assessment using an error matrix, to identify priorities to improve the systems. Problems in the IRMs were identified because of discrepancies in years of initial reference data and Landsat satellite data for mapping. In the experiments, additional reference data improved the LLGC results even when the data were prepared quickly in a very rough scale. The result of adding rough reference data indicated that the system should have a function to collect reference data of human settlement areas missed in the LLGC results.

We developed an on-demand system, which processes human settlement maps according to users' preferences. The developed system used the WPS, which enabled user input, such as target date for an output of settlement mapping. Because the WPS is a standard to implement this sort of service for other web mapping standards, such as WMS and WFS, the developed system is expected to be easily extended to end-user interface with WebGIS.

In conclusion, the following future works are suggested:

1. Improvement of the methodology by analysis of consistency among time-series outputs
2. Data collection system for error reports of missing settlement areas
3. Development of a user-friendly interface using WPS with integration of WMS and WFS, especially for non-GIS users and non-web developers
4. Combinations of other data resources that recently became publicly available, such as ASTER and Sentinel series, for better quality and higher frequency
5. Applications of high-frequency satellite data for finer time scales, such as using data from constellation satellites for monthly data products

Through the further development of the system, more practical applications of the satellite-based human settlement mapping will be expected, such as contributions to the achievements of the Sustainable Development Goals.

ACKNOWLEDGMENTS

This work was supported by the Green Network of Excellence Environmental Information project (GRENE-ei, 2011–2016) of Japan's Ministry of Education, Culture, Sports, Science, and Technology.

REFERENCES

Angel, Shlomo, Stephen C. Sheppard, and Daniel L. Civco. 2005. *The Dynamics of Global Urban Expansion*. Washington D.C.: The World Bank.
Balk, Deborah, Adam Storeygard, Marc Levy, Joanne Gaskell, Manohar Sharma, and Rafael Flor. 2005. Child hunger in the developing world: An analysis of environmental and social correlates. *Food Policy* 30 (5):584–611.
Bartholome, Etienne, and Allan S. Belward. 2005. GLC2000: A new approach to global land cover mapping from Earth observation data. *International Journal of Remote Sensing* 26 (9):1959–1977.
Bicheron, Patrice, Mireille Huc, Caroline Henry, Sophie Bontemps, and Globcover Partners. 2008. GLOBCOVER: Products Description Manual.

Brooker, Simon, Archie C. A. Clements, Peter J. Hotez, Simon I. Hay, Andrew J. Tatem, Donald A. P. Bundy, and Robert W. Snow. 2006. The co-distribution of *Plasmodium falciparum* and hookworm among African schoolchildren. *Malaria Journal* 5 (1):99–99.

Center for International Earth Science Information Network, University Columbia, International Food Policy Research Institute, The World Bank, and Centro Internacional de Agricultura Tropical. 2004. Global Rural–Urban Mapping Project (GRUMP), Alpha Version: Urban Extents. Socioeconomic Data and Applications Center (SEDAC), Columbia University.

Dasgupta, Susmita, Benoit Laplante, Craig Meisner, David Wheeler, and Jianping Yan. 2009. The impact of sea level rise on developing countries: A comparative analysis. *Climatic Change* 93 (3):379–388. doi: 10.1007/s10584-008-9499-5.

Doocy, Shannon, Yuri Gorokhovich, Gilbert M. D. Burnham, Deborah Balk, and Courtland Robinson. 2007. Tsunami mortality estimates and vulnerability mapping in Aceh, Indonesia. *American Journal of Public Health* 97 (S1):S146–S146.

Duan, Yulin, Xiaowei Shao, Yun Shi, Hiroyuki Miyazaki, Koki Iwao, and Ryosuke Shibasaki. 2015. Unsupervised global urban area mapping via automatic labeling from ASTER and PALSAR satellite images. *Remote Sensing* 7 (2):2171–2192.

Esch, Thomas, Hubert Asamer, Martin Boettcher, Fabrice Brito, Andreas Hirner, Mattia Marconcini, Emmanuel Mathot, Annekatrin Metz, Hans Permana, Tomas Soukop, Filip Stanek, Stepan Kuchar, Julian Zeidler, and Jakub Balhar. 2016. Earth observation-supported service platform for the development and provision of thematic information on the built environment—The TEP-Urban Project. *Int. Arch. Photogramm. Remote Sens. Spatial Inf. Sci.*, XLI-B8, 1379–1384. doi: 10.5194/isprs-archives-XLI-B8-1379-2016.

Esch, Thomas, Mattia Marconcini, Andreas Felbier, Achim Roth, Wieke Heldens, Martin Huber, Maximilian Schwinger, Hannes Taubenbock, Andreas Muller, and Stefan Dech. 2013. Urban footprint processor—Fully automated processing chain generating settlement masks from global data of the TanDEM-X mission. *Geoscience and Remote Sensing Letters, IEEE* 10 (6):1617–1621. doi: 10.1109/LGRS.2013.2272953.

Esch, Thomas, Mattia Marconcini, Dimitrios Marmanis, Julian Zeidler, Sherif Elsayed, Annekatrin Metz, Andreas Müller, and Stefan Dech. 2014. Dimensioning urbanization—An advanced procedure for characterizing human settlement properties and patterns using spatial network analysis. *Applied Geography* 55:212–228. doi: http://dx.doi.org/10.1016/j.apgeog.2014.09.009.

European Commission Joint Research Centre. 2014. Global Human Settlement Layer. Accessed October 29. http://ghslsys.jrc.ec.europa.eu/.

Foley, Jonathan A., Ruth DeFries, Gregory P. Asner, Carol Barford, Gordon Bonan, Stephen R. Carpenter, F. Stuart Chapin, Michael T. Coe, Gretchen C. Daily, Holly K. Gibbs, Joseph H. Helkowski, Tracey Holloway, Erica A. Howard, Christopher J. Kucharik, Chad Monfreda, Jonathan A. Patz, I. Colin Prentice, Navin Ramankutty, and Peter K. Snyder. 2005. Global consequences of land use. *Science* 309 (5734):570–574. doi: 10.1126/science.1111772.

Foody, Giles M. 2002. Status of land cover classification accuracy assessment. *Remote Sensing of Environment* 80 (1):185–201.

Friedman, Jerome, Trevor Hastie, and Robert Tibshirani. 2000. Special invited paper. Additive logistic regression: A statistical view of boosting. *The Annals of Statistics* 28 (2):337–374.

Global Land Cover Facility. 2014. Global Land Survey (GLS).

Haykin, Simon. 1998. *Neural Networks: A Comprehensive Foundation*. 2nd ed. Delhi: Prentice Hall.

Ju, Junchang, and David P. Roy. 2008. The availability of cloud-free Landsat ETM+ data over the conterminous United States and globally. *Remote Sensing of Environment* 112 (3):1196–1211.

Koeln, Gregory T., Thomas B. Jones, and Jeannine E. Melican. 2000. GeoCover LC: Generating global land cover from 7600 frames of Landsat TM data. In *Proceedings of ASPRS 2000 Annual Conference*, Washington, D.C., USA.

Koike, Toshio, Petra Koudelova, Patricia Ann Jaranilla-Sanchez, Asif Mumtaz Bhatti, Cho Thanda Nyunt, and Katsunori Tamagawa. 2015. River management system development in Asia based on Data Integration and Analysis System (DIAS) under GEOSS. *Science China Earth Sciences* 58 (1):76–95. doi: 10.1007/s11430-014-5004-3.

Land Processes Distributed Active Archive Center. 2014. Land Cover Type Yearly L3 Global 500 m SIN Grid.

Mease, David, and Abraham Wyner. 2008. Evidence contrary to the statistical view of boosting. *Journal of Machine Learning Research* 9:131–156.

Miyazaki, Hiroyuki, Masahiko Nagai and Ryosuke Shibasaki. 2016. An automated method for time-series human settlement mapping using Landsat data and existing land cover maps. In *2016 IEEE International Geoscience and Remote Sensing Symposium (IGARSS)*, Beijing, China.

Miyazaki, Hiroyuki, Xiaowei Shao, Koki Iwao, and Ryosuke Shibasaki. 2014. Development of a global built-up area map using ASTER satellite images and existing GIS data. In *Global Urban Monitoring and Assessment through Earth Observation*, 121–142. CRC Press.

Montgomery, Mark R. 2008. The urban transformation of the developing world. *Science* 319 (5864):761–764.

Nelson, Gerald Charles, and Richard D. Robertson. 2007. Comparing the GLC2000 and GeoCover LC land cover datasets for use in economic modelling of land use. *International Journal of Remote Sensing* 28 (19):4243–4262.

Ng, Andrew Y., Michael I. Jordan, and Yair Weiss. 2002. On spectral clustering: Analysis and an algorithm. *Advances in Neural Information Processing Systems 14 (NIPS 2001)*.

Omumbo, Judy A., Carlos A. Guerra, Simon I. Hay, and Robert W. Snow. 2005. The influence of urbanisation on measures of *Plasmodium falciparum* infection prevalence in East Africa. *Acta Trop.* 93 (1):11–21.

Open Geospatial Consortium. 2006. OpenGIS Web Map Service (WMS) Implementation Specification. Accessed June 17. http://portal.opengeospatial.org/files/?artifact_id=14416.

Open Geospatial Consortium. 2010. OpenGIS Web Feature Service 2.0 Interface Standard (also ISO 19142). Accessed June 17. http://portal.opengeospatial.org/files/?artifact_id=39967.

Open Geospatial Consortium. 2015. OGC® WPS 2.0 Interface Standard. Accessed 2 February. http://docs.opengeospatial.org/is/14-065/14-065.html.

Pesaresi, Martino, Daniele Ehrlich, Stefano Ferri, A. Florczyk, Sergio Freire, Matina Halkia, Andreea Julea, Thomas Kemper, Pierre Soille, and Vasileios Syrris. 2015. Operating procedure for the production of the Global Human Settlement Layer from Landsat data of the epochs 1975, 1990, 2000, and 2014. JRC Technical Report.

Potere, David, Annemarie Schneider, Shlomo Angel, and Daniel L. Civco. 2009. Mapping urban areas on a global scale: Which of the eight maps now available is more accurate? *International Journal of Remote Sensing* 30 (24):6531–6558.

Schneider, Annemarie, Mark A. Friedl, and David Potere. 2009. A new map of global urban extent from MODIS satellite data. *Environmental Research Letters* 4 (4):44003–44003.

Schneider, Annemarie, Mark A. Friedl, and David Potere. 2010. Mapping global urban areas using MODIS 500-m data: New methods and datasets based on 'urban ecoregions'. *Remote Sensing of Environment* 114 (8):1733–1746. doi: 10.1016/j.rse.2010.03.003.

Schneider, Annemarie, Karen C. Seto, Douglas R. Webster, Jianming Cai, and Binyi Luo. 2003. Spatial and temporal patterns of urban dynamics in Chengdu, 1975–2002. Stanford, California: Asia Pacific Research Center, Stanford University.

Schneider, Annemarie, and Curtis E. Woodcock. 2008. Compact, dispersed, fragmented, extensive? A comparison of urban growth in twenty-five global cities using remotely sensed data, pattern metrics and census information. *Urban Studies* 659–692. doi: 10.1177/0042098007087340 45.

Small, Christopher. 2005. A global analysis of urban reflectance. *International Journal of Remote Sensing* 26 (4):661–681.

Taubenböck, Hannes, Michael Wiesner, Andreas Felbier, Mattia Marconcini, Thomas Esch, and Stefan Dech. 2014. New dimensions of urban landscapes: The spatio-temporal evolution from a polynuclei area to a mega-region based on remote sensing data. *Applied Geography* 47 (0):137–153. doi: http://dx.doi.org/10.1016/j.apgeog.2013.12.002.

United Nations. 2015. Sustainable Development Goal 11—Make cities and human settlements inclusive, safe, resilient and sustainable. Accessed April 3. https://sustainabledevelopment.un.org/sdg11.

US Geological Survey. 2015a. Landsat Bulk Metadata Service.

US Geological Survey. 2015b. Landsat Data Access.

von Luxburg, Ulrike. 2007. A tutorial on spectral clustering. *Statistics and Computing* 17 (4):395–416. doi: 10.1007/s11222-007-9033-z.

Yang, Yuanyuan, Shuwen Zhang, Yansui Liu, Xiaoshi Xing, and Alex de Sherbinin. 2017. Analyzing historical land use changes using a historical land use reconstruction model: A case study in Zhenlai County, northeastern China. *Scientific Reports* 7:41275. doi: 10.1038/srep41275.

Zhou, Dengyong, Olivier Bousquet, Thomas Navin Lal, Jason Weston, and Bernhard Schölkopf. 2003. Learning with local and global consistency. *Advances in Neural Information Processing Systems* 16:321–328.

3 Morphological Building Index (MBI) and Its Applications to Urban Areas

Xin Huang and Tao Zhang

CONTENTS

3.1 INTRODUCTION

The precise location and extraction of buildings is one of the most important information sources for urban management, population estimation, and environment assessment. The advent of high-resolution imagery provides more potential for automatic and accurate building detection. Although the high-resolution data contain abundant information in the spatial domain for building identification, it poses great challenges to the traditional information extraction methods. Because of the complicated urban scenes in high-resolution imagery, the conventional spectral-based methods are inadequate for discrimination between buildings and other spectrally similar classes, such as roads, parking lots, and open areas. In this context, a large number of methods have been proposed considering the spatial information (e.g., structure and context) as a supplement for building detection and verification. The previous studies focusing on building detection can be divided into the following two main categories. Supervised machine-learning methods are among the most widely used approaches that integrate spectral and spatial information for building extraction. Some commonly used spatial features include differential morphological profiles

(DMPs) (Pesaresi and Benediktsson 2001), pixel shape index (Zhang et al. 2006), and the gray-level co-occurrence matrix (Haralick 1979). It should be mentioned that the supervised methods for building detection are subject to training samples collection and the time-consuming process of machine learning. In view of this, some automated methods were proposed for building detection from high-resolution data, using spectral–spatial priors (Sirmacek and Unsalan 2009), marked point process models (Benedek et al. 2010), and semantic-based building/shadow symbiosis (Ok 2013). In the meantime, a number of building indexes have also been proposed to indicate the presence of buildings, which can support the automatic processing of massive high-resolution data. For instance, a procedure for the calculation of a texture-based built-up presence index (PanTex) (Pesaresi et al. 2008) was proposed based on the assumption that buildings and their spatially adjacent shadows can lead to high local contrast. Two improved versions of PanTex (Pesaresi and Gerhardinger 2011) have been presented by considering the vegetation components or highlighting the morphological characteristics of built-up structures. Nevertheless, 5-m spatial resolution is considered optimal for the discrimination of built-up areas using the textural method.

More recently, Huang and Zhang (2011) proposed a morphological building index named MBI for automatic building detection, which has been demonstrated to be an effective morphological metric to extract building structures from high-resolution imagery. It aims to describe the spectral–spatial characteristics of buildings (e.g., brightness, contrast, size, and directionality) based on a set of morphological operations (e.g., top-hat by reconstruction, granulometry, and directionality). Nevertheless, the original MBI is subject to false alarms related to bright soils and open areas since they are usually brighter than their surroundings and present similar spectral properties as buildings. In addition, some omission errors can be found with respect to the heterogeneous and dark roofs. To deal with such problems, the morphological shadow index (MSI) considering the building/shadow symbiosis, as well as a dual-threshold method (Huang and Zhang 2012a), has been proposed as an improved version of the original MBI to alleviate both commission and omission errors. The integration of MBI and MSI performs well in urban areas with high building density. To extend its applicability to large-area high-resolution imagery with complex scenes (e.g., suburban, rural, mountainous, and agricultural areas), recently an enhanced building index (EBI) has been developed to improve MBI detector by simultaneously considering the spectral, contextual, and shape information, which can be regarded as a systematic postprocessing framework aligned with MBI to extract buildings in more complex and challenging environments (Huang et al. 2016b). The original MBI and its variants have been verified as effective building detectors without statistical learning and training samples, and they are fast and easy to implement in a large area for applications in both urban and nonurban areas. In current studies, MBI has been widely used in many researches related to urban land-cover classification and change detection (Huang et al. 2014a; Wen et al. 2016). It is noteworthy that MBI, which served as an automatic building extraction method, has a promising potential to be applied further in the field of urban remote sensing associated with building components (e.g., seismic vulnerability of a building) as it can provide accurate building information, which is one of the most important and basic geographic data sources for urban management. In this chapter, we will summarize three versions of MBI that have been published in international journals and discuss their applications to urban areas.

3.2 THE ORIGINAL MBI

The basic idea of MBI is to build the relationship between the intrinsic properties of the buildings (e.g., brightness, contrast, size, and directionality) and a set of morphological operators (Huang and Zhang 2011). The construction of MBI considers the following characteristics of buildings.

1. *Local Contrast:* The buildings with relative high reflectance and the spatially adjacent shadows lead to a high local contrast. Therefore, the white top-hat (WTH) transformation, which has the ability to highlight bright structures with high local contrast, is used to represent the characteristic of buildings.
2. *Size and Directionality:* A major challenge for designing a building index is to automatically remove roads that have similar spectral properties to buildings. Unlike roads, which are generally elongated structures, buildings present as more isotropic with a limited size. Accordingly, MBI is implemented based on a series of linear structural elements, which can characterize the size and directionality of buildings.

The principle of the MBI is shown in Figure 3.1, and the calculation of the MBI is briefly introduced as follows. First, the maximum value of the visible bands for each pixel is recorded as the brightness image. The brightness image is focused on as buildings often have relatively higher reflectance than their surrounding pixels. Second, multiscale and multidirectional WTH transformations are implemented to generate the DMPs, which can characterize the spatial patterns of buildings in different scales and directions. Finally, the DMP-WTH profiles are integrated by

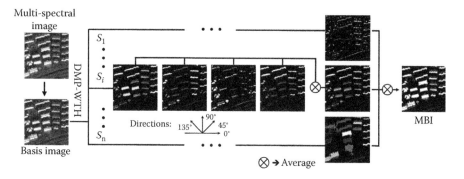

FIGURE 3.1   The flowchart of the MBI calculation.

averaging operators to emphasize the presence of buildings. Therefore, the formulation of the MBI is defined as

$$\text{MBI} = \frac{\displaystyle\sum_{s \in S} \sum_{d \in D} (\text{DMP-WTH}(s,d))}{N_s \times N_D}, \tag{3.1}$$

where DMP-WTH is the WTH-based DMPs, s and d represent the scale and direction of the WTH transformation, and N_S and N_D are the total number of scales and directions, respectively. Since the buildings present relative isotropic characteristic, that is, show high local contrast in different directions, they often correspond to larger MBI feature values.

In order to evaluate the effectiveness of MBI, the GeoEye-1 imagery with four spectral bands of 2-m spatial resolution was used in this study. The image covers the main part of Wuhan in central China and includes a set of representative urban land cover types such as buildings, roads, lakes, bare soils, forests, and grasslands. Four subregions with detailed field surveying are selected for validation. For comparison purposes, another built-up index, namely, PanTex (Pesaresi et al. 2008), as well as a multilevel object-based approach (Tian and Chen 2007), is implemented in the study. Four quantitative measures including omission error (OE), commission error (CE), overall accuracy (OA), and kappa coefficient (Congalton 1991), were used to evaluate the results of building extraction. The number of reference samples and the accuracies of building detection for different algorithms are shown in Table 3.1. It can be obviously seen that the MBI achieves more accurate results than PanTex and the object-based approach. In the meantime, the extracted building maps of the four validation areas are shown in Figure 3.2 for a visual inspection. The results obtained

TABLE 3.1
Accuracies of Building Extraction for Different Methods

| Region | # Test Samples | | Method | Accuracy | | | |
	Buildings	Background		OE (%)	CE (%)	OA (%)	Kappa
A	11,198	11,051	PanTex	41.6	18.8	72.2	0.446
			Object-based	40.1	23.8	70.3	0.407
			MBI	11.4	2.4	93.1	0.862
B	14,332	14,068	PanTex	41.4	19.4	71.9	0.440
			Object-based	38.8	6.0	78.4	0.569
			MBI	6.9	2.1	95.5	0.910
C	15,650	15,301	PanTex	18.4	6.7	87.6	0.753
			Object-based	37.6	7.4	78.4	0.569
			MBI	7.6	2.0	95.1	0.903
D	12,428	14,690	PanTex	67.5	33.0	61.6	0.196
			Object-based	34.0	13.6	79.6	0.581
			MBI	18.2	6.4	89.0	0.777

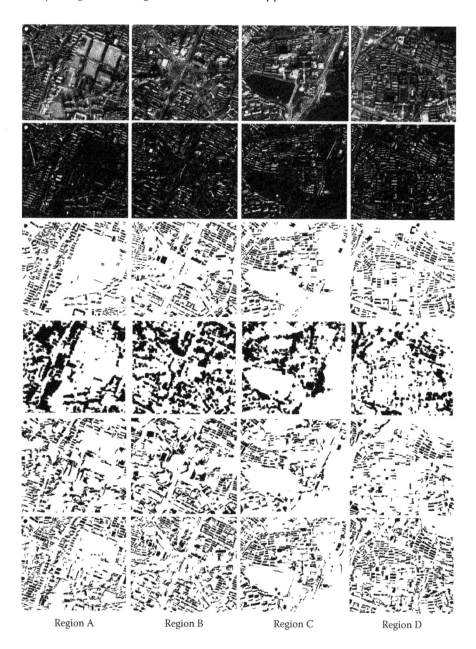

Region A   Region B   Region C   Region D

FIGURE 3.2 Results of building detection for the four validation areas (white = backgrounds and black = buildings). Rows 1, 2, and 3 represent brightness, MBI, and the manually delineated ground truth maps, respectively. Rows 4, 5, and 6 represent the building maps extracted by PanTex, object-based approach, and MBI, respectively.

by the object-based approach are acceptable. However, the inaccurate segmentation (e.g., over- and under-segmentation) may make it difficult to choose suitable shape attributes for identification of a building object. Although PanTex is able to indicate the presence of buildings, it fails to accurately locate their boundaries because of the predefined window size, which is a key parameter for the construction of the index, and may not favor all the building structures with different scales. In general, the MBI algorithm provides the most accurate results according to the visual inspection. Because of the multiscale and multidirectional profiles considered in the MBI calculation, buildings with different scales are effectively detected.

3.3   MORPHOLOGICAL BUILDING/SHADOW INDEX

Although the original MBI performs well in building detection, it is also subject to some commission and omission errors. The commission errors are associated with bright soils and open areas since they often show similar spectral reflectance to buildings and present higher values than their neighborhoods. On the other hand, the omission errors mainly refer to heterogeneous and dark roofs. To alleviate both commission and omission errors for the original MBI, an MSI (Huang and Zhang 2012a) is proposed to detect shadows that are used as a spatial constraint of buildings. MSI can be regarded as a twinborn index of MBI since shadows generally show darker spectral reflectance but similar spatial characteristics to buildings. Accordingly, the principle of MSI is also based on the relationship between the spectral–structural characteristics of shadows, and the calculation of MSI can be extended from MBI by replacing the WTH with the black top-hat (BTH), which has ability to represent the local contrast of shadows. Accordingly, Equation 3.1 can be converted to

$$\text{MSI} = \frac{\sum_{s \in S} \sum_{d \in D} (\text{DMP-BTH}(s,d))}{N_s \times N_D}. \tag{3.2}$$

In Figure 3.3, it can be seen that MBI mainly represents the buildings and the shadows are highlighted in MSI feature image. By integrating the information provided by MBI and MSI, the co-occurrence relationship of buildings and shadows can be characterized, which leads to more effective building extraction from high-resolution imagery. Shadows are spatially adjacent or close to buildings, and hence, the distance to shadows is used as a spatial constraint of buildings. As a result, some commission errors such as bright soils and open areas can be removed. To reduce omission errors caused by heterogeneous and dark roofs, a dual-threshold method is designed to deal with such problem. Specifically, the MBI feature image is separated into low-MBI and high-MBI regions. With respect to low-MBI regions, a relatively strong constraint (small threshold) on the distance between buildings and shadows is applied while a relatively weak constraint (large threshold) is imposed on the high-MBI regions. This dual-threshold processing is able to alleviate omission errors corresponding to buildings with dark roofs.

The experiment was performed on a WorldView-2 image over Hangzhou located in east of China to test the effectiveness of the improved MBI (i.e., the integration of

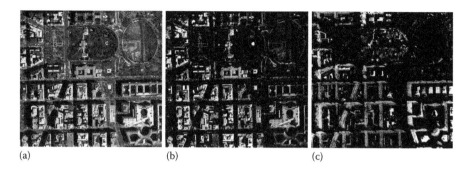

(a)   (b)   (c)

FIGURE 3.3   (a), (b), and (c) are the original image, MBI, and MSI feature images, respectively.

MBI and MSI). The accuracies derived by different methods are reported in Table 3.2, and the generated building maps are displayed in Figure 3.4. It can be seen that a small MBI threshold (T) leads to small omission but large commission errors, and a large value results in small commission but large omission errors. It is usually difficult to find a simple threshold to simultaneously achieve small omission and commission errors. However, the proposed dual-threshold approach effectively addresses this problem with the spatial constraint of shadows. Specifically, the supervised machine-learning method is also conducted for comparison. In this case, two categories of spatial features are fed into the SVM classifier to discriminate between building and non-buildings. The first is the multiscale and multidirectional DMPs that are used to generate the MBI and MSI. The second is a recently developed multiscale urban complexity index (MUCI) (Huang and Zhang 2012b), which is built on the 3D wavelet transform and can be used to discriminate different land covers such as buildings, roads, and vegetation. In this case, 50 building pixels and 300 non-building pixels are randomly selected from the reference data for SVM training. Comparing the results, the DMP-SVM does not perform well, while the MUCI-SVM achieves acceptable results since it outperforms all the simple-threshold MBI method. Overall, it can be stated that the integration of MBI and MSI gives satisfactory performances and is a more suitable option for building detection without training samples, especially for the large-area high-resolution remote sensing imagery.

TABLE 3.2

Accuracies of Building Extraction for Different Methods

# Test Samples			Accuracy			
Buildings	**Background**	**Method**	**OE (%)**	**CE (%)**	**OA (%)**	**Kappa**
49,184	43,439	MBI ($T = 0.5$)	15.6	8.5	87.5	0.751
		MBI ($T = 2$)	56.7	0.4	69.8	0.415
		DMP-SVM	36.7	3.7	79.2	0.592
		MUCI-SVM	18.9	4.6	87.9	0.760
		MBI-MSI	16.1	0.8	91.1	0.823

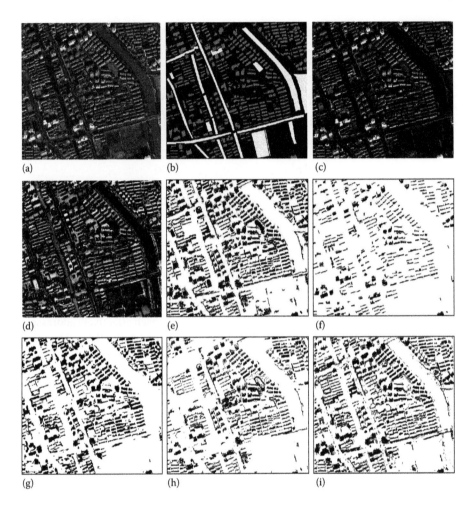

FIGURE 3.4 Results of building extraction for WorldView-2 Hangzhou data set: (a) and (b) are the RGB image and the reference buildings; (c) and (d) show the MBI and MSI feature images, respectively; (e) and (f) are generated by the original MBI algorithm with $T_B = 0.5$ and 2, respectively; (g) is the result of the proposed method; (h) and (i) present the building maps extracted by DMP-SVM and MUCI-SVM, respectively.

3.4 ENHANCED BUILDING INDEX

The effectiveness of the MBI has been verified by a lot of researches (Huang et al. 2014b; Tang et al. 2013) in high building density urban areas. Nevertheless, the aforementioned methods based on the MBI did not consider the difficulty of building detection in suburban, mountainous, agricultural, and rural areas. When dealing with images composed of complex scenes, MBI is mainly subject to commission errors when bright soil and wetland present a higher intensity than their neighborhoods.

The phenomena can be attributed to the fact that the traditional MBI does not make full use of the spectral and contextual information to verify the confusing objects with similar spectral and morphological properties. To address such problems, a systematic postprocessing framework is proposed to describe the characteristics of buildings by simultaneous consideration of the spectral, geometrical, and contextual information, which can be successfully applied to large high-resolution imagery (Huang et al. 2016b). The proposed framework aims to construct an EBI, which improves the MBI detector by a set of constraints (e.g., spectral, shadow, and shape).

1. *Spectral Constraints:* Bright vegetation and soil are the main source of commission errors for building detection. In the proposed framework, the normalized difference vegetation index (NDVI), and the hue component (H) of the images are integrated to reduce such false alarms. Note that some buildings with blue roofs also show relatively high NDVI values. Therefore, both NDVI and hue are jointly considered to remove the bright vegetation and soil while the blue buildings with high NDVI value can be preserved (Figure 3.5). In addition, some water bodies due to eutrophication can also result in commission errors, which can be removed by the normalized difference water index (NDWI) (Figure 3.6).
2. *Shadow Verification and Shape Constraints:* Shadow and shape (e.g., the area and length–width ratio) can serve as constraints for building detection to remove false alarms, such as soil, parking lots, and roads. If the candidate

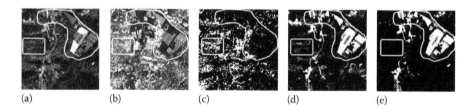

(a)   (b)   (c)   (d)   (e)

FIGURE 3.5   An example of vegetation/soil elimination: (a) RGB; (b) NDVI; (c) binary hue image; (d) original MBI; and (e) MBI feature after vegetation/soil elimination.

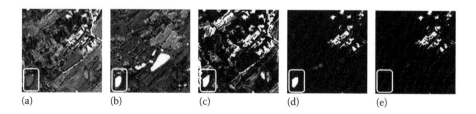

(a)   (b)   (c)   (d)   (e)

FIGURE 3.6   An example of water body elimination: (a) RGB; (b) NDWI; (c) MBI; (d) MBI with vegetation/soil elimination; and (e) MBI with spectral constraint (including vegetation/soil and wetland elimination).

buildings are spatially adjacent to the shadows they cast, the detected buildings can be retained; otherwise, they will be removed. Similarly, the small errors such as noises and narrow and elongated roads can be removed by shape constraint (Figure 3.7).

To quantitatively evaluate the EBI for building detection, the statistical measures, including completeness (Com), correctness (Cor), and quality (Q), were used to assess the accuracies of the building maps. Completeness is the proportion of correctly detected buildings to the total number of reference buildings. Correctness is the percentage of correctly detected buildings to the total number of detected buildings (Song and Haithcoat 2005). An ideal detection should simultaneously have high completeness and correctness. The quality metric provides a trade-off considering both completeness and correctness (Rutzinger et al. 2009). In the meantime, the SVM-DMP method is added for comparison, and the building maps derived by different methods are also displayed in Figures 3.8 and 3.9. The experiments were conducted on two test images including Pangzhihua and Harbin, corresponding to mountainous and rural/agricultural areas, respectively, which pose great challenges to building extraction. Because of the complexity of the challenging scenes, the original MBI is subject to significant false alarms in the non-urban areas. However, according to Table 3.3, EBI can achieve promising performances in these areas and present superiority to the original MBI in both completeness and correctness. With regard to DMP-SVM, although it gives acceptable results

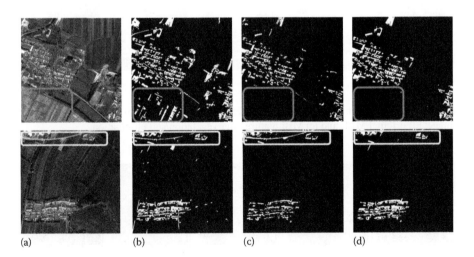

(a)   (b)   (c)   (d)

FIGURE 3.7 Building detection results and the ground reference for the representative test patches: (a) original RGB images with ground reference; (b) the results of the MBI; (c) the results of DMP-SVM; and (d) the results of the EBI. The subregions emphasize the performances of the shadow (first row) and shape constraints (second row).

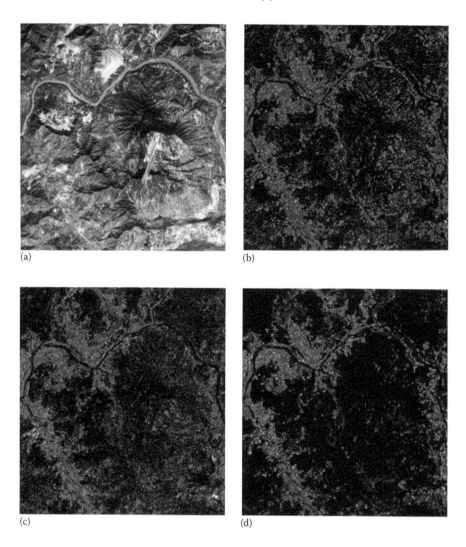

FIGURE 3.8 Building detection results for Pangzhihua: (a) original RGB images; (b) building maps with the result of the MBI; (c) building maps with the result of DMP-SVM; and (d) building maps with the result of the proposed framework. The green color represents the boundary between the high-density and low-density areas.

for building detection, it requires intensive manual labor since additional training samples need to be manually collected. To further compare EBI and DMP-SVM, the computational efficiency (computation time) is also recorded in Table 3.3. It can obviously be seen that EBI can be implemented much more rapidly than DMP-SVM despite ignoring the time of training sample collection. Overall, the results confirm that the EBI is effective for building detection from high-resolution imagery composed of complex scenes.

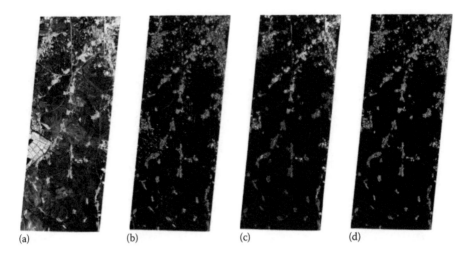

(a)   (b)   (c)   (d)

FIGURE 3.9 Building detection results for Harbin: (a) original RGB images; (b) building maps with the result of the MBI; (c) building maps with the result of DMP-SVM; and (d) building maps with the result of the proposed framework. The green color represents the boundary between the high-density and low-density areas.

TABLE 3.3
Reference Information, Accuracy (%), and Computation Times (Seconds) of the Study Data Sets

Study Area	# Test Samples		Method	Measures			
	Buildings	Background		Com	Cor	Q	Time
Pangzhihua	112,780	916,871	MBI	83.65	72.74	63.69	140.6
			DMP-SVM	87.81	73.91	67.03	348.5
			EBI	89.40	91.49	82.53	174.1
Harbin	77,317	365,455	MBI	86.74	80.25	71.48	44.5
			DMP-SVM	86.45	83.78	74.06	130.2
			EBI	89.58	96.07	86.41	52.6

3.5   APPLICATIONS OF MBI

In the current research community, MBI, as well as its improved versions, has been extensively applied in various fields associated with building components (e.g., building extraction, change detection, image classification, and urban environment application) since it can provide accurate building information with high efficiency when applied to high-resolution remote sensing imagery (Hou et al. 2016; Rubeena and Tiwari 2016). The main applications of MBI in existing literature can be summarized as follows.

3.5.1 BUILDING EXTRACTION

MBI, regarded as an effective building indicator, has been applied to building detection from high-resolution remote sensing images in many studies. Syrris et al. (2015) investigated the sensitivity of contrast-based textural measurements (PanTex) and morphological characteristics (MBI) using remote sensing imagery when diverse image enhancement techniques are implemented. The article pointed out that MBI is a quite accurate building detection index and not based on statistical learning and training samples. In the experiments, MBI achieves results comparable to or even better than those of PanTex. Furthermore, PanTex is strongly dependent on the contrast of the image while MBI performs better with a moderate contrast level that demonstrates the effectiveness of MBI in building detection. Recently, some researchers also improve the original MBI to extract buildings more effectively. Zhang et al. (2016) proposed an efficient framework for building detection over urban areas using morphological technique. The preliminary result of buildings is derived by MBI, which is able to provide potential building structures. Then, a morphological component classification (e.g., core, branch, and bridge) approach is used to reduce some commission errors such as small noisy patches. Moreover, the shape of the buildings can also be regularized by removing the branches, and the holes in the building objects can be identified and filled, leading to more accurate building detection results. In addition, with respect to MBI and MSI, the computational complexity increases as a result of the use of multiscale and multidirectional morphological operators. A computationally efficient implementation of MBI/MSI using graphic processing units was proposed, which significantly reduces the computational burden when dealing with building/shadow detection in large-area high-resolution remote sensing images (Jiménez et al. 2016).

3.5.2 CHANGE DETECTION

Currently, MBI has been successfully applied in the field of change detection. Tang et al. (2013) proposed a building change detection method from high-resolution images over urban areas, which is able to suppress the errors caused by the geometrical differences in multi-temporal images. The interest points of buildings were extracted by integrating MBI and the Harris detector, and the corresponding buildings in the multi-temporal images are matched within a local domain. The proposed method aims to compare buildings and detect potential changes by a local building matching, which is fault tolerant to the geometrical differences of buildings in multi-temporal images. Huang et al. (2014c) proposed a set of novel building change indexes by using MBI and slow feature analysis for building change detection from remote sensing images, in which MBI is served to focus the change detection on building structures. Xiao et al. (2017) adopted MBI and spectral features to calculate the magnitude of the difference in multi-temporal images that can represent the change feature of buildings for subsequent building change detection. In addition, combining MBI and other semantic indexes (e.g., MSI, NDVI, and NDWI), the complicated high-resolution scenes can be represented by low-dimensional data space. Inspired by such idea, an innovative automatic change detection method based on the multi-index image representation has been proposed for urban high-resolution remote sensing imagery (Wen et al. 2016). Specifically, urban primitives, for example, buildings, vegetation, and water, can be automatically extracted by MBI, NDVI, and NDWI, respectively. In this way, change detection can be implemented by measuring the similarity of multi-index histograms (i.e., frequency and spatial arrangement of the urban primitives) between multi-temporal images.

3.5.3   IMAGE CLASSIFICATION

MBI, which served as an effective indicator for building components, has been widely used in remote sensing image classification. Huang et al. (2014a) proposed a novel multi-index learning (MIL) method for high-resolution image classification over urban areas. Two categories of indexes, including primitive indexes (MBI, MSI, and NDVI) and variation indexes based on 3D wavelet transformation, are extracted to describe urban landscapes, which significantly reduces the dimensionality of feature space. The MIL method achieves promising results with a low-dimensional feature space and provides a practical strategy for large-area high-resolution image processing. Recently, Huang et al. (2015b) proposed a novel method for automatic sample selection and labeling for image classification in urban areas. Considering that some automatic information indexes can successfully indicate different land-cover classes, it is possible to develop an automatic method for labeling the training samples rather than manual interpretation. In this study, MBI and MSI are used to generate high-quality training samples for buildings and shadows, respectively. Combined with other information indexes such as NDVI and NDWI, the proposed method can provide a large number of reliable samples with high efficiency and achieve satisfactory results for urban land-cover classification. In addition, Liao et al. (2015) reported the outcomes of the 2014 Data Fusion Contest organized by IEEE Geoscience and Remote Sensing Society (IEEE GRSS). Participants aim to conduct accurate land-cover classification using remote sensing imagery acquired at 20-cm resolution and thermal hyperspectral data at 1-m resolution. The winners of the contest use MBI as one of important features to feed into a classifier, and with some postprocessing steps, the classification results achieve the most accurate performance.

3.5.4   URBAN ENVIRONMENT

MBI has the ability to automatically extract accurate buildings from high-resolution images, which can be further applied in urban environment applications. Recently, the National Agency for Research, a French institution tasked with funding scientific research, released a report related to the evaluation of the seismic vulnerability of the buildings (Chiancone et al. 2013). In the report, MBI plays an important role and provides the accurate parameters of the buildings (e.g., locations and boundaries), which helps characterize the seismic vulnerability of the buildings. Huang et al. (2015a) focused on the urban villages (villages engulfed by ever-expanding urban areas) in mega city regions of China. In general, urban villages are composed of high-density buildings with much less open spaces compared to formal settlements. The features of buildings in urban villages can be characterized by MBI. Based on such idea, a novel multi-index scene model was proposed to analyze the spatiotemporal patterns of urban villages, in which MBI was used to describe the proportion and configuration of buildings in urban villages. In addition, building information is very closely related to population distribution. Wang et al. (2016) used MBI for building extraction and MSI for shadow detection and building height retrieval. Then, building location and height information can be combined to reconstruct 3D models to estimate population.

3.6 CONCLUSION

Building information, one of the most important geographic data sources, is being used for many different applications, for example, real estate, disaster monitoring, and population estimation. The automatic building extraction from high-resolution images is an active research topic and remains a big challenging task in the fields of computer vision and remote sensing owing to the complexity of the buildings (e.g., shape, size, and brightness) and their surrounding environment (e.g., road, parking lot, and open area). In this chapter, a systematic framework is introduced in detail for effective building detection, that is, the MBI and its improved versions. MBI is able to describe spectral–spatial properties of buildings (e.g., contrast, size, and directionality) by a set of morphological operators and can be applied to large-area high-resolution remote sensing images for automatic building extraction. Because of its high efficiency and accuracy, MBI has been extensively employed in different fields of remote sensing and help other researchers to conduct related works. In addition, recently developed generalized differential morphological profiles (GDMPs) may have the potential to construct enhanced MBI by considering the across-scale DMPs, which are more appropriate to describe the multiscale characteristics of buildings (Huang et al. 2016a). By combining multi-source data sets such as socioeconomic data, more in-depth applications of MBI, for example, building density estimation (Zhang et al. 2017) and city function recognition (Zhang and Du 2015), are strongly expected in future studies.

REFERENCES

Benedek, C., Descombes, X., and Zerubia, J., 2010. Building detection in a single remotely sensed image with a point process of rectangles. Paper presented at the International Conference on Pattern Recognition.

Chiancone, A., Mura, M.D., and Chanussot, J., 2013. Sismologie urbaine: Évaluation de la vulnérabilité et des dommages sismiques par méthodes innovantes. http://urbasis.osug.fr/IMG/pdf/m5-urbasis_report-gipsalab-light.pdf.

Congalton, R.G., 1991. A review of assessing the accuracy of classifications of remotely sensed data. *Remote Sensing of Environment* 37 (1):35–46.

Haralick, R.M., 1979. Statistical and structural approaches to texture. *Proceedings of the IEEE* 67 (5):786–804.

Hou, B., Wang, Y., and Liu, Q., 2016. A saliency guided semi-supervised building change detection method for high resolution remote sensing images. *Sensors* 16 (9):1377.

Huang, X., Han, X., Zhang, L., Gong, J., Liao, W., and Benediktsson, J.A., 2016a. Generalized differential morphological profiles for remote sensing image classification. *IEEE Journal of Selected Topics in Applied Earth Observations and Remote Sensing* 9 (4):1736–51.

Huang, X., Liu, H., and Zhang, L., 2015a. Spatiotemporal detection and analysis of urban villages in mega city regions of China using high-resolution remotely sensed imagery. *IEEE Transactions on Geoscience and Remote Sensing* 53 (7):3639–57.

Huang, X., Lu, Q., and Zhang, L., 2014a. A multi-index learning approach for classification of high-resolution remotely sensed images over urban areas. *ISPRS Journal of Photogrammetry and Remote Sensing* 90:36–48.

Huang, X., Weng, C., Lu, Q., Feng, T., and Zhang, L., 2015b. Automatic labelling and selection of training samples for high-resolution remote sensing image classification over urban areas. *Remote Sensing* 7 (12):16024–44.

Huang, X., Yuan, W., Li, J., and Zhang, L., 2016b. A new building extraction postprocessing framework for high-spatial-resolution remote-sensing imagery. *IEEE Journal of Selected Topics in Applied Earth Observations and Remote Sensing* 10 (2):654–68.

Huang, X., and Zhang, L., 2011. A multidirectional and multiscale morphological index for automatic building extraction from multispectral GeoEye-1 imagery. *Photogrammetric Engineering & Remote Sensing* 77(7):721–32.

Huang, X., and Zhang, L., 2012a. Morphological building/shadow index for building extraction from high-resolution imagery over urban areas. *IEEE Journal of Selected Topics in Applied Earth Observations and Remote Sensing* 5 (1):161–72.

Huang, X., and Zhang, L., 2012b. A multiscale urban complexity index based on 3D wavelet transform for spectral–spatial feature extraction and classification: An evaluation on the 8-channel WorldView-2 imagery. *International Journal of Remote Sensing* 33 (8):2641–56.

Huang, X., Zhang, L., and Zhu, T., 2014b. Building change detection from multitemporal high-resolution remotely sensed images based on a morphological building index. *IEEE Journal of Selected Topics in Applied Earth Observations and Remote Sensing* 7 (1):105–15.

Huang, X., Zhu, T., and Tang, Y., 2014c. A novel building change index for automatic building change detection from high-resolution remote sensing imagery. *Remote sensing letters* 5 (8):713–22.

Jiménez, L.I., Plaza, J., and Plaza. A., 2016. Efficient implementation of morphological index for building/shadow extraction from remotely sensed images. *The Journal of Supercomputing*:1–13.

Liao, W., Huang, X., Coillie, F.V., Gautama, S., Pižurica, A., Philips, W., Liu, H., Zhu, T., Shimoni, M., and Moser, G., 2015. Processing of multiresolution thermal hyperspectral and digital color data: Outcome of the 2014 IEEE GRSS data fusion contest. *IEEE Journal of Selected Topics in Applied Earth Observations and Remote Sensing* 8 (6):2984–96.

Ok, A.O., 2013. Automated detection of buildings from single VHR multispectral images using shadow information and graph cuts. *ISPRS Journal of Photogrammetry and Remote Sensing* 86:21–40.

Pesaresi, M., and Benediktsson, J.A., 2001. A new approach for the morphological segmentation of high-resolution satellite imagery. *IEEE Transactions on Geoscience and Remote Sensing* 39 (2):309–20.

Pesaresi, M., and Gerhardinger, A., 2011. Improved textural built-up presence index for automatic recognition of human settlements in arid regions with scattered vegetation. *IEEE Journal of Selected Topics in Applied Earth Observations and Remote Sensing* 4 (1):16–26.

Pesaresi, M., Gerhardinger, A., and Kayitakire, F., 2008. A robust built-up area presence index by anisotropic rotation-invariant textural measure. *IEEE Journal of Selected Topics in Applied Earth Observations and Remote Sensing* 1 (3):180–92.

Rubeena, V., and Tiwari, K.C., 2016. Multisensor multiresolution data fusion for improvement in classification. Paper presented at Proceedings of SPIE.

Rutzinger, M., Rottensteiner, F., and Pfeifer. N., 2009. A comparison of evaluation techniques for building extraction from airborne laser scanning. *IEEE Journal of Selected Topics in Applied Earth Observations and Remote Sensing* 2 (1):11–20.

Sirmacek, B., and Unsalan, C., 2009. Urban-area and building detection using SIFT keypoints and graph theory. *IEEE Transactions on Geoscience and Remote Sensing* 47 (4):1156–67.

Song, W., and Haithcoat, T.L., 2005. Development of comprehensive accuracy assessment indexes for building footprint extraction. *IEEE Transactions on Geoscience and Remote Sensing* 43 (2):402–4.

Syrris, V., Ferri, S., Ehrlich, D., and Pesaresi, M., 2015. Image enhancement and feature extraction based on low-resolution satellite data. *IEEE Journal of Selected Topics in Applied Earth Observations and Remote Sensing* 8 (5):1986–95.

Tang, Y., Huang, X., and Zhang. L., 2013. Fault-tolerant building change detection from urban high-resolution remote sensing imagery. *IEEE Geoscience and Remote Sensing Letters* 10 (5):1060–4.

Tian, J., and Chen, D.M., 2007. Optimization in multi-scale segmentation of high-resolution satellite images for artificial feature recognition. *International Journal of Remote Sensing* 28 (20):4625–44.

Wang, S., Tian, Y., Zhou, Y., Liu, W., and Lin. C., 2016. Fine-scale population estimation by 3D reconstruction of urban residential buildings. *Sensors* 16 (10):1755–81.

Wen, D., Huang X., Zhang, L., and Benediktsson. J.A., 2016. A novel automatic change detection method for urban high-resolution remotely sensed imagery based on multi-index scene representation. *IEEE Transactions on Geoscience and Remote Sensing* 54 (1):609–25.

Xiao, P., Yuan, M., Zhang, X., Feng, X., and Guo. Y., 2017. Cosegmentation for object-based building change detection from high-resolution remotely sensed images. *IEEE Transactions on Geoscience and Remote Sensing.* DOI:10.1109/TGRS.2016.2627638.

Zhang, L., Huang, X., Huang, B., and Li. P., 2006. A pixel shape index coupled with spectral information for classification of high spatial resolution remotely sensed imagery. *IEEE Transactions on Geoscience and Remote Sensing* 44 (10):2950–61.

Zhang, Q., Huang, X., and Zhang. G., 2016. A morphological building detection framework for high-resolution optical imagery over urban areas. *IEEE Geoscience and Remote Sensing Letters* 13 (9):1388–92.

Zhang, T., Huang, X., Wen, D., and Li. J., 2017. Urban building density estimation from high-resolution imagery using multiple features and support vector regression. *IEEE Journal of Selected Topics in Applied Earth Observations and Remote Sensing.* DOI: 10.1109 /JSTARS.2017.2669217.

Zhang, X., and Du, S., 2015. A linear dirichlet mixture model for decomposing scenes: Application to analyzing urban functional zonings. *Remote Sensing of Environment* 169:37–49.

Section II

Assessing and Modeling Urban Landscape Compositions, Patterns, and Structures

4 Stereo-Based Building Roof Mapping in Urban Off-Nadir VHR Satellite Images: Challenges and Solutions

Alaeldin Suliman and Yun Zhang

CONTENTS

4.1 INTRODUCTION

Urban areas are one of the most rapidly growing and continuously changing land covers on the Earth. In 2000, around 40% of the world population lived in urban areas. According to the United Nations, this urban population will continue to increase substantially over the coming decades (Weng and Quattrochi 2007). Because of the search for employment, education, business, and health care environments, most of the urbanization related to the population increase will occur in the developing countries (Hussain and Shan 2016).

Remote sensing images produced by satellite sensors provide a worldwide leading technology for monitoring urban areas even on large scales. Since urban areas are dynamic, structured, and complex, very high resolution (VHR) satellite images are the ideal geo-data for mapping such challenging environments. This is due to the advantages of the VHR satellite images that include submeter ground resolution, fast acquisition, high availability, broad coverage, relatively low prices, and rich information content necessary for mapping complex urban areas (Mishra and Zhang 2012).

Buildings are the predominant object class in dense urban areas. Hence, continuously updated building information, in Geographic Information Systems (GIS), is important for many urban applications including, but not limited to, city planning and management (Nielsen and Ahlqvist 2014), city modeling (Singh et al. 2014), population estimation (Xie et al. 2015), and urban growth monitoring (Sugg et al. 2014). Consequently, building detection in VHR satellite images has become an active research topic in the remote sensing community.

Off-nadir images with along-track (inline) and/or across-track angles are usually acquired by almost all the currently working VHR satellite sensors (e.g., WorldView-4, TripleSat, and Pleiades-1A). This can be clearly observed in the available online databases of VHR satellite image vendors (e.g., DigitalGlobe, *ImageFinder*) where most of the archived VHR images are acquired off-nadir. This is because the off-nadir capability provides fast image acquisition with various modes including mainly the inline stereo mode that rapidly captures stereo images that allow stereo-based information to be extracted. However, despite the high availability of VHR satellite images, the off-nadir ones are usually avoided in the applications of building detection in dense urban areas as reflected in the relevant research publications. This is due to the resulting severe building lean in off-nadir VHR images, which poses a difficulty for the currently available building detection and mapping methods.

Building detection methods in VHR satellite images exploit various types of image information. These information types include textural, spectral, contextual, and morphological image information (Salehi et al. 2012). Recently, a comprehensive survey of the available object detection methods in remote sensing images has been provided by Cheng and Han (2016). They categorized these methods into four groups: template matching based, machine learning based, knowledge based, and OBIA based (Object-Based Image Analysis). However, these methods usually have a limited ability to cope with the inter-similarity among different impervious surfaces in VHR images. This is because these methods rely on image-based information that lacks the geometric information of the imaged 3D surface. For instance, it is very difficult, based on image information only, to automatically distinguish building roofs from parking lots with the same spectral and spatial properties (Ghaffarian and Ghaffarian 2014). Therefore, ancillary information is needed.

Elevation information provides a critical key feature for accurate building detection and reliable differentiation from other similar traffic areas since buildings are inherently elevated objects. Different sources of remote sensing data can be used to generate elevation information including stereo images and LiDAR (Light Detection and Ranging) data. Among these sources, the readily available stereo off-nadir VHR satellite images allow generating elevation data, to support building detection, at a lower cost than the other remote sensing sources. However, because of the apparent building lean resulting from off-nadir acquisition mode, the incorporation of stereo-based geometric information with the off-nadir VHR satellite images to support detecting buildings still encounters a few challenges and difficulties.

The incorporation of stereo-based information for generating surfaces includes three steps: stereo information generation, co-registration, and normalization. Two types of stereo-based information can be generated from stereo images: digital elevation models and disparity (x-parallax) maps. Hence, the third dimension used for stereo-based building detection may be either type of information. Both of the stereo-based information types need to be accurately co-registered with image data to support identifying the elevated objects. However, when orthographic elevation models are integrated with perspective images, problematic misregistration is presented. Additionally, because stereo matching techniques are used to generate both information types, large data gaps are expected as a result of inherent occlusion effects particularly in the case of off-nadir stereo pairs. Furthermore, since the stereo-based information is normally generated for the imaged visible surface, the effect of the terrain relief is included. This effect makes identifying the aboveground (off-terrain) objects not directly applicable by thresholding the generated stereo-based information. Hence, this information needs to be normalized by eliminating or minimizing the terrain relief influence. Accordingly, each of the three incorporation steps introduces some challenges for building detection applications, especially for the case of employing off-nadir VHR satellite images.

Hence, this chapter describes, in Section 4.2, the principle of the stereo-based building detection method adopted for this research. Then, three challenges associated with the use of off-nadir VHR images are identified in Section 4.3. Correspondingly, three solutions are described briefly in Section 4.4. The developed concepts are proven with experimental results and discussions in Section 4.5. Finally, the conclusions drawn are given in Section 4.6.

4.2 PRINCIPLE OF STEREO-BASED BUILDING DETECTION

The building detection principle adapted in our research is based on thresholding the third dimension that identifies the elevated objects. This dimension is the stereo-based information that may represent either elevation or disparity. The detection methodology includes three stages: stereo image processing, building detection, and performance evaluation. Figure 4.1 shows these stages along with the steps involved. Further description of these stages is provided in the following sections.

4.2.1 STEREO IMAGE PROCESSING

This stage includes the preparation steps of the stereo-based information to be incorporated and applied for building detection. The incorporation steps of the stereo-based information, along with a notion about the expected challenges, are described below.

4.2.1.1 Stereo-Based Information Generation

Elevation information can be extracted from stereo satellite images using the well-established digital photogrammetric approaches. These approaches usually start with epipolar rectification to generate image pairs with eliminated y-parallax (e.g., Zhao, Yuan, and Liu 2008). These image pairs allow efficient implementation of dense matching by restricting the search area to be linear. After that, a dense image matching technique is executed to find accurate point matches. Then, a forward space intersection process is applied. In this process, the image coordinates of the matched points are combined with the relevant sensor model in a bundle block adjustment process (Grodecki and Dial 2003) to extract elevation data that represent the digital surface model (DSM) of the imaged area.

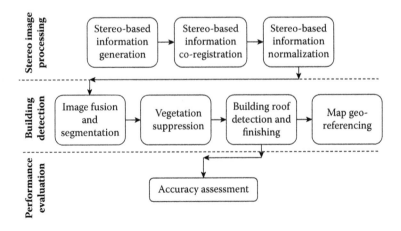

FIGURE 4.1   Principle of the stereo-based building detection methodology.

Disparity information, on the other hand, is measured within the image spaces of a stereo pair. Disparity maps are generated from epipolar rectified stereo pairs using epipolar-based matching algorithms. On the basis of these maps, typically, the elevation data are photogrammetrically derived as described earlier. Hence, if the disparity information is used, the space intersection and bundle adjustment processes are bypassed.

In our research, the adopted epipolar rectification model is the projection reference plane (PRP) epipolarity model introduced by Wang et al. (2011). This model is selected, based on our preliminary testing, because it provides simple and efficient epipolar rectification in the object space for the satellite images acquired by line array sensors. For generating disparity maps, the adopted epipolar-based matching algorithm is the semiglobal matching (SGM)—developed by Hirschmüller (2008)—since it is very successful even in dense urban areas as concluded by Alobeid et al. (2010).

The occlusion created by the building lean within stereo pairs poses a problem for all matching algorithms. It prevents the matching from being made and hence produces data gaps in both types of stereo-based information. When off-nadir VHR satellite images are used, these gaps cannot be filled by interpolation because they are very large owing to the severe building lean in such images. Therefore, this is one challenge that needs to be mitigated.

4.2.1.2 Stereo-Based Information Co-Registration

To identify the elevated objects correctly in the off-nadir satellite images, the height information (elevation or disparity) needs to be co-registered accurately with the image data. Hence, after generating the stereo-based information, an accurate co-registration technique should be executed.

In the case of disparity maps, perfect image-disparity co-registration is directly achieved without any effort. This is because these maps share the same reference frame of the selected reference image as per the definition of disparity maps. These maps simply represent the shift in pixels of all image pixels in the right image relative to the left stereo mate. Therefore, by using disparity information, the steps of space intersection and co-registration are bypassed.

In the case of DSM elevation data, on the other hand, image-elevation co-registration can be accurately achieved by any of the techniques reviewed by Mishra and Zhang (2012) especially for nadir images. However, as concluded in this review, all of these techniques have limitations when off-nadir images acquired over dense urban areas are employed. Therefore, the co-registration challenge of employing off-nadir images with either pre-existing or photogrammetrically-derived DSM data is another challenge that needs to be overcome.

4.2.1.3 Stereo-Based Information Normalization

Before applying the thresholding operation to identify the elevated objects, the terrain effect should be eliminated or minimized. This process is referred to as normalization, which relates the stereo-based information to the datum representing the varying level of the bare earth. Typically, the normalization is conducted in two steps: first, extracting the terrain height information (elevation or disparity) from the registered stereo-based data, and then subtracting the extracted terrain information from the original registered data. In the case of elevation data, this process is commonly expressed in the following equation:

$$nDSM = DSM - DTM, \qquad (4.1)$$

where nDSM is the normalized DSM elevation model representing only the above ground objects and DTM is the extracted Digital Terrain Elevation Model representing only the bare-earth elevations.

In our research, the local-minima technique introduced by Zhang et al. (2004) is selected. This technique is based on interpolating the local minima elevations identified within a predefined size of a moving window. It is designed to work successfully in different terrain relief and structured environments without any presumptions. However, in the case of stereo-based information derived from off-nadir satellite stereo images acquired over dense urban areas, this technique faces a big challenge in interpolating the resulting data gaps because they are quite large when off-nadir stereo pairs are used. Therefore, this is the third challenge that needs to be resolved.

4.2.2 BUILDING DETECTION

After processing the stereo images and preparing the derived stereo-based information, the stereo-based building detection procedure can be started. This procedure comprises four steps as discussed below. The last step is required when geo-referenced results are needed.

4.2.2.1 Image Fusion and Segmentation

VHR satellite images usually have four multispectral bands and one panchromatic band. While the color information is provided in the multispectral bands, these bands usually have a resolution of one-fourth of that of the panchromatic one. Therefore, an image fusion/pan-sharpening technique needs to be utilized to combine these bands and take the advantages of both types. The UNB pan-sharpening technique, which was introduced by Zhang (2004), is selected in our research because it is designed specifically for VHR images acquired by the new satellite sensors.

To reduce the complexity of VHR images, image segmentation is required to be executed for the image in which buildings will be detected. This reduction is made by dividing the image into small segments based on a homogeneity measure of the color information. Multi-resolution segmentation, introduced by Baatz and Schäpe (2000), is one of the most appropriate techniques for segmenting VHR images in urban areas (Dey 2013). Thus, our research adopts this multi-resolution technique for the image segmentation step.

4.2.2.2 Vegetation Suppression

Since the elevations are going to have a crucial role in building detection, other elevated features, such as trees, must be removed to avoid confusion with the building objects. Fortunately, urban vegetation and trees can be effectively detected and removed by using vegetation indices. In our research, the Normalized Deference Vegetation Index (NDVI) is selected since it is one of the most accurate and popular vegetation indices.

4.2.2.3   Building Detection and Finishing

To delineate elevated objects, a thresholding operation for the normalized stereo-based information needs to be applied. The off-terrain tree objects are already suppressed in the previous step. The best thresholding value can be selected and tested empirically.

To enhance the representation of the detected building objects, postprocessing procedures should be applied. These procedures generally comprise steps such as including the segments circumfluent by the detected building objects in the building class, merging the building segments of relatively short border length to nonbuilding segments to enlarge their areas, and excluding the segments of small areas after the merge. Last, morphological operations and final editing could be performed to enhance the final shapes.

Because the detected rooftops have offsets from their corresponding footprints owing to the building lean in off-nadir images, it is worth mentioning that the detection performance should be evaluated before correcting these rooftops' offsets if the reference data are going to be manually digitized from the off-nadir image. Conversely, if the reference data available are geo-referenced maps, the evaluation will be after correcting the offsets of the mapped rooftops and geo-referencing the detection map.

4.2.2.4   Map Geo-Referencing

After detecting the building objects, enhancing their shapes, and validating the achieved results (in the case of manually generated reference data), these objects should be geo-referenced to their correct orthographic locations. This allows easy input and integration with other existing GIS layers. Therefore, these objects need to be shifted and registered to their correct geo-locations.

Each building rooftop has an offset distance from its footprint relative to the building elevation. Hence, to shift the detected rooftop objects to their correct geo-locations, the elevation information for these rooftop objects needs to be available in order to calculate the distances of these varying offsets. Therefore, the missed information poses an additional problem that needs to be solved for geo-referencing the detection roof map using direct registration techniques.

4.2.3   Performance Evaluation

The building detection performance is evaluated by comparing the mapped buildings' roofs against a reference data set. Completeness, correctness, and overall quality are three widely used measures to assess the detection performance. Completeness is the percentage of the entities in the reference data that are correctly detected. Correctness indicates how well the detected entities match the reference data. In contrast, as a compound performance metric, the overall-quality measure balances completeness and correctness (Potůčková and Hofman 2016; Rutzinger et al. 2009). Hence, this measure will be used in this chapter as a general indication of the building detection quality. The formulas of the three performance measures are expressed as follows:

$$\text{Completeness (Comp.)} = \text{TP}/(\text{TP} + \text{FN}) \qquad (4.2)$$

$$\text{Correctness (Corr.)} = \text{TP}/(\text{TP} + \text{FP}) \qquad (4.3)$$

$$\text{Overall Quality (OQ)} = \text{TP}/(\text{TP} + \text{FP} + \text{FN}), \qquad (4.4)$$

where true positive (TP) is the number of correctly identified building roof segments. False negative (FN) is the number of building roof segments in the reference data set that are not detected or are wrongly labeled as not roofs. False positive (FP) represents the number of building roof segments that are detected but do not correspond to the reference data set. These measured entities in our research are selected to represent the total number of pixels that are related to the building-roof class (Potůčková and Hofman 2016). Additionally, the reference (ground truth) data used for the evaluation are generated manually from the original detection image since they are not usually available. Consequently, the result evaluation is proposed to be before the map geo-referencing step. If a geo-referenced map or an orthoimage is available for the building roofs of the imaged area, the geo-referencing accuracy can be evaluated.

4.3   IDENTIFIED CHALLENGES

The challenges identified in this research are mentioned briefly during the description, in Section 4.2, of the building detection procedure. The challenges mainly result from the severe building lean in off-nadir satellite images when incorporated with stereo-based information. Since off-nadir images are a 2D perspective projection of the 3D real world, the sides of the elevated objects will be captured as leaning objects and hence they add more complexity to the acquired scenes. Consequently, the building roofs will be separated from their footprints. This separation creates offsets between the building roofs and their correct orthographic locations.

Both the building roof offsets and building façades create additional problems in the incorporation of stereo-based information with off-nadir images for mapping building rooftops. On the one hand, the varying offsets of the roofs prevent both efficient image-to-elevation co-registration and direct map geo-referencing. On the other hand, building façades create occluded areas that prevent successful matching from being made and hence produce large data gaps that complicate the normalization process. All of these problems associated with off-nadir images, as described in the following subsections, reduce the quality of the stereo-based building detection results.

4.3.1   Efficient Image-Elevation Co-Registration

To support building detection, stereo-based information is required to be co-registered accurately with the corresponding image data. The challenge results from the offsets of the roofs in off-nadir images. In the case of disparity information, the measured disparity maps created by matching techniques are co-registered perfectly, as per disparity definition, to the selected reference stereo image. Hence, no effort is required to achieve pixel-level co-registration even with off-nadir images.

In contrast to the disparity maps, accurate co-registration of elevation models with off-nadir VHR satellite images acquired over dense urban areas is very challenging. This results mainly from the dissimilarity of the projections of the two data types being co-registered. While the 3D elevation models typically have an orthographic projection, the 2D images have perspective projections. In the elevation data, consequently, the building roofs will occupy the same locations as their corresponding footprints. However, this is not the case for off-nadir images because the building roofs are shifted from their corresponding footprints.

Almost all categories of the currently available methods for image-elevation co-registration are reviewed in Mishra and Zhang (2012) and in Suliman and Zhang (2015). It was concluded that none of these methods can provide accurate co-registration in the case of off-nadir images unless a true orthorectification technique is implemented. However, the implementation of this technique is expensive (in terms of input data and computation) and the seamless true orthoimages are difficult to achieve. This makes true orthorectification an inefficient and unfeasible technique. Hence, the first concern of the current research is to address the problem of efficient image-elevation co-registration for off-nadir VHR satellite images. The proposed solution should provide pixel-level co-registration accuracy and reduce the required implementation cost (e.g., input data, calculation steps, and computation cost) in order to be considered as a more efficient technique than what is currently available.

4.3.2 DISPARITY GAPS AND NORMALIZATION

The second challenge associated with off-nadir VHR satellite images is the severe occlusion effects resulting from building façades in off-nadir stereo images acquired over dense urban areas. Since these images are 2D perspective projections of the 3D real world, the elevated objects, mainly the building façades, will hide large areas behind. These occluded areas will appear in only one of the stereo image pairs and thus no successful matching can be made. As a result, many data gaps will be created.

The stereo-based information needs to be normalized before it can be applied in the building detection process. Extracting the terrain relief information is essential for the normalization step. However, the large data gaps must be filled before applying the extraction and normalization processes.

There are two methods to fill the resulting data gaps: filling gaps by surface interpolation (internal data source) and filling gaps by additional data (external data source). The first method (interpolation-based filling) is the typical solution when only one stereo pair is available for deriving stereo-based information in either case of elevation or disparity data. The second method (additional data–based filling) is applicable when several stereo pairs are available for stereo-based information extraction.

In the case of one off-nadir stereo pair being available, filling the resulting large gaps by surface interpolation techniques in dense urban areas is very risky. It reduces the quality of the generated stereo-based data by producing misleading surface information. This wrong information negatively affects the subsequent process of normalizing the surface stereo-based information required to remove the terrain influence. Therefore, this research attempts to mitigate the problem of the disparity gaps and the process of stereo-based information normalization by measuring directly the aboveground (i.e., normalized) disparity information to bypass the steps of interpolating the occlusion disparity gaps and removing the terrain relief influence.

When multiple stereo pairs are available, extra stereo-based elevation data can be easily generated and directly applied for both detecting the outliers and filling the existing gaps. This is because the derived elevations for the same ground point are equal regardless of the stereo pairs used. However, this advantage of generating supplementary data (i.e., additional and consistent data) is not directly available in the case of disparity information measured from different stereo pairs. This is because the disparity value—which is the relative location shift of a shared point within a stereo pair—is a relative co-relation between two stereo images. Hence, different stereo pairs produce different disparity values for the same object space point. Thus, our research attempts to address the problem of generating supplementary disparity data from different inline stereo pairs that are applicable for enriching disparity maps by filling the large occlusion-produced gaps in off-nadir stereo satellite images.

4.3.3 DIRECT MAP GEO-REFERENCING

The offset between a building rooftop and its footprint is created by the off-nadir viewing mode during image acquisition. Furthermore, this offset varies from one building to another based on the elevation of each building individually. As a result, any building detection method that uses image-based information and cues cannot shift the detected rooftops to their correct corresponding footprints unless the elevation information is available. Hence, to close the cycle of building detection by map geo-referencing, this problem needs to be addressed.

4.4   DEVELOPED SOLUTIONS

In the current research, three challenges are identified in the process of incorporating elevation or disparity data for the purpose of stereo-based building detection in off-nadir VHR satellite images. The first challenge is the efficient elevation co-registration with off-nadir VHR satellite images. The second challenge is related to the problems associated with occlusion-produced disparity gaps. On the basis of the availability of the stereo pairs, this challenge can be addressed in two ways: (1) mitigating the negative effects of the occlusion-produced disparity gaps to bypass the normalization process in the case of only one stereo pair being available, and (2) generating consistent supplementary disparity data to fill the resulting disparity gaps in the case of more than one stereo pair being available. The third challenge is correcting the varying offsets of the building rooftops from their corresponding footprints in the detection map to allow direct map geo-referencing.

The following subsections describe the developed solutions corresponding to the three identified challenges. While the first problem is addressed in the first solution, the first case of the second challenge is addressed in the second solution. However, the third solution covers the remaining challenges.

4.4.1 LINE-OF-SIGHT DSM SOLUTION

The image-elevation co-registration problem is mainly caused by the differences in the projections of the two data types being co-registered. While the DSMs are commonly in orthographic projections, off-nadir VHR images are acquired with perspective projections that create building lean. Hence, the proposed solution is to incorporate the sensor model information of the off-nadir image being co-registered to project the DSM elevations back to the image space and create an elevation layer representing the so-called Line-of-Sight DSM (LoS-DSM). This elevation model has the same perspective projection as the off-nadir image being overlaid in order to efficiently achieve accurate co-registration and preserve the original image information without implementing sophisticated algorithms (e.g., algorithms for occlusion detection and compensation) as in the case of true orthorectification.

The development and testing of the LoS-DSM solution is detailed in Suliman and Zhang (2015). In this study, this solution was generated using the sensor model information of the image being co-registered, as shown in Figure 4.2, in two major phases: (1) deriving DSM photogrammetrically from stereo images, and then (2) re-projecting the derived DSM back to one of the stereo images for accurate co-registration. For the case of pre-existing DSM data, more than one elevation value is expected to compete for the same pixel location because of the occlusion effect. Hence, the LoS-DSM is developed by selecting the highest elevation value. Finally, the data set achieved after this image-elevation co-registration forms the basis for mapping off-terrain features even in off-nadir VHR images.

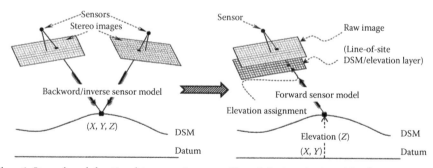

Phase 1: Stereo-based elevation data generation using inverse sensor model

Phase 2: Back projection of the elevations using forward sensor model

FIGURE 4.2   The two major phases of the developed LoS-DSM solution. (Reprinted, with permission, from Suliman, Alaeldin, and Yun Zhang. 2015. Development of line-of-sight digital surface model for co-registering off-nadir VHR satellite imagery with elevation data. *IEEE Journal of Selected Topics in Applied Earth Observations and Remote Sensing* 8 (5):1913–1923. © 2015 IEEE.)

4.4.2 REGISTRATION-BASED MAPPING OF THE ABOVEGROUND DISPARITIES

With the aim of decreasing the computational steps required for detecting build-ings, the proposed solution replaces photogrammetrically derived elevation data with matching-based disparity information to reduce the computational steps and costs. Moreover, since the generated disparity maps fit exactly the same reference frame of the selected reference image, no effort is required for image-disparity co-registration.

Therefore, in the case of only one stereo pair being available, the proposed solu-tion is registration-based mapping of the aboveground disparities (RMAD) for the off-terrain features. This solution starts with epipolar rectification of the two origi-nal images to eliminate the y-direction disparity (y-parallax) of all corresponding pixels and then the co-registration of these epipolar images based on the corre-sponding ground-level objects (e.g., road intersection points) to eliminate the dispar-ity in the x-direction (x-parallax) of the terrain (after minimizing the terrain relief variation as illustrated in Figure 4.3). Hence, the remaining measurable disparity, which can be mapped by any epipolar-based matching technique, should represent only the off-terrain objects (i.e., normalized surface disparity map [nSDM]). This is achievable for reasonably nonflat terrains as it is in the case of modern dense urban areas because they follow the geometric design standards of city roads (e.g., AASHTO 2001).

The development and testing of the RMAD method is published in Suliman et al. (2016). The method attempts to bypass the steps of interpolating the gaps and nor-malizing the surface information by co-registering the corresponding terrain-level features in an epipolar stereo pair to minimize the disparity values of the terrain-level features when measured by an epipolar-based matching technique.

By this registration-based minimization, the mapped disparities from the con-structed epipolar stereo pair will be close to zero values for the terrain features.

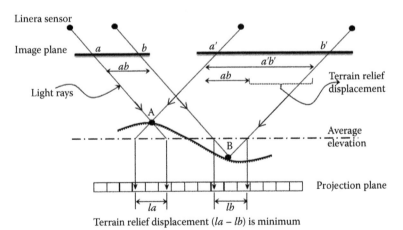

Terrain relief displacement ($la - lb$) is minimum

FIGURE 4.3 Terrain relief minimization in the RMAD method. (Reproduced with permis-sion from the American Society for Photogrammetry and Remote Sensing, Bethesda, Maryland, www.asprs.org.)

Hence, only the disparities of the elevated/aboveground objects will be measured by all epipolar-based matching techniques. The achieved data set of the co-registered and normalized disparity map forms the basis for mapping off-terrain features in off-nadir VHR stereo images acquired over dense and reasonably nonflat urban areas.

4.4.3 DISPARITY-BASED ELEVATION CO-REGISTRATION

When multiple inline stereo pairs are available, several disparity maps can be generated. However, direct fusion of these maps is not applicable as is the case with elevation models. This is due to the fact that the disparity value represents a co-relation between two stereo images. This value will be different for the same object space point when measured from a different stereo pair. This problem will hence prevent filling the occlusion-produced gaps by supplementary disparity data. Therefore, the solution to this challenge is to derive the conversion formula that helps in transferring the disparity values among different domains.

In addition to this problem, the varying offsets between the building roofs and their corresponding footprints in off-nadir images prevent direct geo-referencing of the generated detection maps of building-roofs. These offsets cannot be corrected unless the elevation data of the shifted roofs are available. Hence, the solution to this challenge is to derive efficiently the required elevation data from the co-registered disparity information and then calculate the shift distances to correct the varying roof offsets.

A method for filling the occlusion-produced gaps by supplementary disparity data is introduced in Suliman and Zhang (2017). This method is further extended to derive co-registered elevation data and to address the challenge of varying rooftop-to-footprint offsets of the building objects. The development and testing of the combined and extended solution are provided in Suliman and Zhang (2018). This extended solution is based on developing a scale relationship among the corresponding disparities from different epipolar domains and with their corresponding ground elevation. To achieve such a relationship, it is required to project the inline satellite images employed onto a horizontal plane in the relevant object space at the average terrain elevation of the imaged area. A second projection onto a horizontal plane parallel to the first one is required to derive the transferring scales among the corresponding disparities, the disparity-to-elevation conversion scales, and the correction shift distances of the varying roof offsets. This solution leads to generating elevation data based on a disparity map that are co-registered perfectly to the image data. Hence, this solution is called disparity-based elevation co-registration (DECR).

The DECR solution produces a disparity map enriched from different inline stereo pairs. Additionally, this map can be scaled to represent elevation data in the form of LoS-DSM that is co-registered perfectly to the reference image. The achieved data set after generating either the co-registered disparity or elevation data forms the basis for 3D-supported building detection even in off-nadir VHR images. Furthermore, the DECR solution provides the correction distances for each building roof individually (the varying rooftops offsets). Accordingly, the applicability of the DECR method can be demonstrated through building detection in off-nadir VHR satellite imagery followed by detection map geo-referencing based on the corrected rooftops' offsets. The major phases involved in this solution are described briefly in the following subsections.

4.4.3.1   Image Projection and Rectification

This phase involves two main steps: a two-plane projection process and epipolar recti-fication. The two-plane projection process utilizes two object space horizontal planes at two different elevations for projecting the employed inline images. On the one hand, the first projection is made from a horizontal object-space plane at an elevation (Z_1; the aver-age terrain elevation of the imaged area) to the image-space pixel location. The aim of this first projection process is to generate images in the object space with minimized ter-rain relief distortions. On the other hand, the second projection process is executed from the calculated image-space location in the first projection process back to a horizontal object-space plane at a different elevation (Z_2; the datum elevation). The aim of this is to generate a 3D point in the object space that is collinear to the light ray connecting the corresponding image pixel location to the same point at the average terrain elevation (Z_1). This collinear 3D point at Z_2 allows calculating shift information (ΔX, ΔY) in the object space that can be used for deriving the conversion disparity and elevation scales.

The epipolar rectification is required to generate images with eliminated y-parallax. Since the employed images are already projected in the object space, the rectification process needs to be made in the relevant virtual object space too. In the case of line array satellite sensors, the rigorous epipolar rectification of the acquired images produces non-straight epipolar lines (Habib et al. 2005). However, these epi-polar lines can be approximated by straight lines if the line array satellite images are projected onto the relevant virtual object space. On the basis of this conclusion, Wang et al. (2011) developed the PRP epipolarity model. Therefore, this model is adopted in the DECR solution to epipolar rectify the employed inline images.

To derive the conversion scales required later in the DECR solution, the same image rotation process used to create the epipolar images in the object space is applied to calcu-late the shift values of all ground pixels of the selected reference image. Hence, if the epi-polar rotation angle is θ, the computed shift information (ΔX, ΔY) can be easily rotated to determine the corresponding shift information in the epipolar plane ($\Delta X'$, $\Delta Y'$).

4.4.3.2   Scale-Based Formulas Derivation

By now, the required shift information to calculate the conversion scales is avail-able. The core concept of the DECR solution is illustrated in Figure 4.4. This figure, which is in the epipolar direction, shows four inline stereo satellite images that are projected on two different horizontal planes in the object space. While the first pro-jection plane is at elevation Z_1, the second projection plane is at elevation Z_2. The light rays projected onto the two horizontal object-space planes are all represented in the epipolar plane, $X'Z$, along the same epipolar line. The intersection point of these light rays is the actual point imaged in the object space. The photogrammetric triangulation approaches solve for this point (i.e., space intersection) to reconstruct the imaged object space. The intersections of the rays with the horizontal average level plane (Z_1) are the ground locations of the projected pixels. Hence, the distances between these projected pixels are the object-space disparities (D).

From Figure 4.4, three scale-based formulas can be derived based on the basic geometric rules: domain-to-domain disparity formula (DDF), disparity-to-elevation formula (DEF), and correct distance formula (CDF). While DDF provides the

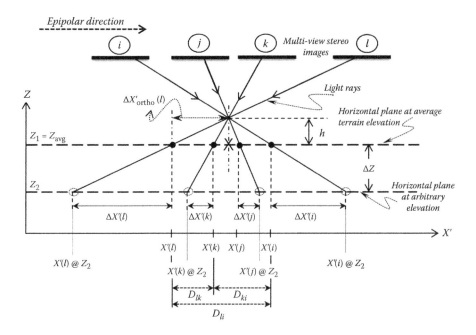

FIGURE 4.4 The concept of the DECR solution. (Reproduced with permission from the Scientific Research Publishing, Irvine CA 92619-4821, USA. www.scirp.org.)

corresponding disparity values among different stereo domains, DEF derives the elevation value from its corresponding disparity value. In contrast, CDF calculates the distance between a projected image pixel and its correct orthographic location in the epipolar direction. The three derived formulas of DDF, DEF, and CDF are expressed in Equations 4.5, 4.6, and 4.7, respectively.

$$D_{lk} = \left(\frac{\Delta X'(l) - \Delta X'(k)}{\Delta X'(k) - \Delta X'(i)} \right) \cdot D_{ki} = \left(\frac{\Delta X'(l) - \Delta X'(k)}{\Delta X'(l) - \Delta X'(i)} \right) \cdot D_{li}, \tag{4.5}$$

where $D_{l,k}$ is the object-space disparity value between the image (l) and image (k), which is transferred from other object-space disparity values of other stereo domains such as $D_{k,i}$ or $D_{l,i}$; $\Delta X'(i)$, $\Delta X'(l)$, and $\Delta X'(k)$ are the computed offsets in the epipolar direction of the same point from the images (i, l, and k) respectively.

$$h = \left(\frac{\Delta Z}{\Delta X'(l) - \Delta X'(k)} \right) \cdot D_{lk} = \left(\frac{\Delta Z}{\Delta X'(l) - \Delta X'(i)} \right) \cdot D_{li}, \tag{4.6}$$

where h is the disparity-based derived elevation value above the selected projection plane; ΔZ is the difference between the selected two parallel object-space planes (if $Z_2 = 0$, then $\Delta Z = Z_{avg}$). Thus, the absolute elevation value from the zero level equals the calculated h value in addition to the elevation of the first projection plane (i.e., $h + Z_{avg}$).

$$X'_{\text{ortho}}(l) = X'(l) - \Delta X'_{\text{ortho}}(l) = X'(l) - \left(\frac{h}{\Delta Z}\right) \cdot \Delta X'(l), \tag{4.7}$$

where $\Delta X'_{\text{ortho}}(l)$ is the distance between the ground pixel $X'(l)$ of the epipolar image (l) to its correct orthographic location $X'_{\text{ortho}}(l)$ in the epipolar direction. Similar to those in Figure 4.4, the distances for the projected image (i), (j), and (k) are $\Delta X'_{\text{ortho}}(i)$, $\Delta X'_{\text{ortho}}(j)$, and $\Delta X'_{\text{ortho}}(k)$, respectively. The correct orthographic location of all these pixels is the same (indicated in the figure as X) since they all represent the same imaged object-space point.

4.4.3.3 Applying the Derived Formulas

Once the epipolar images are created, several disparity maps can be generated from different stereo pair combinations by executing a matching algorithm (e.g., SGM algorithm). To exclude any mismatches that may occur in the matching process, a visibility check should be executed. The most common visibility and consistency check techniques are reviewed by Egnal and Wildes (2002). In this solution, the left–right checking (LRC) technique is selected because of its efficiency and simplicity.

After excluding the false matches from the disparity maps, these maps need to be fused together. However, the disparity conversion formula (DDF) must be applied to every map before fusing them to make the disparity values equivalent to those of the selected reference map (i.e., supplementary disparity maps). After that, the data gaps of the reference maps are filled by implementing a fusion technique to fuse the supplementary disparity maps together. The 3D-median filter fusion technique described by Zhang et al. (2014) is recommended in this study for its efficiency and accuracy. After fusing all the available maps, some small areas may still be without data. Because these gaps are relatively small, an interpolation technique can be used to fill them with minimal risk of inaccuracy.

Once the enriched disparity map is achieved, the DEF can be applied to convert the disparity values to ground elevations. These elevations are co-registered to the selected reference image with pixel-level accuracy. The created elevation model created through these disparity data represents a kind of LoS-DSM since it has the same perspective view of the reference image.

Once the elevation data are available for each rooftop in the reference image, the CDF is applied to correct the varying offsets between the building roofs and their corresponding footprints. The polygon object of each detected roof is shifted by the computed CDF value using its representative point (i.e., the inner centroid of the polygon shape) to the correct footprint location. After that, a geo-referencing process can be applied directly using a 2D-polynomial transformation formula. This step completes the cycle of mapping building roof objects in off-nadir images by providing a geo-referenced detection map.

4.5 EXPERIMENTAL RESULTS AND ANALYSIS

In this section, the experimental results obtained through the developed solutions are described. All the experiments include building detection results in off-nadir VHR satellite images. The data sets used to validate each solution are described in the following subsection.

4.5.1 DATA SETS

Three data sets from two satellite sensors are used to validate the developed solutions. Each solution is validated through experiments on one of these three data sets. Table 4.1 shows the details of these test data.

4.5.2 INTERMEDIATE RESULTS AND DISCUSSION

In this section, the implementation results of the three developed solutions are presented. The last subsection includes the building detection results achieved based on stereo information incorporated using the three developed solutions. While the challenges of elevation co-registration and disparity normalization are addressed in the LoS-DSM and RMAD solutions, respectively, the challenges of disparity gap filling and varying rooftops' offsets are addressed in the DECR solution.

TABLE 4.1
Details of the Data Set Used to Validate Each of the Developed Solutions

No.	Test Data Description	Sol.
Dataset-1	A subset of tri-stereo VHR satellite images acquired by the Pleiades-1A sensor over the urban area of Melbourne, Australia, on February/2012. The product ground resolutions are 0.5 m/pixel and 2 m/pixel for the panchromatic and the four multispectral bands, respectively.	LoS-DSM solution
	The off-nadir acquisition angles of the forward and backward images are approximately +15 and −15, respectively.	
	The stereo pairs are used to derive photogrammetrically digital surface model elevations to be co-registered with one of the off-nadir images.	
	A set of ground control points (GCPs), of accuracy within 10–15 cm, was used to validate the quality of the stereo-based derived data.	
	The area contains more than 100 buildings. This test area was selected specifically to represent a dense urban environment with a variety of building shapes, sizes, and heights including 15 high-rise ones.	
Dataset-2	The two off-nadir stereo images of Dataset-1 (one stereo pair).	RMAD solution
Dataset-3	A subset of five stereo VHR satellite images acquired by the WorldView-2 sensor over the urban area of Rio de Janeiro, Brazil, on January 2010.	DECR solution
	Each of the five MIS-VHR satellite images has eight spectral bands of 1000×1000 pixels and a panchromatic band of 4000×4000 pixels with 2-m and 0.5-m resolutions, respectively.	
	The off-nadir acquisition angles of these images are approximately +45°, +33°, +8°, −30°, and −45°.	
	The imaged area is of a modern city and has many buildings with different shapes and sizes. Most of the buildings are high-rise elevations where building lean and façades 165 are prominent.	

4.5.2.1   LoS-DSM Results and Discussion

The LoS-DSM solution, as introduced in Suliman and Zhang (2015), was developed to resolve the challenge of image-elevation co-registration in off-nadir VHR satellite images. In the solution algorithm, elevation data were generated photogrammetrically. The product of this step was a DSM with an orthographic projection. Since direct co-registration with the image data was not directly possible for off-nadir images, the DSM elevations were then back projected to one of the employed off-nadir images using the forward sensor model information. This back projection led to an elevation layer that has a perspective projection the same as that of the off-nadir image being co-registered. This elevation layer represents the LoS-DSM.

The developed solution was also compared against the traditional 2D image registration. Table 4.2 shows the image-elevation co-registration results achieved based on both the traditional 2D image registration and the LoS-DSM solution. In both methods, the original image data were preserved. While the co-registration achieved based on this LoS-DSM solution was of subpixel accuracy (in terms of root mean square error [RMSE]), the co-registration achieved based on traditional 2D image registration was about 10 times worse than that achieved by the LoS-DSM solution. This quantitative result can be seen visually in the example shown in the same table.

More details and evaluations are provided in Suliman and Zhang (2015). The severe misregistration effect was investigated in the same study through implementing a stereo-based building detection procedure based on both the LoS-DSM solution and the conventional 2D image registration method. The percentage of improvement was identified.

4.5.2.2   RMAD Results and Discussion

The RMAD solution, as introduced in Suliman et al. (2016), was developed to mitigate the normalization challenge owing to the occlusion-produced gaps in the case of only one stereo pair acquired off-nadir over a dense urban area. As described earlier, the stereo pair images were projected onto a horizontal plane in the object space at the average terrain elevation of the imaged area. Then, these two images were epipolar rectified and then co-registered based on terrain-level points (road intersection points). The result of that is minimizing the disparities of the terrain-level objects and maximizing those for the off-terrain-level features (aboveground) in this epipolar co-registered image pair.

Figure 4.5 shows the achieved co-registration results of a severe off-nadir stereo pair. In this figure, the right image of the epipolar pair was co-registered to the left one by applying a 2D translation calculated from a set of ground-level points. The quality of the co-registration, in our implementation, is shown by four points defined by intersections of road centers.

After that, the SGM matching algorithm was executed. The generated disparity map from this epipolar stereo pair was representing the off-terrain objects. Figure 4.6 shows part of the generated disparity map.

More details and evaluations are provided in Suliman et al. (2016). The severe interpolation effects were investigated in the same study through executing the traditional normalization steps (extracting and then subtracting the terrain influence from the surface data) and then implementing a stereo-based building detection procedure based on both the RMAD solution and the conventional techniques. The percentage of improvement was determined.

TABLE 4.2

Quantitative and Qualitative Results Achieved from Both 2D Image Registration and LoS-DSM Methods

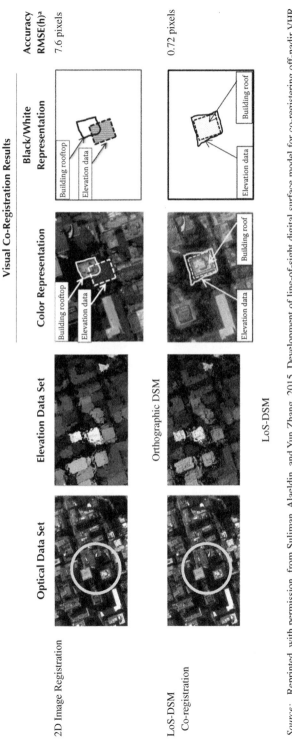

Source: Reprinted, with permission, from Suliman, Alaeldin, and Yun Zhang. 2015. Development of line-of-sight digital surface model for co-registering off-nadir VHR satellite imagery with elevation data. *IEEE Journal of Selected Topics in Applied Earth Observations and Remote Sensing* 8 (5):1913–1923. © 2015 IEEE.

a   RMSE (h), the planimetric/horizontal root mean square error.

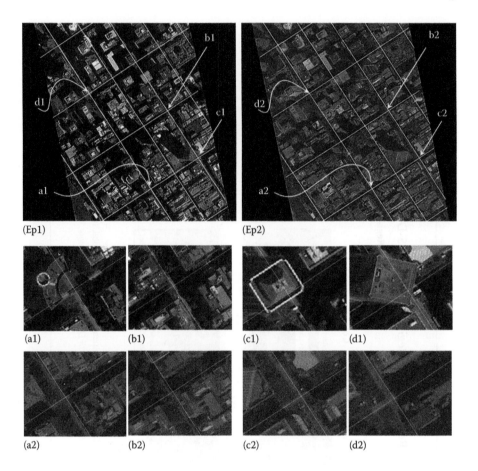

FIGURE 4.5 The result of co-registereing road center lines (straight white lines) in the right epipolar (Ep1) and the left epipolar (Ep2) images. The sub-images are a magnification of the four road intersections in the left (a1, b1, c1, d1) and in the right (a2, b2, c2, d2) images. VHR satellite images. (Reproduced with permission from the American Society for Photogrammetry and Remote Sensing, Bethesda, Maryland, www.asprs.org.)

4.5.2.3 DECR Results and Discussion

As explained in Suliman and Zhang (2018), the DECR method was developed to generate an enriched disparity map based on fusing supplementary disparity data measured from multiple inline stereo satellite images. Additionally, the method was developed to generate a co-registered LoS-DSM, more efficiently than the original algorithm, based on an elevation-to-disparity proportionality relationship. Furthermore, the detection map geo-referencing is included in the DECR solution by correcting the varying offsets of the building roofs in off-nadir images.

As described in major DECR phases earlier, the employed inline stereo satellite images were projected onto a horizontal object-space plane at the average terrain elevation and also onto a second plane at the datum of the imaged area. Then,

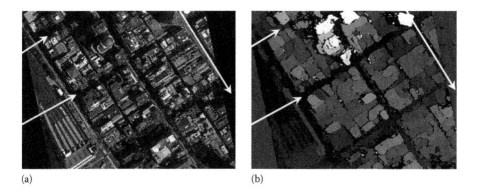

(a)   (b)

FIGURE 4.6   Aboveground disparity map. (a) The left VHR epipolar image, and (b) the resulting aboveground disparity map generated by RMAD technique using SGM algorithm. (Reproduced with permission from the American Society for Photogrammetry and Remote Sensing, Bethesda, Maryland, www.asprs.org.)

the epipolar images were created and the required shift information in the epipolar direction was calculated. Figure 4.7 shows the achieved disparity proportionality among different stereo domains.

This proportionality allows transferring the disparity values from different stereo pairs to the reference image, using DDF as in Equation 4.5, to generate the supplementary disparity maps. These maps are then fused using the 3D-median filter to generate an enriched disparity map. Figure 4.8 shows the generated supplementary disparity maps and the fusion result.

Finally, EDF as in Equation 4.6 was applied to convert the fused disparity data into elevation data that are co-registered to the selected reference image (in our case, an off-nadir image is selected intentionally). These disparity-based co-registered elevation data represent a LoS-DSM. The quality of this elevation co-registration is shown visually in Figure 4.9 through a 3D-rendered representation. More details of the quantitative evaluation of the generated DECR-derived elevations are provided in Suliman and Zhang (2018). The conclusion of the implemented testing and evaluation indicates the high success of the DECR method.

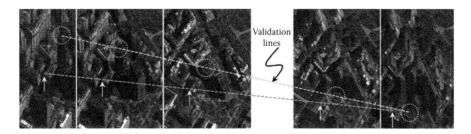

FIGURE 4.7   Disparity inter-proportionality validation among all epipolar images. (Reproduced with permission from the Scientific Research Publishing, Irvine CA 92619-4821, USA. www.scirp.org.)

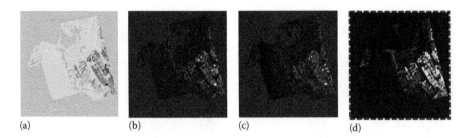

(a)   (b)   (c)   (d)

FIGURE 4.8 The fusion of the supplementary disparity data. The enhanced disparity map (d) fused from supplementary disparity data (a, b, c) generated from different epipolar stereo pairs. (Reproduced with permission from the Scientific Research Publishing, Irvine CA 92619-4821, USA. www.scirp.org.)

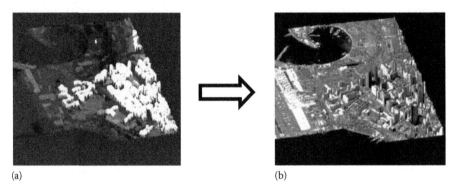

(a)   (b)

FIGURE 4.9 The co-registered LoS-DSM validation. (a) 3D isometric view of the generated LoS-DSM based on the DECR method. (b) 3D rendered representation of the LoS-DSM co-registered to the selected reference image. (Reproduced with permission from the Scientific Research Publishing, Irvine CA 92619-4821, USA. www.scirp.org.)

4.5.3 RESULTS OF STEREO-BASED BUILDING DETECTION

For the three developed solutions described in this chapter, the stereo-based building detection procedure described in Section 4.2 was implemented to evaluate the performance and the improvements resulting from the proposed solutions. The following three subsections describe the achieved building detection results based on the developed solutions.

4.5.3.1 Building Detection Results Based on LoS-DSM Solution

A stereo-based building detection procedure based on elevation data was implemented on an off-nadir image from Dataset-1 (in Table 4.1). This procedure was executed on co-registered elevations based on 2D image registration and the LoS-DSM solution. A representative part from the achieved result is shown in Figure 4.10.

In this figure, the detected rooftops using LoS-DSM were distinguished from each other even when they were very close and the segmentation result was not

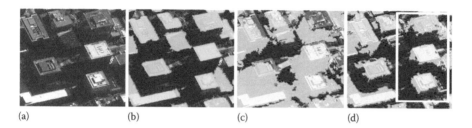

(a)   (b)   (c)   (d)

FIGURE 4.10   A part of the study area contains a few adjacent high-rise buildings. (a) Original off-nadir imagery, (b) reference data of the building roofs, (c) the building detection results achieved using LoS-DSM, and (d) the building detection results achieved by 2D image registration. (Reprinted, with permission, from Suliman, Alaeldin, and Yun Zhang. 2015. Development of line-of-sight digital surface model for co-registering off-nadir VHR satellite imagery with elevation data. *IEEE Journal of Selected Topics in Applied Earth Observations and Remote Sensing* 8 (5):1913–1923. © 2015 IEEE.)

optimized or enhanced. In addition to that, the rooftops of the three high-rise buildings were detected correctly despite the large relief displacements. This is due to the accurate registration of the elevation data achieved by developing the LoS-DSM, which is able to lean with the buildings' lean. In contrast, all of these advantages are missed in the result achieved based on the conventional 2D image registration method. The quality of the building detection results for the two methods under comparison is provided in Table 4.3.

From this table, the detection overall quality achieved by the LoS-DSM method was found to be almost 12% better than that for the conventional registration method while all other parameters used in the building detection process were set exactly the same. Hence, this amount of improvement is mainly attributed to the registration accuracy that assisted in correctly detecting the high-rise buildings' roofs even with severe relief displacements.

TABLE 4.3
Comparison of the Building Detection Results Achieved by Using 2D Image Registration and the LoS-DSM Co-Registration Solution

Method	Overall Quality (%)
2D image registration	58.9
LoS-DSM solution	70.7
Improvement of the LoS-DSM solution	+11.8

Source: Reprinted, with permission, from Suliman, Alaeldin, and Yun Zhang. 2015. Development of line-of-sight digital surface model for co-registering off-nadir VHR satellite imagery with elevation data. *IEEE Journal of Selected Topics in Applied Earth Observations and Remote Sensing* 8 (5):1913–1923. © 2015 IEEE.

4.5.3.2   Building Detection Results Based on RMAD Solution

A stereo-based building detection procedure based on disparity data was implemented on an off-nadir image from Dataset-2 (in Table 4.1). This procedure was
executed on the achieved RMAD-based disparity data as described briefly in Section
4.4.2. Additionally, the same building detection procedure was executed based on a
disparity map that was interpolated and normalized using the traditional normalization steps. Figure 4.11 shows the achieved disparity maps based on the RMAD
method and the traditional interpolation and normalization techniques. These maps
were used directly to identify the elevated objects in the off-nadir image employed.
The detection results are shown in the same figure.

As a quantitative evaluation for the achieved results, Table 4.4 shows a comparison between the developed RMAD technique and the conventional normalization
technique. The RMAD-based detection overall quality of 75%, as in Table 4.4, is
affected directly by the missed roofs from the detection. However, the achieved quality is still promising for building roof detection using only disparity information.
Moreover, there are no building façades detected in the results due to having VHR
stereo images of opposite backward and forward viewing angles. Unlike the currently available techniques, this viewing geometry represents an advantage in our
solution.

As in Table 4.4, the quality of the detection achieved based on the conventional
interpolation and normalization techniques was only 43%. This low value was
affected directly by the high false detection achieved (i.e., high completeness). This
proves the severe negative effects of the interpolation process in dense urban areas.
In contrast, the developed RMAD technique provided an efficient solution for such
dense areas. The degree of the improvement in the tested area in this study was 32%.
Obviously, this value mainly depends on the nature of the urban area. However, it
still shows an attractive degree of improvement for the proposed disparity normalization technique based on co-registration.

■ Dark gray: true positive (TP)
▨ Light gray: false positive (FP)

(a)   (b)   (c)   (d)

FIGURE 4.11   Comparison between the developed RMAD and conventional normalization
techniques. (a) The achieved aboveground disparity map generated by RMAD, and (b) the
surface disparity map normalized by conventional steps. Results based on these two normalized disparity maps are shown in (c) and (d), respectively. (Reproduced with permission from
the American Society for Photogrammetry and Remote Sensing, Bethesda, Maryland, www
.asprs.org.)

TABLE 4.4
Comparison of the Building Detection Results Achieved
by the RMAD Technique and the Conventional
Normalization Technique

Method	Overall Quality (%)
RMAD technique	75
Conventional technique	43
Improvement of the RAMD solution	+32

Source: Reproduced with permission from the American Society for Photogrammetry and Remote Sensing, Bethesda, Maryland, www .asprs.org.

4.5.3.3 Building Detection Results Based on DECR Solution

A stereo-based building detection procedure based on disparity and elevation was implemented on an off-nadir image from Dataset-3 (in Table 4.1). This procedure was executed on the achieved DECR elevation data as described in Section 4.4.3. However, a visibility checking technique (the LRC) was executed during the detection process for two reasons: (1) to eliminate the building facades by generating an occlusion mask and (2) to identify the unstable/false matches. This LRC technique was executed after suppressing the vegetation and thresholding the DECR elevations, to identify building roofs, based on empirical threshold values. Figure 4.12 shows the

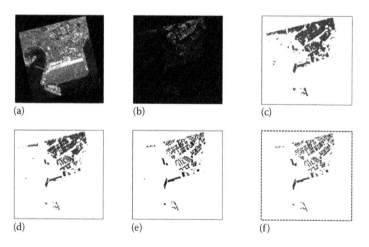

(a)   (b)   (c)

(d)   (e)   (f)

FIGURE 4.12 The intermediate building detection result based on DECR method. (a) Epipolar off-nadir VHR reference image (Ep-2). (b) Disparity-based co-registered LoS-DSM using DECR method. (c) Detected off-terrain objects based on a thresholding operation of a value close to Z_{avg}. (d) Resulting objects after suppressing vegetation objects based on an NDVI bitmap. (e) Resulting objects after removing the building façades. (f) The manually generated reference data for comparison. (Reproduced with permission from the Scientific Research Publishing, Irvine CA 92619-4821, USA. www.scirp.org.)

intermediate results and the final enhanced building detection results along with the manually generated reference data.

When the results were evaluated quantitatively against manually generated reference data, the overall-quality measure was found to be 92%. This high-quality result is attributed to the perfect image-elevation co-registration since the DECR exploits the advantages of the perfectly co-registered disparity maps. Additionally, the accuracy of the stereo-based extracted information (both disparities and elevations) is reflected directly in the detection quality. This indicates that the derived data were complete and almost all the occlusion-produced gaps were filled correctly.

Finally, to close the cycle of building detection by allowing a direct geo-reference of the detection map, the varying roof offsets were then corrected. First, the mapped rooftops were all abstracted by their inner-centroid points and the elevations at these points were extracted. Then, based on Equation 4.7, the mapped building roofs in the epipolar image were all shifted—based on their inner centroids—to the correct footprint location in the epipolar image. Then, a 2D-polynomial function was used to geo-reference the detection roof map. Figure 4.13 shows examples of the result before and after geo-reference. An ortho-rectified image was generated for the test data set to be used as a reference data to validate the geo-referencing result.

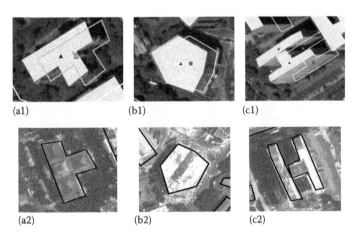

▲ The location of RP of detected roof object in the epipolar image.
■ The location of the RP of the moved roof object to the correct location in the epipolar image.

FIGURE 4.13 A few examples of the geo-referenced roof objects for visual assessment. The upper row subfigures (a1, b1, c1) show the detection result of three building roofs as represented with the filled polygons. The white line shapes represent the correct orthographic locations of the detected roofs after removing the lean effect. The shift distance to correct the lean effect is the distance between the inner centroids of each filled polygon and its corresponding line shapes. The lower row subfigures (a2, b2, c2) illustrate the geo-referencing result after applying the calculated shift distance. The geo-referencing quality is shown through the achieved alignment between the black borders and building roofs. (Reproduced from the Scientific Research Publishing, Irvine CA 92619-4821, USA. http://www.scirp.org. With permission.)

Object-based evaluation of the building roof geo-referencing indicates that almost 100% of the roofs were geo-referenced correctly.

4.6 CONCLUSIONS

For the application of building detection, height information that includes either elevation or disparity is important for reliable building identification since buildings are inherently elevated objects. However, the traditional 3D-assisted building detection procedures usually avoid using off-nadir VHR images, despite their high availability, as a source for the required height information.

In this chapter, the principle of stereo-based building detection was described. The challenges of incorporating stereo-based information with off-nadir VHR satellite images were identified. Accordingly, three solutions were developed and validated.

The challenges associated with building lean in off-nadir images include building roof offsetting from its footprint, and object occlusion and building façades, in addition to the variation of the roofs' offsets relative to the building heights. While the offsets of building roofs create difficulties in the co-registration between image and elevation data, the building façades and occlusions create large data gaps when stereo matching is executed automatically in off-nadir image pairs. Furthermore, because of the variation in building-roof offsets, the mapped roofs extracted from off-nadir images cannot be directly geo-referenced to existing maps for effective information integration.

In this chapter, three efficient solutions for the identified problems were presented. The applicability of all the developed solutions was demonstrated through 3D-assisted building detection in off-nadir VHR satellite images.

First, the LoS-DSM solution was developed to provide more efficient image-elevation co-registration than the currently available techniques. An improvement of about 12% in building detection results was achieved over the traditional 2D registration method. This solution is recommended when either pre-existing DSM data or multiple overlapped images are available.

Second, the RMAD solution was developed to reduce the computation cost and allow bypassing a few time-consuming steps required by the traditional normalization methods. An improvement of about 30% in building detection result was achieved over the traditional normalization technique. This technique is recommended when only one off-nadir stereo pair of VHR satellite images is available. However, the identified limitation of the RMAD method is expected to appear in hilly areas. This means that the minimization of the terrain variation by defining a PRP at the average elevation for the area will not be enough to eliminate the terrain disparities.

Third, the DECR solution was introduced to generate and fuse supplementary disparity data and to efficiently derive elevation data that are co-registered accurately to one of the employed off-nadir images. Additionally, this solution, based on the derived building elevations, provides the shift distances required to correct the building roofs' offsets and geo-reference the generated detection map. This solution is recommended when multi-view off-nadir VHR stereo satellite images acquired via inline acquisition capability over a dense urban area are available. On the basis of the DECR solution, an overall quality of 92% was achieved in the building detection result. The geo-referencing result of the building roofs based on the developed correct-distance formula was found to be almost 100%. However, the DECR method was developed to derive elevations based on a scale relationship with the corresponding disparity data, and it was tested using inline stereo satellite images. Hence, the relationship requires further investigation and the method requires testing using across-track stereo VHR satellite images.

On the basis of the outcome achieved, this research proved the possibility of using off-nadir VHR satellite images for accurate urban building detection. It significantly increases the data source scope for building detection since most (>95%) VHR satellite images are off-nadir and traditional methods cannot effectively handle off-nadir images.

ACKNOWLEDGMENTS

The research was supported in part by funding from Libyan Ministry of Higher Education and Scientific Research and the Canada Research Chair Program. The authors would like to thank Airbus Defence and Space and Digital Globe for providing the data used in the research.

REFERENCES

American Association of State Highway and Transportation Officials (AASHTO). 2001. *A Policy on Geometric Design of Highways and Streets 2001*. Washington, D.C.: The Association.

Alobeid Abdalla, Karsten Jacobsen, and Christian Heipke. 2010. Comparison of matching algorithms for DSM generation in urban areas from Ikonos imagery. *Photogrammetric Engineering and Remote Sensing* 76 (9):1041–1050.

Dey, Vivik. 2013. Image segmentation techniques for urban land cover segmentation of VHR imagery: Recent developments and future prospects. *International Journal of Geoinformatics* 9 (4):15–35.

Egnal, Geoffrey, and Richard P. Wildes. 2002. Detecting binocular half-occlusions: Empirical comparisons of five approaches. *Pattern Analysis and Machine Intelligence, IEEE Transactions on* 24 (8):1127–1133.

Ghaffarian, Salar, and Saman Ghaffarian. 2014. Automatic building detection based on supervised classification using high resolution Google Earth images. *ISPRS Technical Commission III* Symposium, Zurich, Switserland, 5–7 September.

Gong, Cheng, and Junwei Han. 2016. A survey on object detection in optical remote sensing images. *arXiv preprint arXiv:1603.06201*.

Grodecki, Jacek, and Gene Dial. 2003. Block adjustment of high-resolution satellite images described by rational polynomials. *Photogrammetric Engineering and Remote Sensing* 69 (1):59–68.

Habib, Ayman F., Michel Morgan, Soo Jeong, and Kyung-Ok Kim. 2005. Analysis of epipolar geometry in linear array scanner scenes. *The Photogrammetric Record* 20 (109):27–47. doi: 10.1111/j.1477-9730.2005.00303.x.

Hirschmüller, Heiko. 2008. Stereo processing by semiglobal matching and mutual information. *IEEE Transactions on Pattern Analysis and Machine Intelligence* 30 (2):328–341. doi: 10.1109/TPAMI.2007.1166.

Hussain, Ejaz, and Jie Shan. 2016. Urban building extraction through object-based image classification assisted by digital surface model and zoning map. *International Journal of Image and Data Fusion* 7 (1):63–82. doi: 10.1080/19479832.2015.1119206.

Martin, Baatz, and Arno Schäpe. 2000. Multiresolution segmentation: An optimization approach for high quality multi-scale image segmentation. *Angewandte Geographische Informationsverarbeitung XII*:12–23.

Mishra, Rakesh, and Yun Zhang. 2012. A review of optical imagery and airborne LiDAR data registration methods. *The Open Remote Sensing Journal* 5:54–63.

Nielsen, Michael Meinild, and Ola Ahlqvist. 2014. Classification of different urban categories corresponding to the strategic spatial level of urban planning and management using a SPOT4 scene. *Journal of Spatial Science* 60 (1):99–117. doi: 10.1080/14498596.2014.943309.

Potůčková, Markéta, and Petr Hofman. 2016. Comparison of quality measures for building outline extraction. *The Photogrammetric Record* 31 (154):193–209.

Rutzinger, Martin, Franz Rottensteiner, and Norbert Pfeifer. 2009. A comparison of evaluation techniques for building extraction from airborne laser scanning. *IEEE Journal of Selected Topics in Applied Earth Observations and Remote Sensing* 2 (1):11–20.

Salehi, Bahram, Yun Zhang, Ming Zhong, and Vivik Dey. 2012. A review of the effectiveness of spatial information used in urban land cover classification of VHR imagery. *International Journal of Geoinformatics* 8 (2):35–51.

Singh, Surendra Pal, Kamal Jain, and V. Ravibabu Mandla. 2014. Image based 3D city modeling: Comparative study. *The International Archives of Photogrammetry, Remote Sensing and Spatial Information Sciences* 40 (5):537–546. doi:10.5194/isprsarchives-XL-5-537-2014.

Sugg, Zachary P., Tobias Finke, David C. Goodrich, M. Susan Moran, and Stephen R. Yool. 2014. Mapping impervious surfaces using object-oriented classification in a semiarid urban region. *Photogrammetric Engineering and Remote Sensing* 80 (4):343–352. doi: 10.14358/PERS.80.4.343.

Suliman, Alaeldin, and Yun Zhang. 2015. Development of line-of-sight digital surface model for co-registering off-nadir VHR satellite imagery with elevation data. *IEEE Journal of Selected Topics in Applied Earth Observations and Remote Sensing* 8 (5):1913–1923.

Suliman, Alaeldin, and Yun Zhang. 2017. Double projection planes method for generating enriched disparity maps from multi-view stereo satellite images. *Photogrammetric Engineering and Remote Sensing* 83 (11): 749–760.

Suliman, Alaeldin, and Yun Zhang. 2018. Disparity-based generation of line-of-sight DSM for image-elevation co-registration to support building detection in off-nadir VHR satellite imagery. *Journal of Geographic Information System* 10 (1): In Press.

Suliman, Alaeldin, Yun Zhang, and Raid Al-Tahir. 2016. Registration-based mapping of aboveground disparities (RMAD) for building detection in off-nadir VHR stereo satellite imagery *Photogrammetric Engineering and Remote Sensing* 82 (7):535–546.

Wang, Mi, Fen Hu, and Jonathan Li. 2011. Epipolar resampling of linear pushbroom satellite imagery by a new epipolarity model. *ISPRS Journal of Photogrammetry and Remote Sensing* 66 (3):347–355. doi: http://dx.doi.org/10.1016/j.isprsjprs.2011.01.002.

Weng, Qihao, and Dale A. Quattrochi. 2007. *Urban Remote Sensing*. Boca Raton: CRC Press.

Xie, Yanhua, Anthea Weng, and Qihao Weng. 2015. Population estimation of urban residential communities using remotely sensed morphologic data. *IEEE Geoscience and Remote Sensing Letters* 12 (5):1111–1115. doi: 10.1109/LGRS.2014.2385597.

Zhang, Jing, Yang Cao, Zhigang Zheng, Changwen Chen, and Zengfu Wang. 2014. A new closed loop method of super-resolution for multi-view images. *Machine Vision and Applications* 25 (7):1685–1695.

Zhang, Yun. 2004. Understanding image fusion. *Photogrammetric Engineering and Remote Sensing* 70 (6):657–661.

Zhang, Yun, C. Vincent Tao, and J. Bryan Mercer. 2004. An initial study on automatic reconstruction of ground DEMs from airborne IfSAR DSMs. *Photogrammetric Engineering and Remote Sensing* 70 (4):427–438.

Zhao, Dan, Xiuxiao Yuan, and Xin Liu. 2008. Epipolar line generation from IKONOS imagery based on rational function model. *International Archives of the Photogrammetry, Remote Sensing and Spatial Information Sciences* 37 (B4):1293–1297.

5 Beyond Built-Up: The Internal Makeup of Urban Areas

Benjamin Bechtel, Martino Pesaresi,
Aneta J. Florczyk, and Gerald Mills

CONTENTS

5.1 INTRODUCTION

The Anthropocene Working Group of the International Union of Geological Sciences recently concluded that, as a concept, the Anthropocene is geologically real and it identified the mid-twentieth century as the optimal beginning of this Epoch.* This date also represents the beginnings of a second "wave" of urbanization that has accompanied global population growth; while the global population was 2.5 billion (of which 750 million were urbanites) in 1950, by now, more than half of a population of 7.3 billion live in cities, and among 80% of today's humanity live in areas with a population density of more than 300 persons per square kilometer (Pesaresi et al. 2016b). While global population will continue to increase (by 2030, there will be an estimated 8.5 billion), this growth will

---

* https://phys.org/news/2016-08-anthropocene-scientists.html

become concentrated in urban areas as the rural population declines (UN 2015). Cities are simultaneously places that host most of the prominent historical–cultural artifacts and political and economic infrastructure; permit the development of advanced labor market and occasions for human exchange and cultural and social development; and arguably offer the most spatially efficient solution for managing human occupation of the planet. At the same time, they are also places of intense landscape change and pollutant emissions (including greenhouse gases) that profoundly modify natural systems at all scales. It is difficult to escape the conclusion that planetary urbanization is a major driver of environmental change at a hierarchy of scales, including that of the planet. Finally, cities are also exposed to a range of natural hazards, including flooding and storms owing to their location close to coasts, in river valleys, and at low elevation. Therefore, globally consistent data about cities and, more generally, human settlements are absolutely critical for monitoring post-2015 international frameworks, including the UN Third Conference on Housing and Sustainable Urban Development—Habitat III—in 2016, the Sustainable Development Goals, the UN Framework Convention on Climate Change, and the Sendai Framework for Disaster Risk Reduction 2015–2030 (DRR).

Given the scale and rapidity of the urbanization processes, these data are needed urgently. As objects of study, cities pose a conundrum at the scale of the planet. They occupy relatively small and fragmented portions of the global landscape, their edges are complex to describe, and their makeup is extremely heterogeneous in terms of their surface materials and their spatial patterns and arrangements (see Figure 5.1). However, a sufficiently detailed and consistent database on the character and internal composition of cities globally is needed to underpin their scientific study; to permit the development of models, test hypotheses, and monitoring of trends; and to explore possible alternative urban development scenarios.

Ideally, data on the internal makeup of cities worldwide would inform us of aspects of urban form and function and be available at a "useful" scale to support decision making. Urban form describes three aspects of the physical landscape: the surface cover (the proportion of the landscape that is paved and vegetated), the material composition of the surface (the character of the built fabric), and the geometry (the dimensions of buildings and streets). Function describes the metabolism of the city—that is, the inputs of food, water, materials, and energy that are needed to sustain and change the urban landscape and support its occupation. The metabolism includes the release of waste energy and water into the overlying atmosphere (the anthropogenic heat flux) (Oke et al. 2017). The relative roles of form and function differ from city to city depending on background climate (warm or cool), economic structure (e.g., industrial sector), and urban layout (e.g., urban extent and transportation infrastructure). Moreover, these two aspects of cities are correlated: for example, the evidence suggests that in the same circumstances, compact, densely inhabited and built cities use less resources per capita.

Urban data to support decision making at a global scale should permit comparison between cities, allow the assessment of exposure to hazard, and guide appropriate adaptation and mitigation strategies. Given the extraordinary spatial heterogeneity within and among cities, the challenge is how best to describe urban landscapes in a meaningful manner. One approach is to decompose the urban landscape into neighborhoods (~1 km²) that are relatively homogenous in terms of physical layout

(a)

(b)

FIGURE 5.1   Shades of urban—diversity of urban structures within the built-up: (a) wooden house in Eastern Finland, (b) center of Dublin. (© Benjamin Bechtel, all rights reserved.)

(Continued)

(c)

FIGURE 5.1 (CONTINUED)   Shades of urban—diversity of urban structures within the built-up: (c) center of Hong Kong. (© Benjamin Bechtel, all rights reserved.)

(e.g., road dimensions, green cover, building types, and placement) and common functions such as residential, industry, and so on. In other words, a consistent description of urban land use and land cover (LULC) is needed but LULC schemes are usually designed for city-specific purposes and are inconsistent in terms of scales and categories. Satellite-based sensors are ideally suited to acquiring consistent land-cover data at a global scale, but the signal from urban landscapes is mixed spectrally owing to the number of distinct entities (trees, buildings, parks, etc.) that are contributing; these entities have a characteristic scale of 10 m (Small 2003, 2005). As a consequence, detailed mapping of urban surfaces would require metric- or decametric-scale sensors but, until recently, the only global urban databases were generated using sensors with spatial resolution ranging from hundreds to thousands of meters (Potere et al. 2009; Schneider et al. 2003, 2009). As a result, the available databases have identified compact and large patches of urbanized surfaces but generally fail to identify low-density, open, and scattered settlement patterns.

Studies that have used high-resolution imagery have, until recently, been used to report on the comparative extent of urbanized areas for selected cities; for example, Angel et al. (2005) mapped 120 cities over 1990 and 2000 and Taubenböck et al. (2012) conducted a systematic analysis of 27 current mega cities using multi-temporal Landsat data from 1975, 1990, and 2000 and TerraSar-X data from 2010. The science is now at a stage where the detailed mapping of the global urban cover is possible. The Global Urban Footprint was generated using TerraSar-X data for the years 2011–2013 as part of the TanDEM-X mission (Marconcini et al. 2013). The Monitoring of Global Land Cover project has produced a detailed global impervious surface cover data set for a 2006 base year (Gong et al. 2013). The 30-m-resolution global land cover (GlobeLand30) data set has been derived using two Landsat data collections (years 2000 and 2010), and GlobeLand30 integrates other internationally available land cover products and relies on large use of manual editing of the final information done by domain experts (Chen et al. 2015). A global urban area map was also produced by Miyazaki et al. (2013) using an automated classification method based on ASTER data. The approach combines Learning with Local and Global Consistency and logistic regression with urban maps. Finally, globally complete and multi-temporal assessment of artificial built-up surfaces were first introduced by the Global Human Settlement Layer (GHSL) project in 2014, using global sets of different Landsat sensor data collected in 1975, 1990, 2000, and 2014 (Pesaresi et al. 2016a).

These databases on the extent of the built-up area are used as a first-order description of human settlements and are useful as baseline data in many studies. For example, satellite-derived data can be used to assess exposure at all scales (Ehrlich and Tenerelli 2013; Ehrlich et al. 2013) and exposure at the global level is mainly derived from human settlement information (Pesaresi et al. 2015a). However, information on the urban land cover alone that does not discriminate between neighborhoods is very limited in its capacity to support more sophisticated analysis of urban risks that includes vulnerability. Moreover, existing modeling tools that are capable of evaluating the urban effect on local climate (e.g., urban heat island [UHI]) and hydrology (e.g., pluvial flooding) and assessing the impact of planning strategies cannot be employed. Satellite data can help decompose the urban land cover into common urban structural types (USTs), but the complexity of urban built forms, the heterogeneity of materials, and the multiplicity of spectral properties have impeded progress. As a result, UST studies to date have focused on individual cities and the methodological approaches are not generic enough to be applied on a global basis (Heiden et al. 2012; Voltersen et al. 2015; Wurm et al. 2009).

Recently, there have been two projects on mapping the spatial character of cities that have the potential for providing a standard approach for a global urban database. The World Urban Database and Portal Tools (WUDAPT) project uses the Local Climate Zone (LCZ) classification scheme (Stewart and Oke 2012), available satellite data, and local expertise to describe the characteristics of different urban neighborhoods in a city landscape (Bechtel et al. 2015). The GHSL project is designed to produce maps of built-up density automatically by integrating several available sources that capture aspects of human settlement including remotely sensed imagery. In addition, it delivers an experimental GHSL-LABEL product that gathers built-up area characteristics stratified by vegetation cover and building height (Pesaresi et al. 2016a).

In this chapter, the WUDAPT-LCZ scheme and the GHSL-LABEL product are compared based on selected cities, and their advantages and disadvantages as a standard for UST classification in urban remote sensing are presented. There is a pressing need across a range of scientific and practical disciplines for detailed and consistent data on urban landscapes for cities worldwide. To illustrate, we start by discussing climate change and cities and overcoming the impediments to linking global-scale change to urban-scale decisions.

5.2   URBAN DATA FOR GLOBAL CLIMATE SCIENCE

The small and fragmented nature of urban areas at a planetary scale makes them difficult to include as spatial entities in climate models. Their aggregate role as a significant driver of global climate change is well known, but until very recently, the science at this scale has not explicitly incorporated cities; rather, they are included as anthropogenic drivers in the sectoral sources of greenhouse gases (mitigation) and as places at risk of projected climate changes (adaptation). Naturally, this is a weak solution as it greatly diminishes the capacity to make informed decisions at urban scales. Improving the global climate science in this context requires three components: technological advancements that will allow cities to be explicitly included in Earth System Models, theoretical developments that integrate the hierarchy of climate effects and integrate urban climate science, and a sufficiently detailed and consistent database on the character of cities (the focus of this chapter) that can support these initiatives.

The urban impact on climates has been studied systematically for more than 50 years and the measurement of the urban effect on climate is more than 200 years old. The best known of these effects is the UHI, which describes the fact that cities are generally warmer than the surrounding areas. The UHI is an outcome of landscape transformation and direct heating by human activities. The replacement of natural cover by paved surfaces means that, generally, urban surfaces are dry and much of the available energy is used to warm the substrate and overlying air and little is expended in evaporation. Typical urban fabrics are dense and have specific thermal and radiative properties, which makes them suited to the absorption and retention of solar energy. Finally, the crenulated urban surface restricts access to solar energy in streets and slows the passage of air, but it also restricts the nighttime loss of energy to the overlying sky. In addition, the anthropogenic heat flux (caused by building heat and vehicle loss mainly) also contributes to the UHI. Typically, the magnitude of the UHI increases from the edge of the city to the city center, and this pattern conforms to typical urban layouts: widely spaced buildings with ample green space in the suburbs and tightly spaced, taller buildings with limited vegetation in the city center. This general description of the UHI illustrates that its magnitude and spatial form are closely related to the character of the urban surface (Oke et al. 2017). In fact, the LCZ scheme was designed specifically to provide a rational and universal description of urban landscapes suitable for UHI studies, but critically, it provides a framework for capturing data on a range of surface characteristics that regulate many climatic and hydrological impacts. The value of these data has been demonstrated in a number of urban modeling and observation studies; for instance, these data have been integrated with the Surface Urban Energy and Water Balance Scheme (Alexander et al. 2015, 2016) and the Weather Research Forecasting model (Brousse et al. 2016) to explore the urban impact for selected cities.

However, the potential for such urban data is much greater. While the UHI (and other urban phenomena) was largely seen as a local-scale issue, sustained and accelerating global urbanization and the recognition of the impact of cities on global climate change (and vice versa) have changed this view. Improvements in atmospheric modeling capacity now permit multi-scalar approaches that can incorporate

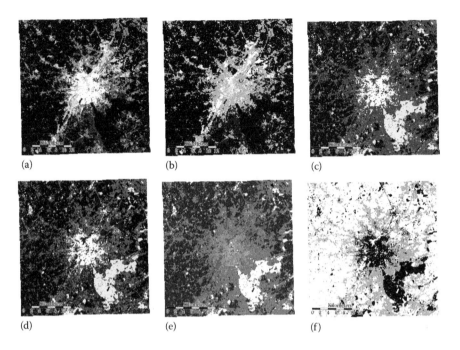

(a)   (b)   (c)

(d)   (e)   (f)

FIGURE 5.2 Urban canopy parameters estimated from the Local Climate Zone map for Brussels: (a) built fraction, (b) impervious fraction, (c) height of roughness elements, (d) aspect ratio, (e) surface roughness, and (f) albedo.

urban-scale processes into global climate models (Jackson et al. 2010). These developments are central to addressing the drivers of climate change and assessing the impacts of projected changes for the places where humanity is concentrated. This gap is recognized in the latest IPCC assessment reports on both adaptation and mitigation (Pachauri et al. 2014).

Previous studies found that a discrimination into urban neighborhoods is suitable for modeling purposes and that even a decomposition into just three classes might be sufficient for some modeling applications (Lee et al. 2011; Loridan et al. 2010). Here, we investigate whether the GHSL-LABEL and WUDAPT-LCZ products could be unified into a semantically aggregated scheme that can fulfill this need. Moreover, the LCZ scheme is linked to a number of physical descriptors and properties of the respective neighborhoods, which readily can be used for modeling purposes. Figure 5.2 shows some of these so-called urban canopy parameters derived from LCZs for Brussels, Belgium.

5.3 GHSL-LABEL

The GHSL-LABEL data were produced in 2014 during the first experimental test of the GHSL processing workflow using global satellite data records collected by the Landsat satellite platform in the past 40 years (Pesaresi et al. 2016a). The remote sensing data used in the Landsat GHSL experiment of 2014 consisted of 32,808 scenes organized in four collections corresponding to 1975, 1990, 2000, and 2014, including 7588, 7375, 8756, and 9089 satellite scenes, respectively. The collections for 1975, 1990, and 2000 were made by the Global Land Survey data preprocessed by the Maryland University and available for public use since 2008 (Gutman et al. 2013). The collection for 2014 was made by a direct download of 9089 Landsat 8 scenes from the USGS platform, mainly targeting data collected in high vegetation growing season.

In the standard GHSL paradigm, supervised satellite data classification techniques are applied in order to produce a systematic assessment of the presence of artificial built-up surfaces on the whole Earth's landmass. The training set data used as input of the supervised classification are derived from available global information layers containing an approximate representation (in both spatial and thematic domains) of the target class abstraction under analysis in the satellite data input. These classification techniques have been recently introduced as Symbolic Machine Learning (SML) (Pesaresi et al. 2016d) and are based on data sequencing and associative analysis similarly to applications in other data mining domains such as bioinformatics and genome characterization. The SML classification techniques showed good performances in data processing scenarios with large volume of input data, large noise in the training set, and ill-defined target class abstraction in the input image data feature space (Pesaresi et al. 2015b, 2016c).

In the GHSL, the built-up area class abstraction is defined as the *union of all the spatial units collected by the specific sensor and containing a roofed built structure or any portion of it* (Pesaresi et al. 2013). The GHSL-LABEL variant was introduced in the GHSL experimental workflow of 2014 in order to test the possibility to generate automatically global, fine-scale multi-class abstractions improving the capacity to describe the internal physical characteristics of the human settlements (Pesaresi et al. 2016a). The LABEL experimental product was generated by processing the 2014 Landsat GHSL, made exclusively by Landsat 8 (LS8) data scenes. The process was based on two main distinct analytical reasoning phases: (a) *inductive*, supported by SML techniques, and (b) *deductive*, supported by expert knowledge.

In the inductive phase, the confidence in assigning 1 of 12 land-cover classes was evaluated for each spectral data sequence recorded by the satellite sensor, using SML techniques. The set of class abstractions included in the evaluation were (a) a simplification (aggregation, logical union) of the GlobCover (GLC) classification schema (Bontemps et al. 2011) to 10 land cover classes, (b) the *road surface* class abstraction as derived from the OpenStreetMap data repository (www.openstreetmap.org), and (c) the *built-up area* class abstraction as derived from the standard GHSL workflow. At the end of this inductive phase, each satellite image data record (pixel) was assigned to the class abstraction associated to the maximum of confidence according to the SML assessment of each individual class hypothesis against the set of all the others. In the subsequent deductive phase, the resulting data records labeled with the *built-up* class abstraction during the inductive phase were subdivided into eight subclasses according to a rule-based hierarchical combination of the three criteria listed below:

i. The presence of high-reflectance surfaces
ii. The presence of vegetation cover
iii. The estimated volume of the built-up structures

The first criterion was assessed from the information included in the quality (QA) band of the LS8 data. Specifically, morphological filtering (open by reconstruction) was applied for small isolated patches of data classified as *cirrus* within the built-up area. The Automatic Cloud Cover Assessment algorithm supporting the production of the QA band of the LS8 data is known to produce false positives for highly reflecting materials (Irish et al. 2006) (e.g., dry soils, silicosis rocks, and large concrete surfaces)—modern, large prefabricated buildings (such as commercial buildings) fit this criterion. The second criterion was assessed by the normalized vegetation index (NDVI) calculated from the top-of-atmosphere radiometrically calibrated image data using the calibration parameters included in the LS8 metadata package. The third criterion was assessed by a 3D level-contrast index (3Dr) calculated from morphological and textural filtering of an available global digital surface model (DSM) according to the open and free data access policy of the GHSL project. Specifically, the shuttle radar topographic mission (NASA 2015) and the ASTER GDEM project (https://asterweb.jpl.nasa.gov/gdem.asp) were processed in this phase.

The deductive phase of the classification of the built-up areas followed a set of rules. For all the image data records (pixels) classified as built-up area in the inductive phase, first check if they are *high reflectance* and label the output class correspondingly. If not, check for different cutoff values of the estimated vegetation cover considered as an inverse proxy of the intensity of imperviousness. In this phase, four subclasses of built-up areas are identified as *very light imperviousness* (NDVI > 0.4), *light imperviousness* (0.3 < NDVI ≤ 0.4), *medium imperviousness* (0.2 < NDVI ≤ 0.3), and *strong imperviousness* (NDVI ≤ 0.2), characterizing the built-up class by different levels of the vegetation cover as estimated from the NDVI measures. Subsequently, only if the data record falls inside the *strong imperviousness* case should one check for different cutoff values of the estimated volumetric index from DSM data. Specifically, four different subclasses are identified in this phase: *average low-rise built-up structures* (3Dr ≤ 25 m), *average medium-rise built-up structures* (25 m < 3Dr ≤ 50 m), *average high-rise built-up structures* (50 m < 3Dr ≤ 100 m), and *average very high rise built-up structures* (3Dr > 100 m). The 3Dr index expresses the estimation of the average difference between terrain and top roof surface elevation values (meters) in a spatial neighborhood of circa 150 m of radius.

Table 5.1 shows the final classification schema of 19 classes (11 classes belonging to the *not built-up* and 8 to *built-up* abstractions) produced in the GHSL-LABEL experimental output. It reports about the GHSL-LABEL class code and nomenclature, the source used as training set in the inductive learning phase, and the rule set applied to the "built-up" class abstraction in order to produce the final classification schema.

5.4 WUDAPT-LCZ

The LCZ classification scheme was originally developed to provide a standardized approach for describing and reporting meteorological field sites used in UHI studies (Stewart 2011). It provides a physical description of 17 different landscape types (that are largely culturally neutral), which can be used to explain the thermal response of the near-surface atmosphere (Stewart and Oke 2012) (see Table 5.2). Each type is

TABLE 5.1

The GHSL-LABEL Classification Schema, the Sources Used as Training Set in the Inductive Learning Phase, and Rule Set Applied to the "Built-Up" Class Abstraction

	Description		Inductive (SML *Training Set Source*)	Deductive *(Rule Set)*
1	Other		GLC	
2	Ice and snow		GLC	
3	Bare soil and rocks		GLC	
4	Shrubs and grassland		GLC	
5	Mosaic croplands and forest		GLC	
6	Rain cropland		GLC	
7	Irrigated cropland		GLC	
8	Forest		GLC	
9	Water	Occasionally water/ land–water interface	GLC	
10		Surface water	GLC	
11	Roads		OSM	
12	Built-up	Highly reflecting roof	GHSL, various	Filtering from Landsat QA
13		Very light impervious		NDVI > 0.4
14		Light impervious		$0.3 < \text{NDVI} \leq 0.4$
15		Medium impervious		$0.2 < \text{NDVI} \leq 0.3$
16		Strong impervious	Low rise		NDVI ≤ 0.2 and 3Dr ≤ 25 m
17			Medium rise		NDVI ≤ 0.2 and $25 < 3\text{Dr} \leq 50$ m
18			High rise		NDVI ≤ 0.2 and $50 < 3\text{Dr} \leq 100$ m
19			Very high rise		NDVI ≤ 0.2 and 3Dr > 100 m

Source: Adapted from Bechtel, Pesaresi et al. (2016).

associated with recognizable urban forms and is linked to typical ranges of numerical values associated with surface cover, fabric, geometry, and anthropogenic heat flux. Mapping LCZ types across a city is therefore also a mapping of the numerical descriptors (Bechtel et al. 2015); for example, an LCZ map can be used to derive spatial coverages of impermeable surface cover, of approximate building height, and of anthropogenic heat flux, which can be incorporated in climate models at urban, regional, and even global scales (Alexander et al. 2015; Ching 2013).

Multiple schemes for mapping LCZs have been proposed and evaluated in the course of developing the WUDAPT project, including (i) manual sampling of individual grid cells using a Geo-Wiki and subsequent digitization of homogenous LCZs, (ii) a GIS-based approach using building data (Geletič and Lehnert 2016; Lelovics

TABLE 5.2
Classes of the Local Climate Zones Scheme

Built Zones

High ← Building Density[b] → Low

	ID	Zone Name	ID	Zone Name	ID	Zone Name
Low ← Building	1	Compact high rise	4	Open high rise	8	Large low rise
Height[a] → High	2	Compact midrise	5	Open midrise	9	Sparsely built
	3	Compact low rise	6	Open low rise	10	Heavy industry
	7	Lightweight low rise				

		Natural Zones		[a]*Building Height*
	A	Dense trees		"High rise" >25 m
	B	Scatter trees		"Midrise" 10–25 m
Low ← Veg.	C	Bush/scrub		"Low rise" 3–10 m
Height → High	D	Low plant cover		[b]*Building Fraction*
	E	Paved/bare rock		"Compact" >40%
	F	Bare soil/sand		"Open" 20%–40%
	G	Water		"Sparse" 10%–20%

Source: Adapted from Bechtel, Pesaresi et al. (2016).

et al. 2014), (iii) object-based image analysis (Gamba et al. 2012), and (iv) supervised pixel-based classification (Bechtel 2011; Bechtel and Daneke 2012). To achieve the aims of universality and transferability demanded by WUDAPT, (iv) was found to be comparably robust, sufficiently objective, computing efficient, and based on free data (Bechtel et al. 2015). To enable an operator without specific knowledge in image processing, a simplified version of the original scheme was implemented in the open source software SAGA GIS (Conrad et al. 2015). The appearance of the zones differs within and especially between cities, and as a result of differing cultural construction practices, materials, and background climate, a particular urban LCZ type will exhibit different spectral properties in different parts of the world (Schneider et al. 2009). Thus, site-specific training data are needed, which can be collected using a template in Google Earth. As standard data sources, multi-spectral and thermal Landsat data from different seasons were chosen, which implies that the discrimination is based on urban cover and fabric rather than on structure and metabolism. However, other input data such as Sentinel 1 (Bechtel et al. 2016b), Sentinel 2 (Kaloustian et al. 2017), and ASTER (Xu et al. 2017) have been successfully applied as well. The classification is conducted using a supervised Random Forest classifier (Breiman 2001) on a 100-m-resolution grid, which is coarser then the size of the objects (buildings, trees, roads, etc.) but finer than the size of the neighborhoods in order to obtain comparable spectral signals for all pixels (Bechtel et al. 2015). Since the LCZ is defined as hundreds of meters to kilometers in scale, a second version is produced using a spatial postprocessing filter. In the following, LCZ refers to zones derived by the WUDAPT method if nothing else is specified. Manuals for the full procedure are provided in www.wupdapt.org.

5.5 COMPARISON LABEL-LCZ

A first comparison between the GHSL-LABEL and WUDAPT-LCZ data sets was conducted in Bechtel et al. (2016a). Summarizing, it revealed a good agreement at the city level and at the kilometer scale, while the agreement at the pixel scale was limited because of the mismatch in grid scale and typology. Generally, GHSL-LABEL was found to preserve more detail owing to the higher resolution, while for the WUDAPT-LCZs, the internal makeup seemed somewhat clearer. For the example of Milan, the built-up areas showed very good agreement at the 1-km scale in terms of spatial pattern and radial distribution as a function of distance from town. Further, aggregated class set pairs that represent open and compact (light/medium and strong built) showed substantial agreement for the case of Milan, but the respective commercial classes (LCZ 8: *large low rise*, LABEL 12: *highly reflecting roof*) did not correspond (Bechtel et al. 2016a). It was concluded that the initial results were promising, but the study should be supplemented with different cities and, at later stages, updated versions of both data sets. Because of the mismatch in resolution, projection, and typology, the comparison at aggregated level in terms of spatial and semantic resolution was found appropriate.

This study now presents an extended comparison between GHSL-LABEL and WUDAPT-LCZ based on 50 cities on five continents (Africa, America, Asia, Australia, and Europe) using the prescribed methodology. The cities, presented in Figure 5.3, are Amsterdam, Aracaju, Arnhem, Blanca, Barcelona, Berlin, Bogor, Bologna, Brussels, Caracas, Changsha, Chicago, Cork, Dublin, Glasgow, Hamburg, Hangzhou, Houston, Jinan, Khartoum, Kolkata, Lisbon, London, Madrid, Manchester, Matsuyama, Melbourne, Milan, Paris, Pearl River Delta, Qingdao, Rio de Janeiro, Sao Paulo, Seoul, Shenyang, Suzhou, Sydney, Tainan, Tianjin, Toulouse, Vancouver, Venice, Vienna, Vitoria, Warsaw, Washington, DC, Wuhan, Wuxi, Xi'an, and Xiamen.

The LCZ data sets for these cities were taken from the WUDAPT pre-release database, which means some of the cities are still under revision. However, part of our interest is to investigate if problematic LCZ data sets of lower quality can be identified by the cross-comparison. The cities and the pre-release versions of the data sets are listed in Appendix A.

5.5.1 COMPARISON METHODS

The comparison was conducted as follows. First, the WUDAPT-LCZ maps were projected to the GHSL-LABEL Coordinate System (WGS84/Pseudo-Mercator, EPSG 3857) and the corresponding tiles from GHSL-LABEL were selected and mosaicked to the same target grid. Both data sets were then reclassified using six sets for LCZ (*all built LCZ, no sparse, compact and open, compact, open, and commercial LCZ*) and four sets for GHSL-LABEL (*all built GHSL, strong built, light-medium, and commercial GHSL*). The members of each set are given in Table 5.3.

Subsequently, the reclassified sets were resampled to 1000 m and the areal fraction occupied by each set in each grid cell was calculated. The correspondence between the new WUDAPT-LCZ and GHSL-LABEL pair sets was then evaluated visually and quantitatively. In particular, the following accordance measures were

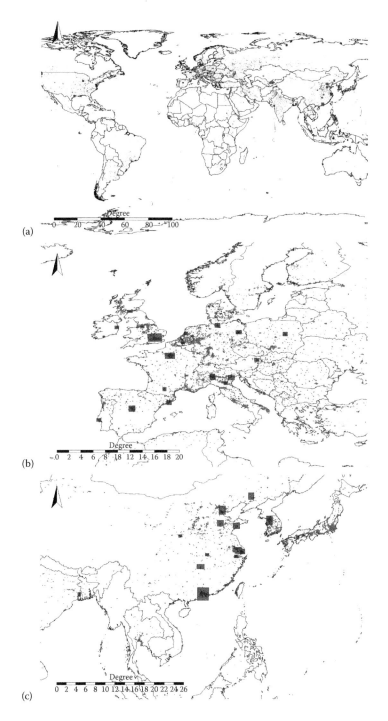

FIGURE 5.3 Cities included in comparison study. (a) Worldwide and (b) European and (c) Asian subsets.

TABLE 5.3

Built-Up Classes in GHSL-LABEL and WUDAPT-LCZ Typologies; Sets Used for Comparison

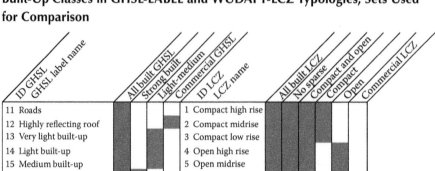

ID GHSL / GHSL label name	ID LCZ / LCZ name
11 Roads	1 Compact high rise
12 Highly reflecting roof	2 Compact midrise
13 Very light built-up	3 Compact low rise
14 Light built-up	4 Open high rise
15 Medium built-up	5 Open midrise
16 Strong built-up, low rise	6 Open low rise
17 Strong built-up, midrise	7 Light eight low rise
18 Strong built-up, high rise	8 Large low rise
19 Strong built-up, very high rise	9 Sparsely built
	10 Heavy industry

computed: correlation coefficient (R), mean distance (MD) equal to the LCZ set fraction minus GHSL-LABELS set fraction, mean absolute distance (MAD), and a linear fit using linear regression. Since the aggregated grids represent areal fractions, the differences MD and MAD are displayed in percentages.

Before the final evaluation, we tested the sensitivity of the results to selected parameters, particularly scale. The results were insensitive to the resolution of the fine grid in the tested range; 40 m was selected as the appropriate scale, which is close to the GHSL-LABEL layer and a good compromise between accuracy and processing and storage costs. However, the results were slightly sensitive to the resolution of the resampled coarser grid with slightly better results for smaller resolutions. To guarantee a good sample within the cells but preserve the urban structure, 1000 m (in Pseudo-Mercator) was selected. Finally, the WUDAPT-LCZ is routinely produced in two versions: the raw classification output in 100 m and a spatially filtered version using a majority filter of radius 3 pixels that is more consistent with the local scale definition of the zones. Here, we used the unfiltered LCZ, since better agreement with GHSL-LABEL was found. This is not surprising, since the filtering increases the effective scale mismatch between both datasets.

5.5.2 VISUAL COMPARISON

For visual comparison, similar color maps were defined for both data sets. Selected examples are shown in Figure 5.4.

From the shown examples, it can be seen that the settlement patterns are generally similar. For Brussels, the GHSL-LABEL shows more differentiation within the compact areas while the discretization in compact, open, and warehouse types seems a

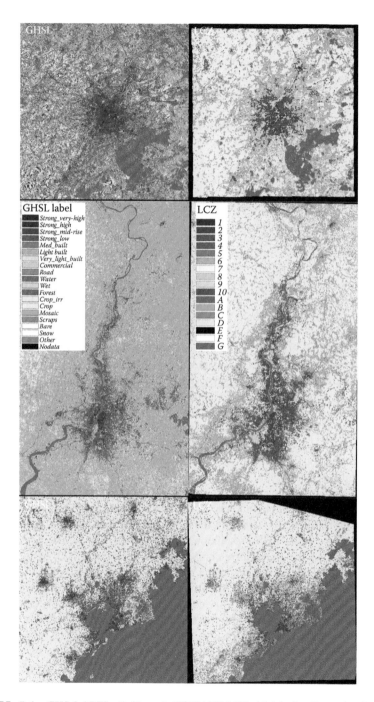

FIGURE 5.4 GHS-LABEL (left) and WUDAPT-LCZ (right) for Brussels, Belgium, Kolkata, India, and Qingdao, China.

bit clearer for the WUDAPT-LCZ map. Further, it can be seen that most of the roads are not classified as such in GHSL-LABEL but as medium and strong built-up types. For Kolkata, the LCZ map contains more small settlements, which are composed of water, trees, and buildings. In the west, some Landsat orbit effect can be seen in the natural classes of GHSL-LABEL. The different water coverage is explained by the seasonal changes of rice paddies. For Qingdao, more of the small settlement structures are preserved by GHSL-LABEL. The LCZ map shows large warehouse areas in the north and northwest, which cannot be found in GHSL-LABEL.

5.5.3 SET COMPARISONS

Figure 5.5 shows scatterplots for all GHSL-LABEL and WUDAPT-LCZ set combinations for the example of Sao Paulo, Brazil. Almost all sets show strong correlations. This essentially reflects the shape of the urban area that determines the occurrence of all classes, and accordingly, different sets of the same data source are also correlated. However, there are distinct differences. The best match is found between all built (GHSL) and no sparse (LCZ), which indicates very good agreement of the built-up. The worse agreement of all built (LCZ) is related to LCZ type 9 (sparsely built), which has a built fraction of only 10%–20% and therefore is dominantly natural cover. The compact and open (LCZ) set generally shows very similar patterns to no sparse (LCZ) but is somehow smeared with partly lower fractions than the no-sparse (LCZ) set. This is mostly attributed to the missing LCZs 8 and 10 (large low rise and heavy industry), which are mostly contained in the strong built GHSL types. From the remaining combinations, the commercial (GHSL) set shows much lower fractions than any LCZ set and hardly any relevant correlation. The compact LCZ set shows the best agreement with the strong built GHSL-LABEL set and the open LCZ set agrees well with the light-medium GHSL-LABEL set. Low fractions of open (LCZ) are associated with either low or very high fractions of strong built (GHSL) with some co-occurrence at medium fractions. Likewise, medium fractions of compact LCZ co-occur with medium fractions of light-medium (GHSL) and both very high and very low compact (LCZ) fractions with low light-medium (GHSL) fractions. Generally, the light-medium (GHSL) fractions are mostly below 50%, since at the GHSL resolution they are mixed with natural types.

Table 5.4 shows the quantitative comparison (R, MD, and MAD) of all set combinations. The presented values are the median values of the 50 cities. Generally, the results confirm the findings of Sao Paulo. The highest correlation (R = 0.89) was found between the all-built (GHSL) and no-sparse (LCZ) sets. This combination also showed a small (absolute) MD (−0.65%) and the smallest MAD (9.25%) from all set combinations of all built sets with all respective sets from the other data set. The strong built (GHSL) showed the highest correlation within the no sparse (LCZ) (R = 0.73), but if MD and MAD are taken into account, the best accordance was with the compact (LCZ) set (R = 0.66, MD = −6.09%, MAD = 4.18%). The light-to-medium built (GHSL) set showed the highest correlation with the compact and open and open (LCZ) sets. The smallest (absolute) MD and MAD were found for the open (LCZ) sets (MD = 0.10%, MAD = 8.15%). For the commercial (GHSL) set,

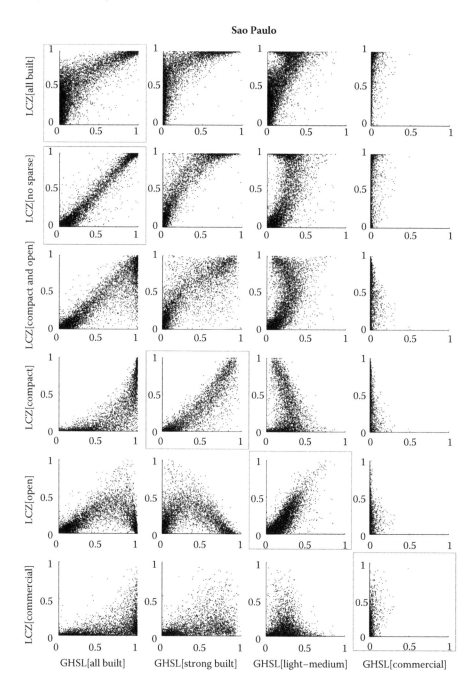

FIGURE 5.5 Co-occurrence of all GHSL-LABEL and WUDAPT-LCZ sets at 1 km scale for Sao Paulo, Brazil.

TABLE 5.4

Agreement between WUDAPT-LCZ (Rows) and GHSL-LABEL (Columns) Sets

R (Median)	All Built	Strong Built	Light-Medium	Commercial
All built	0.86	0.65	0.75	0.30
No sparse	0.89	0.73	0.77	0.34
Compact and open	0.81	0.65	0.78	0.12
Compact	0.53	0.66	0.31	0.11
Open	0.71	0.41	0.79	0.11
Commercial	0.50	0.53	0.30	0.55
MD (Median)	All Built	Strong Built	Light-Medium	Commercial
All built	5.02	18.15	11.99	24.04
No sparse	−0.65	13.62	5.24	18.96
Compact and open	−4.65	9.14	1.54	15.60
Compact	−15.87	−2.97	−8.65	1.56
Open	−7.00	6.09	0.10	12.88
Commercial	−17.00	−2.62	−7.30	2.40
MAD (Median)	All Built	Strong Built	Light-Medium	Commercial
All built	11.35	20.34	15.72	24.04
No sparse	9.25	14.46	12.08	18.97
Compact and open	9.99	11.92	9.38	15.87
Compact	15.97	4.18	9.95	2.56
Open	11.66	11.65	8.15	13.16
Commercial	17.11	4.30	8.84	2.60

Note: Median correlation coefficient R, median mean distance (MD; in %), and median mean absolute distance (MAD; in %) of 50 cities.

the accordance was generally weak; however, the best match was still found with commercial (LCZ) ($R = 0.55$, MD = 2.40%, MAD = 2.40%), while MD and MAD were generally low because of the comparably little occurrence.

Summarizing, the best matches at aggregated level were all built (GHSL) and no sparse (LCZ), strong built (GHSL) and compact (LCZ), light-medium (GHSL) and open (LCZ), and commercial (GHSL and LCZ), which were therefore investigated in more detail. To reveal the effect of LCZ 9 (sparsely built), the all-built (GHSL and LCZ) combination was added, resulting in a total of five set combinations of particular relevance. These set pairs are indicated by boxes in Figure 5.5 and listed in Table 5.5. Figure 5.6 shows the fractions for the preferred sets for Sao Paulo in the same order.

Selected statistics of the comparison of the five selected sets over the 50 cities are presented in Figure 5.7. Again, the best agreement was between the second selected set pair, while the first set pair (including the sparse LCZ class) showed a wider variation in R and even more in MD. The correlations between sets 3 to 5 were also high but showed considerably variation over the 50 cities. This means that the

TABLE 5.5
Selected Set Pairs of GHSL-LABEL and WUDAPT-LCZ for Further Comparison

Comparison ID (Nr)	Full (1)	Built (2)	Compact (3)	Open (4)	Commercial (5)
LCZ set	All built	No sparse	Compact	Open	Commercial
GHSL set	All built	All built	Strong built	Light-medium	Commercial

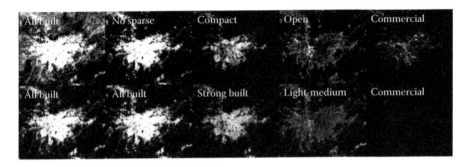

FIGURE 5.6 Fractions of the preferred sets of GHSL-LABEL (upper) and WUDAPT-LCZ (lower) for Sao Paulo.

agreement between the more differentiated structures is less consistent than for the built-up, which could indicate that the class match is not identical for all biophysical backgrounds. The distribution of the MD for set pairs 3 and 4 shows that the GHSL tended to have larger fractions of strong built-up than the LCZ compact types (negative MD), while for the open sets, LCZ had slightly higher fractions.

Interestingly, substantial differences were found between continents (see Figure 5.8). While in general the agreement was slightly higher for the European and American cities than for the Asian cities, the agreement of set pair 3 (compact) was higher for Asian cities than for European cities while the agreement of set pair 4 (open) was particularly low for Asian compared to both European and American cities. African and Australian cities were too few to be compared and the only African city (Khartoum) had no significant correlation for set pair 4, likely since open classes are of minor relevance in a desert environment. The differences between continents might be related to the large complexity in the internal makeup of Asian cities, but it could also indicate that the class borders of GHSL-LABEL vary with the biophysical background (especially vegetation) or that the LCZ training areas are not consistent, which needs further investigation.

In the following, selected examples from different continents are presented and briefly discussed; some additional cities are presented in Appendix C.

The example of Brussels, Belgium, in Figure 5.9 shows a good agreement of the built-up. Also, the accordance of set pairs 3 (compact) and 4 (open) is very good, with both data sets clearly delimiting the urban core. For the latter, GHSL-LABEL has somewhat lower fractions, which is likely related to the resolution mismatch

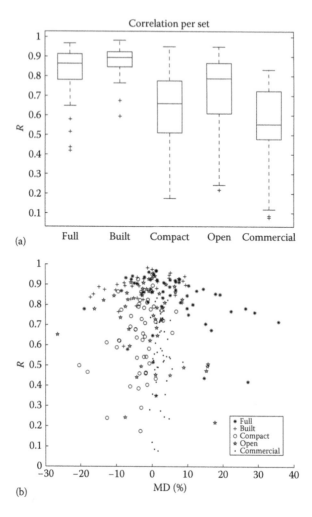

FIGURE 5.7 Correlation for the five selected set pairs: (a) Distribution over 50 cities; (b) comparison with mean distance (MD).

resulting in a higher fraction of natural classes in open built structures. Interestingly, set pair 1 shows a better agreement then set pair 2 ($R = 0.93$ and 0.89, respectively). This indicates that the LCZ 9 areas are also built-up areas in GHSL-LABEL for the respective example. Additionally, the GHSL-LABEL pattern shows more settlement structures in the outskirts. Figure 5.16a shows that this is due to the better recognition of linear isolated structures in the higher-resolution GHSL-LABEL data set. However, GHSL-LABEL also identifies the road network as strong built types.

The example of Caracas, Venezuela (see Figure 5.10), shows a nice agreement of not only the built-up (set pair 2) but also the compact (set pair 3, $R = 0.95$) and open types (set pair 4, $R = 0.90$). While the fractions agree very well for the compact types (MD = 0.1%), the fractions for open types are lower for GHSL-LABEL (MD = 3.9%), which also results in slightly lower fractions for all built classes.

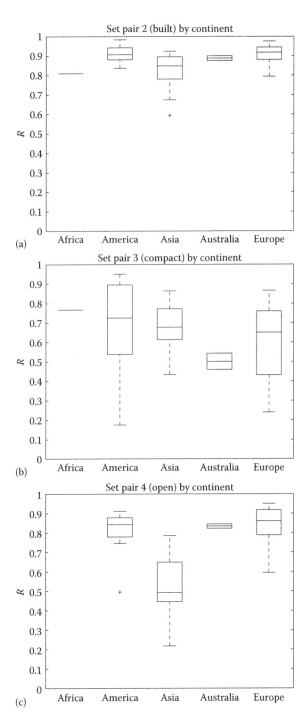

FIGURE 5.8 Agreement of selected set pairs by continent. Correlation for (a) set pair 2 (built), (b) set pair 3 (compact), and (c) set pair 4 (open).

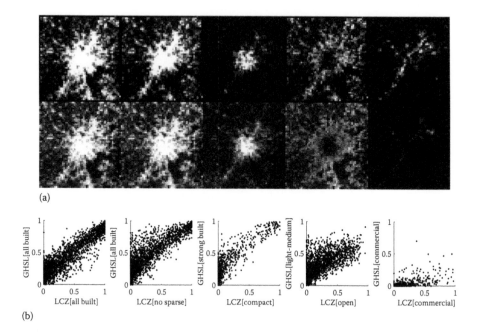

FIGURE 5.9 Fractions of WUDAPT-WUDAPT-LCZ (upper) and GHSL-LABEL (middle) sets full (1), built (2), compact (3), open (4), and commercial (5) for Brussels, Belgium. Lower row: scatterplots between WUDAPT-LCZ and GHSL-LABELS fractions.

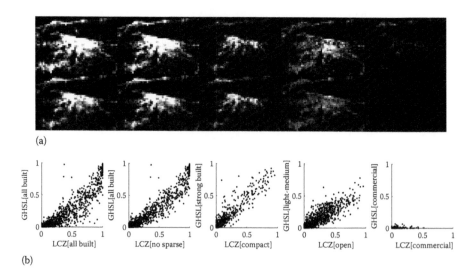

FIGURE 5.10 Fractions of WUDAPT-LCZ (upper) and GHSL-LABEL (middle) sets full (1), built (2), compact (3), open (4), and commercial (5) for Caracas, Venezuela. Lower row: scatterplots between WUDAPT-LCZ and GHSL-LABELS fractions.

The agreement for Dublin, Ireland, shown in Figure 5.11 is generally quite good as well. The built-up (set pair 2) shows a correlation of $R = 0.97$ (MD = 1.1%) and the open classes (set pair 4) shows a correlation of $R = 0.95$ (MD = 0.4%). The agreement of the compact classes (set pair 3) is a bit lower ($R = 0.71$, MD = −2.0%). This is partly due to commercial warehouse areas and roads, and partly due to suburban structures being assigned to strong built classes (cf. Figure 5.16b). As previously seen, the open classes have slightly lower fractions and the compact classes have slightly higher fractions in GHSL-LABEL. The agreement between the commercial sets (pair 5, $R = 0.76$) indicates that the spatial patterns still show relevant similarities, even though the GHSL-LABEL identifies less warehouse areas.

The results for Hamburg, Germany, are displayed in Figure 5.12. The agreement for the built-up (set pair 2) is very good ($R = 0.96$, MD = −0.9%). The open types also show very high agreement ($R = 0.93$), and the fractions in GHSL-LABEL are slightly higher (MD = −0.9%). This is also true for the compact types (set pair 3, $R = 0.78$, MD = −2.2%). However, the linear fit is different, with a slope of the regression line of 1.45 for the compact sets and 0.73 for the open sets. The higher fractions of strong built classes can also be seen in the maps, which delineate a much larger urban core for GHSL-LABEL, containing most of the harbor areas (Figure 5.16c). For LCZ, the narrower parts of the river Elbe in the southeast vanish because of the lower resolution and water areas are partly misclassified.

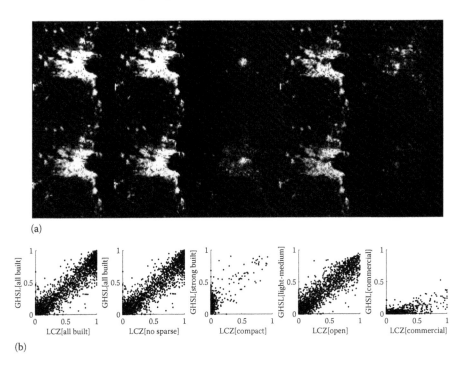

(a)

(b)

FIGURE 5.11 Fractions of WUDAPT-LCZ (upper) and GHSL-LABEL (middle) sets full (1), built (2), compact (3), open (4), and commercial (5) for Dublin, Ireland. Lower row: scatterplots between WUDAPT-LCZ and GHSL-LABELS fractions.

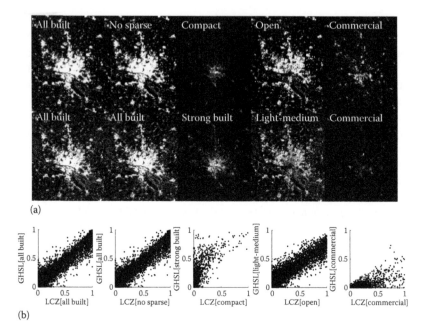

(b)

FIGURE 5.12 Fractions of WUDAPT-LCZ (upper) and GHSL-LABEL (middle) sets full (1), built (2), compact (3), open (4), and commercial (5) for Hamburg, Germany. Lower row: scatterplots between WUDAPT-LCZ and GHSL-LABELS fractions.

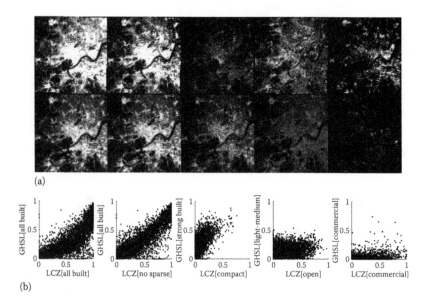

(b)

FIGURE 5.13 Fractions of WUDAPT-LCZ (upper) and GHSL-LABEL (middle) sets full (1), built (2), compact (3), open (4), and commercial (5) for Hangzhou, China. Lower row: scatterplots between WUDAPT-LCZ and GHSL-LABELS fractions.

Hangzhou, China (Figure 5.13), is an example of moderate agreement. While the built-up (set pair 2) has a correlation of 0.90, those of the compact ($R = 0.77$) and even more open ($R = 0.49$) sets are worse. As seen before, the fractions of GHSL-LABEL are higher for the compact sets (MD = −8.6%) and lower for open sets (MD = 8.88%). However, if the commercial set is added to LCZ set compact, the agreement with the GHSL-LABEL strong built set considerably increases ($R = 0.83$, MD = 1.4%). Nevertheless, it must be stated that the discrimination in open and compact classes substantially differs between the two data sets for Hangzhou. This is mainly related to the LCZ type 4 (open high rise), which is typically identified as strong built in GHSL-LABEL.

Another example from Asia is presented in Figure 5.14. For Kolkata, India, the shape of the urban areas mostly agrees (set pair 2: $R = 0.84$), but LCZ shows some speckle in the outskirts, resulting in a very broad point cloud in the scatterplot (set pair 2: MD = 9.9%). This is mainly due to small settlements that consist of houses and trees surrounded by ponds (Figure 5.16d). Since these have a quite

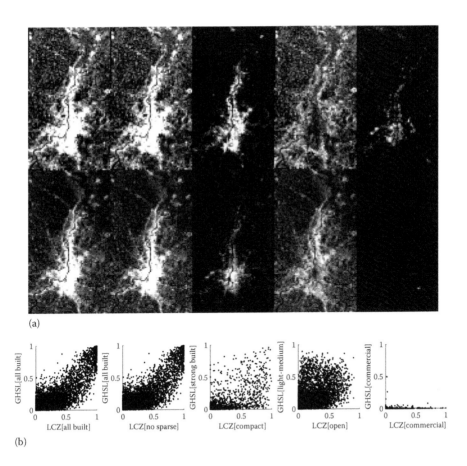

(a)

(b)

FIGURE 5.14 Fractions of WUDAPT-LCZ (upper) and GHSL-LABEL (middle) sets full (1), built (2), compact (3), open (4), and commercial (5) for Kolkata, India. Lower row: scatterplots between WUDAPT-LCZ and GHSL-LABELS fractions.

specific spectral signature, they are better recognized in the LCZ data, but at the same time, the built-up area is overestimated. Regarding the differentiation of compact and open types, both data sets clearly delineate an urban core, which, however, is larger for LCZ. Therefore, the chosen sets do not agree well for this case.

Qingdao, China (see Figure 5.15), shows good agreement on the built-up (set pair 2: $R = 0.92$, MD = −0.4%) but little agreement on the compact ($R = 0.69$, MD = −6.1) and open ($R = 0.35$, MD = 1.1%) set pairs, which is related to the previous findings (Figure 5.15). In particular, the open classes around the coast are mostly open high rise (LCZ 4) and identified as strong built by GHSL-LABEL. Further, the warehouse areas have mostly blue and red roof materials and therefore are less reflective than expected and thus are included in the strong built GHSL-LABEL classes (Figure 5.16e). Accordingly, if LCZ 8 (large low rise) is added to LCZ set compact, the agreement increases considerably (to $R = 0.82$ and MD = −1.7%). The color of the warehouse roof is a good example of the limitation of a global model as for GHSL-LABEL and the benefits of site- or at least region-specific training data.

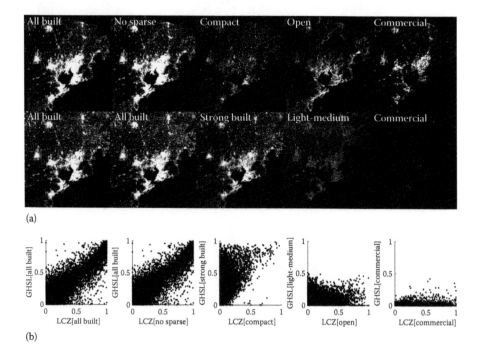

FIGURE 5.15 Fractions of WUDAPT-LCZ (upper) and GHSL-LABEL (middle) sets full (1), built (2), compact (3), open (4), and commercial (5) for Qingdao, China. Lower row: scatterplots between WUDAPT-LCZ and GHSL-LABELS fractions.

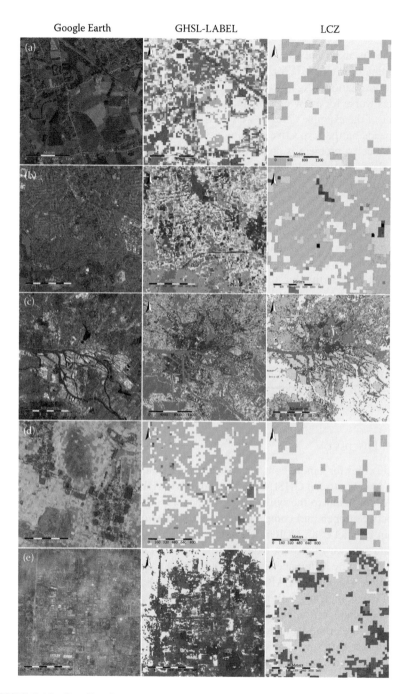

FIGURE 5.16 Details of agreement for selected examples: Google Earth image, GHSL, LCZ. (a) outskirts of Brussels, Belgium; (b) southern Dublin, Ireland; (c) inner city and Harbor area of Hamburg, Germany; (d) small settlements with ponds in Kolkata, India; (e) commercial areas in Qingdao, China. Color maps are the same as in Figure 5.4.

5.5.4 OVERALL AGREEMENT

After the discussion of specific aspects of individual cities, Figure 5.17 shows the distribution of R and MD for the five set pairs for all 50 cities, considering all significant correlations. Again, the best fit is found for set pair 2. Set pair 1 mainly performs worse for a few cities, which indicates that the LCZ 9 problem, identified for some of the example cities as well as Milan in Bechtel et al. (2016a), is not global but depends on the respective training data. Thus, the cross-comparison can easily reveal if such a problem is present for a specific LCZ map. Compact and open sets mostly show considerable agreement as well, but not uniformly for all cities. For the compact set pair and the open set pair, 24% and 20% of the cities, respectively, have a correlation of less than 0.5. For the commercial set pair, these are also only 32% of the cities, which means a large number shows relevant agreement of the patterns despite systematic underestimation by GHSL-LABEL. Generally, the agreement of the open class set is better than that of the compact class set (see also Table 5.4).

The lower row shows the linear fits between WUDAPT-LCZ and GHSL-LABEL sets. It can be seen that for all set pairs except the compact/strong built types, LCZ has higher fractions. This likely reflects the scale differences: At the higher grid resolution of GHSL, the sparse and open classes are partially decomposed into built and natural pixels. Therefore, the effect is strong for set pair 4 but not for set pair 3. The better accordance of no sparse to GHSL-LABEL built-up compared again reflects the problems with LCZ 9, which, by definition, has a build fraction of only 10%–20% and is additionally frequently misclassified. The much higher fraction of LCZ commercial compared to GHSL-LABEL reflects the potential for identifying reflective roofs in the GHSL approach.

5.5.5 DISCUSSION

Using 50 cities for a more comprehensive comparison, good agreement on the built-up was found between GHSL-LABEL and WUDAPT-LCZ. Further, both data sets add relevant detail to the built-up, and at coarser scale, there is substantial agreement between open and compact set pairs for most cities as found for the case of Milan in Bechtel et al. (2016a). However, the correlation and the linear fit between both data sets vary between cities and there is no universal relation. This is most likely affected by differences in the LCZ training data between cities and the cross-comparison with GHSL-LABEL proved useful to identify both doubtful maps and areas of low confidence within the maps.

Second, there are apparently systematic differences in the biophysical background and the urban structures of a specific cultural and historic background that determine the differing class accordance between cities. In particular it was found that the agreement differs between continents. While the agreement of set pair 3 (compact) was relatively high for the Asian cities, the agreement of set pair 4 (open) was particularly low for Asian cities compared to European and American cities. This is most likely related to the frequent occurrence of LCZ type 4 (open high rise), which is an open class in the LCZ scheme but undoubtedly intensively built. Thus, a further comparison should consider new set pairs. This could also account

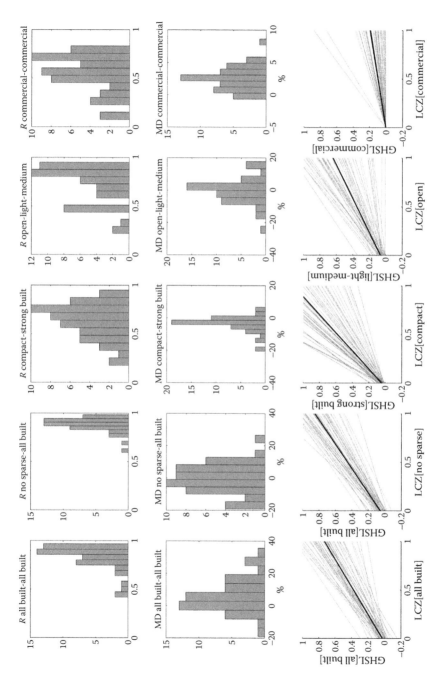

FIGURE 5.17 Overall agreement for all cities. Histograms of *R* and mean difference. Linear fits for all cities.

for the fact that LCZ 8 (large low rise) mostly falls into the strong built GHSL-LABEL classes. On the other hand, the region-specific training data from WUDAPT could be used to improve the discrimination of warehouse areas in GHSL-LABEL. A respective study is currently under way. An improved scheme could also consider making the comparison in the UTM coordinate system, to account for the sensitivity against coarse resolution and to increase consistency.

5.6 CONCLUSIONS

Global decision-making processes need accurate and globally consistent data for evidence-based policy making that includes evaluation of risks and hazard management, development of concepts, monitoring and understanding of trends, transfer knowledge between places, and exploration of alternative scenarios. It also needs these data to support the observational and modeling research that underpins policy making; in the absence of such data, much of our science is mute on the implications of global change for urban places. Consequently, the next generation of global urban mapping products must focus on the acquisition of consistent information on the internal makeup of cities and the spatial character of urban form and function.

WUDAPT-LCZ and GHSL-LABEL represent two approaches for generating better discretization of urban landscapes, both in experimental phase. The GHSL-LABEL is a global product derived by a methodology, developed for big remote sensing data scenarios. The built-up classes are derived by physical characteristics of settlements (i.e., built-up spatial density, height, roof reflectance, and vegetation presence). LCZs are a generic typology of urban structures, which can be mapped using RS data and a supervised classifier. There is considerable evidence that LCZs can discriminate between the climate impact of different types of urbanization, but the scheme itself provides a potentially much wider scope in domains such as planning or emergency response. Both WUDAPT-LCZ and GHSL-LABEL have specific advantages and disadvantages. The typology of the LCZs provides information on a large number of climatic and physical properties, but the current classification procedure needs city-specific training data provided by experts. The GHSL processing workflow is fully automatic (i.e., no human intervention during all processing steps), but it depends greatly on input data resolution and quality of training data (prone to errors). Additionally, the GHSL-LABEL classification schema is driven by physical rather than by functional properties of settlements.

For further comparison of the two data sets, additional spatial metrics and supplementary data (such as soil sealing, building height, LIDAR, and OpenStreetMap) will be considered. In addition, ways to combine both methodologies, such as incorporation of the LABEL 3D roughness into the LCZ classification and refinement of the thresholds in LABEL to increase consistency with the LCZ classes, will be studied. One of the improvements of the GHSL-LABEL workflow can be an additional filtering, which will potentially increase the stability of the output classification. This extension and more reliable spatial and thematic training data will be tested in the future on Sentinel data, which will also improve spatial accuracy. For WUDAPT, cross-comparison is considered to become part of the standard quality assessment protocol. Eventually, better classification algorithms are expected from recent research (Bechtel et al. 2017), including the data fusion contest (Tuia et al. 2017).

ACKNOWLEDGMENTS

We thank all data providers and WUDAPT contributors, i.e., Cai Meng, Chao Ren, Csilla Gal, Daniel Fenner, Debashish Das, Deepak Thapa, Guillaume Dumas, Je Woo, Joan Gilabert Mestre, Julia Hidalgo, Karenia Cordova, Lorenzo Giovannini, Luke Gebert, M. Kottas, Maria Brovelli, Maria De Fatima Andrade, Marie-Leen Verdonck, Max Anjos, Micheal Foley, Monika Tomaszewsk, Natasha Picone, Nial Buckley, Nial Buckley, Oscar Brousse, Paul Alexander, Ran Wang, Rohinton Emmanuel, Taciana Albuquerque, Tandang Yuliadi Dwi Putra, and Yueyi Feng. This work was partly founded by the Cluster of Excellence 'CliSAP' (EXC177), University of Hamburg, funded through the German Science Foundation (DFG).

REFERENCES

Alexander, P. J., B. Bechtel, W. T. L. Chow, R. Fealy, and G. Mills. 2016. Linking urban climate classification with an urban energy and water budget model: Multi-site and multi-seasonal evaluation. *Urban Climate* 17 (September): 196–215. doi:10.1016/j .uclim.2016.08.003.

Alexander, P. J., G. Mills, and R. Fealy. 2015. Using LCZ data to run an urban energy balance model. *Urban Climate* 13 (September): 14–37. doi:10.1016/j.uclim.2015.05.001.

Angel, S., S. Sheppard, D. L. Civco, R. Buckley, A. Chabaeva, L. Gitlin, A. Kraley, J. Parent, and M. Perlin. 2005. *The Dynamics of Global Urban Expansion*. Citeseer. http://citeseerx.ist .psu.edu/viewdoc/download?doi=10.1.1.309.2715&rep=rep1&type=pdf.

Bechtel, B., P. J. Alexander, J. Böhner, J. Ching, O. Conrad, J. Feddema, G. Mills, L. See, and I. Stewart. See, and I. Stewart. 2015. Mapping local climate zones for a world-wide database of the form and function of cities. *ISPRS International Journal of Geo-Information* 4 (1): 199–219. doi:10.3390/ijgi4010199.

Bechtel, B. 2011. Multitemporal Landsat data for urban heat island assessment and classification of local climate zones. In *Urban Remote Sensing Event (JURSE), 2011 Joint*, 129–32. IEEE. doi:10.1109/JURSE.2011.5764736.

Bechtel, B., O. Conrad, M. Tamminga, M. L. Verdonck, F. Van Coillie, D. Tuia, M. Demuzere et al. 2017. Beyond the urban mask: Local climate zones as a generic descriptor of urban areas—Potential and recent developments. In *2017 Joint Urban Remote Sensing Event (JURSE)*, 1–4. doi:10.1109/JURSE.2017.7924557.

Bechtel, B., and C. Daneke. 2012. Classification of local climate zones based on multiple earth observation data. *IEEE Journal of Selected Topics in Applied Earth Observations and Remote Sensing* 5 (4): 1191–1202. doi:10.1109/JSTARS.2012.2189873.

Bechtel, B., M. Pesaresi, L. See, G. Mills, J. Ching, P. J. Alexander, J. J. Feddema, A. J. Florczyk, and I. Stewart. 2016a. Towards consistent mapping of urban structures—Global Human Settlement Layer and Local Climate Zones. *ISPRS—International Archives of the Photogrammetry, Remote Sensing and Spatial Information Sciences* XLI-B8 (June): 1371–1378. doi:10.5194/isprs-archives-XLI-B8-1371-2016.

Bechtel, B., L. See, G. Mills, and M. Foley. 2016b. Classification of local climate zones using SAR and multispectral data in an arid environment. *IEEE Journal of Selected Topics in Applied Earth Observations and Remote Sensing* PP (99): 1–9. doi:10.1109 /JSTARS.2016.2531420.

Bontemps, S., P. Defourny, E. V. Bogaert, O. Arino, V. Kalogirou, and J. R. Perez. 2011. GLOBCOVER 2009—Products Description and Validation Report. http://www.citeulike .org/group/15400/article/12770349.

Breiman, L. 2001. Random forests. *Machine Learning* 45 (1): 5–32.

Brousse, O., A. Martilli, M. Foley, G. Mills, and B. Bechtel. 2016. WUDAPT, an efficient land use producing data tool for mesoscale models? Integration of urban LCZ in WRF over Madrid. *Urban Climate* 17 (September): 116–134. doi:10.1016/j.uclim.2016.04.001.

Chen, J., J. Chen, A. Liao, X. Cao, L. Chen, X. Chen, C. He et al. 2015. Global land cover mapping at 30m resolution: A POK-based operational approach. *ISPRS Journal of Photogrammetry and Remote Sensing* 103: 7–27.

Ching, J. K. S. 2013. A perspective on urban canopy layer modeling for weather, climate and air quality applications. *Urban Climate* 3 (May): 13–39. doi:10.1016/j .uclim.2013.02.001.

Conrad, O., B. Bechtel, M. Bock, H. Dietrich, E. Fischer, L. Gerlitz, J. Wehberg, V. Wichmann, and J. Böhner. 2015. System for Automated Geoscientific Analyses (SAGA) v. 2.1.4. *Geosci. Model Dev.* 8 (7): 1991–2007. doi:10.5194/gmd-8-1991-2015.

Ehrlich, D., T. Kemper, X. Blaes, and P. Soille. 2013. Extracting building stock information from optical satellite imagery for mapping earthquake exposure and its vulnerability. *Natural Hazards* 68 (1): 79–95.

Ehrlich, D., and P. Tenerelli. 2013. Optical satellite imagery for quantifying spatio-temporal dimension of physical exposure in disaster risk assessments. *Natural Hazards* 68 (3): 1271–1289.

Gamba, P., G. Lisini, P. Liu, P. Du, and H. Lin. 2012. Urban climate zone detection and discrimination using object-based analysis of VHR scenes. *Proceedings of the 4th GEOBIA*, 7–9.

Geletič, J., and M. Lehnert. 2016. GIS-based delineation of local climate zones: The case of medium-sized central European cities. *Moravian Geographical Reports* 24 (3): 2–12. doi:10.1515/mgr-2016-0012.

Gong, P., J. Wang, L. Yu, Y. Zhao, Y. Zhao, L. Liang, Z. Niu et al. 2013. Finer resolution observation and monitoring of global land cover: First mapping results with Landsat TM and ETM+ data. *International Journal of Remote Sensing* 34 (7): 2607–2654.

Gutman, G., C. Huang, G. Chander, P. Noojipady, and J. G. Masek. 2013. Assessment of the NASA–USGS Global Land Survey (GLS) datasets. *Remote Sensing of Environment* 134: 249–265.

Irish, R. R., J. L. Barker, S. N. Godward, and T. Arvidson. 2006. Characterization of the Landsat-7 ETM+ Automated Cloud-Cover Assessment (ACCA) algorithm. *Photogrammetric Engineering & Remote Sensing 3* 72: 1179–1188.

Jackson, T. L., J. J. Feddema, K. W. Oleson, G. B. Bonan, and J. T. Bauer. 2010. Parameterization of urban characteristics for global climate modeling. *Annals of the Association of American Geographers* 100 (4): 848–865. doi:10.1080/00045608.2010 .497328.

Kaloustian, N., M. Tamminga, and B. Bechtel. 2017. Local climate zones and annual surface thermal response in a Mediterranean city. In *Urban Remote Sensing Event (JURSE), 2017 Joint*, 1–4. IEEE. http://ieeexplore.ieee.org/abstract/document/7924597/.

Lee, S.-H., S.-W. Kim, W. M. Angevine, L. Bianco, S. A. McKeen, C. J. Senff, M. Trainer, S. C. Tucker, and R. J. Zamora. 2011. Evaluation of urban surface parameterizations in the WRF model using measurements during the Texas Air Quality Study 2006 Field Campaign. *Atmospheric Chemistry and Physics* 11 (5): 2127–2143.

Lelovics, E., J. Unger, T. Gál, and C. V. Gál. 2014. Design of an urban monitoring network based on local climate zone mapping and temperature pattern modelling. *Climate Research* 60 (1): 51–62. doi:10.3354/cr01220.

Loridan, T., C. S. B. Grimmond, S. Grossman-Clarke, F. Chen, M. Tewari, K. Manning, A. Martilli, H. Kusaka, and M. Best. 2010. Trade-offs and responsiveness of the single-layer urban canopy parametrization in WRF: An offline evaluation using the MOSCEM optimization algorithm and field observations. *Quarterly Journal of the Royal Meteorological Society* 136 (649): 997–1019.

Marconcini, M., T. Esch, A. Felbier, and W. Heldens. 2013. Unsupervised high-resolution global monitoring of urban settlements. In *Geoscience and Remote Sensing Symposium (IGARSS), 2013 IEEE International*, 4241–4244. IEEE. http://ieeexplore.ieee.org /abstract/document/6723770/.

Miyazaki, H., X. Shao, K. Iwao, and R. Shibasaki. 2013. An automated method for global urban area mapping by integrating ASTER satellite images and GIS data. *IEEE Journal of Selected Topics in Applied Earth Observations and Remote Sensing* 6 (2): 1004–1019. doi:10.1109/JSTARS.2012.2226563.

NASA. 2015. The Shuttle Radar Topography Mission (SRTM) Collection User Guide. https:// lpdaac.usgs.gov/sites/default/files/public/measures/docs/NASA_SRTM_V3.pdf.

Oke, T. R., G. Mills, A. Christen, and J. A. Voogt. 2017. *Urban Climates*. Cambridge University Press doi:10.1017/9781139016476.

Pachauri, R. K., M. R. Allen, V. R. Barros, J. Broome, W. Cramer, R. Christ, J. A. Church et al. 2014. Climate Change 2014: Synthesis Report. Contribution of Working Groups I, II and III to the Fifth Assessment Report of the Intergovernmental Panel on Climate Change. http:// epic.awi.de/37530/.

Pesaresi, M., D. Ehrlich, S. Ferri, A. Florczyk, S. Freire, F. Haag, M. Halkia, A. M. Julea, T. Kemper, and P. Soille. 2015a. Global human settlement analysis for disaster risk reduction. *The International Archives of Photogrammetry, Remote Sensing and Spatial Information Sciences* 40 (7): 837.

Pesaresi, M., S. Ferri, D. Ehrlich, A. J. Florczyk, S. Freire, M. Halkia, A. Julena, T. Kemper, P. Soille, and V. Syrris. 2016a. *Operating Procedure for the Production of the Global Human Settlement Layer from Landsat Data of the Epochs 1975, 1990, 2000, and 2014.* Vol. EUR 27741 EN. JRC Technical Report. http://bookshop.europa.eu/en /operating-procedure-for-the-production-of-the-global-human-settlement-layer-from -landsat-data-of-the-epochs-1975-1990-2000-and-2014-pbLBNA27741/?AllPersonalA uthorNames=true.

Pesaresi, M., G. Huadong, X. Blaes, D. Ehrlich, S. Ferri, L. Gueguen, M. Halkia et al. 2013. A global human settlement layer from optical HR/VHR RS data: Concept and first results. *IEEE Journal of Selected Topics in Applied Earth Observations and Remote Sensing* 6 (5): 2102–2131. doi:10.1109/JSTARS.2013.2271445.

Pesaresi, M., M. Melchiorri, and T. Kemper. 2016b. *Atlas of the Human Planet 2016. Mapping Human Presence on Earth with the Global Human Settlement Layer.*

Pesaresi, M., V. Syrris, and A. Julea. 2015b. Benchmarking of the symbolic machine learn-ing classifier with state of the art image classification methods—Application to remote sensing imagery. JRC Technical Report.

Pesaresi, M., V. Syrris, and A. Julea. 2016c. Analyzing big remote sensing data via symbolic machine learning. In *Proceedings of the 2016 Conference on Big Data from Space, Santa Cruz de Tenerife, Spain*, 15–17. https://www.researchgate.net/profile/Vasileios _Syrris/publication/299643667_ANALYZING_BIG_REMOTE_SENSING_DATA _VIA_SYMBOLIC_MACHINE_LEARNING/links/5703ba9b08aedbac127087c1.pdf.

Pesaresi, M., V. Syrris, and A. Julea. 2016d. A new method for earth observation data analyt-ics based on symbolic machine learning. *Remote Sensing* 8 (5): 399.

Potere, D., A. Schneider, S. Angel, and D. L. Civco. 2009. Mapping urban areas on a global scale: Which of the eight maps now available is more accurate? *International Journal of Remote Sensing* 30 (24): 6531–6558.

Schneider, A., M. A. Friedl, and D. Potere. 2009. A new map of global urban extent from MODIS satellite data. *Environmental Research Letters* 4 (4): 044003. doi:10.1088/1748-9326/4/4/044003.

Schneider, A., M. A. Friedl, D. K. McIver, and C. E. Woodcock. 2003. Mapping urban areas by fusing multiple sources of coarse resolution remotely sensed data. *Photogrammetric Engineering & Remote Sensing* 69 (12): 1377–1386.

Small, C. 2003. High spatial resolution spectral mixture analysis of urban reflectance. *Remote Sensing of Environment* 88 (1): 170–186.

Small, C. 2005. A global analysis of urban reflectance. *International Journal of Remote Sensing* 26 (4): 661–681.

Stewart, I. D. 2011. A systematic review and scientific critique of methodology in modern urban heat island literature. *International Journal of Climatology* 31 (2): 200–217. doi:10.1002/joc.2141.

Stewart, I. D., and T. R. Oke. 2012. Local climate zones for urban temperature studies. *Bulletin of the American Meteorological Society* 93 (12): 1879–1900. doi:10.1175 /BAMS-D-11-00019.1.

Taubenböck, H., T. Esch, A. Felbier, M. Wiesner, A. Roth, and S. Dech. 2012. Monitoring urbanization in mega cities from space. *Remote Sensing of Environment* 117: 162–176.

Tuia, D., G. Moser, B. Le Saux, B. Bechtel, and L. See. 2017. 2017 IEEE GRSS Data Fusion Contest: Open Data for Global Multimodal Land Use Classification [Technical Committees]. *IEEE Geoscience and Remote Sensing Magazine* 5 (1): 70–73. doi:10.1109 /MGRS.2016.2645380.

Xu, Y., C. Ren, M. Cai, Y. Y. E. Ng, and T. Wu. 2017. Classification of local climate zones using ASTER and Landsat data for high-density cities. *IEEE Journal of Selected Topics in Applied Earth Observations and Remote Sensing.* http://ieeexplore.ieee.org /abstract/document/7891588/.

APPENDIX A

City	Version	City	Version
Amsterdam	Amsterdam_RanWang_MF_20151201	Matsuyama	Matsuyama_DeepakThapa_MF_20170220
Aracaju	Aracaju_MaxAnjos_MF_20170129	Melbourne	Melbourne_LukeGebert_MF_20170220
Arnhem	Arnhem_RanWang_MF_20170129	Milan	Milan_MariaBrovelli_MF_20151016
Blanca	Blanca_NatashaPicone_MF_20161125	Paris	Paris_GuillaumeDumas_MF_20161001
Barcelona	Gilabert_Mestre_MF_20170129	Pearl River Delta	CaiMeng_WangRan_MF_20170115
Berlin	DanielFenner_BB_20151117_v6	Qingdao	Qingdao_ChaoRen_MF_20170228
Bogor	Bogor_Tandang_MF_20161125	Rio	Max_Anjos_MF_20170228
Bologna	Bologna_LorenzoGiovannini_MF_20170307	Sao Paulo	MF_200150930_bb20161128_MF20170228
Brussels	Brussels_Brussels_16-12-2015_Verdonck	Seoul	Seoul_JEWOO_MF_20170228
Caracas	Caracas_KareniaCordova_MF_20170308	Shenyang	Shenyang_ChaoRen_MF_20170228
Changsha	Changsha_ChaoRen_MF_20170309	Suzhou	Suzhou_ChaoRen_MF_20160403
Chicago	Chicago_Chicago_MichealFoley_20160126	Sydney	Sydney_Yueyi_MF_20161124
Cork	Cork_PaulAlexander_MF_20170130	Tainan	Tainan_ChaoRen_MF_20170228
Dublin	Dublin_PaulAlexander_MF_20170130	Tianjin	Tianjin_ChaoRen_MF_20160403
Glasgow	Glasgow_REmmanuel_MF_20161128	Toulouse	Toulouse_Toulouse_JuliaHidalgo_20160119
Hamburg	Kottas_MF_BB_20161014	Vancouver	Vancouver_Vancouver_MichealFoley_20170228
Hangzhou	Hangzhou_ChaoRen_MF_20170217	Venice	Venice_LorenzoGiovannini_MF_20160808
Houston	Houston_NialBuckley_MF_20170214	Vienna	Vienna_17112015_KrisH_OscarB
Jinan	Jinan_ChaoRen_MF_20170307	Vitoria	Vitoria_TacianaAlbuquerque_MF_20151201
Khartoum	Khartoum_MichealFoley_200170320	Warsaw	Warsaw_MonikaTomaszewsk_MF_20160920
Kolkata	DebashishDas_MF_BB_20161121	Washington	DC_Csilla_MF_20170228
Lisbon	Max_Anjos_MF_20170217	Wuhan	Wuhan_ChaoRen_MF_20161129
London	London_NialBuckley_MF_20170213	Wuxi	Wuxi_ChaoRen_MF_20160404
Madrid	Madrid_OscarBrousse_BB_20151122	Xi'an	ChaoRen_MF_BB_20170227
Manchester	Manchester_MichealFoley_20170217	Xiamen	Xiamen_ChaoRen_MF_20160405

APPENDIX B

City	R All Built	R No Sparse–All Built	R Compact–Strong Built	R Open–Light–Medium	R Commercial	MAD All Built	MAD No Sparse–All Built	MAD Compact–Strong Built	MAD Open–Light–Medium	MAD Commercial
Amsterdam	0.88	0.89	0.24	0.79	0.42	9.7	9.1	13.4	11.0	2.3
Aracaju	0.90	0.91	0.81	0.50	0.12	5.2	5.2	3.4	4.5	1.0
Arnhem	0.78	0.83	0.29	0.87	0.49	13.5	9.7	4.3	8.2	1.8
Bahia	0.85	0.85	0.58	0.81	0.28	6.2	6.2	1.9	4.3	0.8
Barcelona	0.91	0.91	0.79	0.83	0.67	8.0	8.0	4.1	8.9	2.6
Berlin	0.92	0.92	0.64	0.87	0.55	9.2	8.2	2.8	8.7	3.3
Bogor	0.78	0.78	0.66	0.65	0.09	21.0	21.0	7.9	28.5	0.7
Bologna	0.42	0.93	0.46	0.85	0.72	27.9	3.3	1.9	3.1	2.3
Brussels	0.93	0.89	0.87	0.74	0.67	11.9	16.7	9.7	13.3	3.8
Caracas	0.93	0.96	0.95	0.90	0.35	8.2	5.9	1.7	5.2	1.2
Changsha	0.67	0.67	0.79	0.51	0.52	17.1	17.1	2.1	16.1	0.6
Chicago	0.91	0.84	0.54	0.78	0.73	9.8	17.7	4.4	17.2	2.8
Cork	0.88	0.95	0.78	0.93	0.72	14.5	5.5	0.8	2.7	5.1
Dublin	0.97	0.97	0.71	0.95	0.76	3.7	3.7	2.1	3.4	2.5
Glasgow	0.76	0.94	0.54	0.93	0.77	29.1	10.5	1.7	9.4	3.4
Hamburg	0.96	0.96	0.78	0.93	0.77	5.8	5.8	2.3	6.3	2.3
Hangzhou	0.85	0.90	0.77	0.49	0.52	20.1	12.4	9.1	12.7	9.0
Houston	0.87	0.89	0.18	0.83	0.52	10.0	9.3	3.3	10.4	3.4
Jinan	0.81	0.85	0.43	0.45	0.25	9.3	7.9	5.7	4.5	3.5
Khartoum	0.71	0.81	0.77	none	0.66	16.3	10.1	9.1	0.4	2.2
Kolkata	0.84	0.84	0.72	0.45	0.08	13.0	13.0	4.4	13.9	1.7
Lisbon	0.92	0.93	0.73	0.89	0.83	7.4	6.1	3.2	7.2	1.5
London	0.87	0.87	0.39	0.79	0.61	11.8	12.3	3.9	11.9	1.4

(Continued)

APPENDIX B (CONTINUED)

City	R All Built	R No Sparse-All Built	R Compact-Strong Built	R Open-Light-Medium	R Commercial	MAD All Built	MAD No Sparse-All Built	MAD Compact-Strong Built	MAD Open-Light-Medium	MAD Commercial
Madrid	0.91	0.91	0.66	0.86	0.71	5.5	4.7	4.0	4.1	2.1
Manchester	0.52	0.85	0.61	0.68	0.79	23.5	17.2	12.8	15.8	2.9
Matsuyama	0.94	0.83	0.86	none	0.48	3.5	6.0	3.2	4.8	0.8
Melbourne	0.79	0.90	0.46	0.85	0.82	14.4	12.0	3.7	13.1	4.4
Milan	0.77	0.94	0.74	0.88	0.63	24.1	7.6	6.2	6.8	4.4
Paris	0.97	0.98	0.72	0.91	0.57	4.3	3.8	2.3	5.2	1.7
Pearl	0.92	0.92	0.80	0.79	0.72	6.3	6.3	4.2	6.4	2.6
Qingdao	0.92	0.92	0.69	0.35	0.42	5.4	5.4	6.4	5.8	4.2
Rio	0.94	0.94	0.90	0.86	0.53	6.5	6.5	3.2	8.2	0.5
Sao Paulo	0.85	0.98	0.93	0.88	0.52	18.5	3.6	4.4	4.3	3.5
Seoul	0.89	0.89	0.68	0.73	0.24	8.3	8.3	5.1	7.2	5.0
Shenyang	0.81	0.85	0.62	0.65	0.36	10.4	9.4	9.2	7.4	3.8
Suzhou	0.87	0.87	0.50	0.47	0.54	13.5	13.5	20.5	17.5	4.8
Sydney	0.93	0.88	0.54	0.82	0.73	9.1	14.1	3.6	14.6	2.8
Tainan	0.86	0.77	0.68	0.44	0.74	7.8	10.8	5.8	6.8	1.2
Tianjin	0.58	0.59	0.59	0.24	0.27	17.7	17.1	10.9	10.4	2.6
Toulouse	0.87	0.96	0.46	0.94	0.71	13.2	4.0	1.8	4.1	1.4
Vancouver	0.83	0.91	0.64	0.91	0.58	12.6	9.4	4.5	8.1	3.3
Venice	0.44	0.79	0.40	0.60	0.57	20.5	7.0	1.7	8.0	2.9
Vienna	0.65	0.90	0.82	0.83	0.82	15.5	9.2	3.4	8.5	1.3
Vitoria	0.75	0.88	0.82	0.75	0.36	10.8	4.2	1.5	4.1	0.2
Warsaw	0.80	0.83	0.40	0.69	0.56	16.1	10.5	6.1	11.8	1.1
Washington	0.72	0.92	0.51	0.86	0.74	36.3	13.0	3.2	12.6	3.2

(Continued)

APPENDIX B (CONTINUED)

City	R All Built	R No Sparse–All Built	R Compact–Strong Built	R Open–Light–Medium	R Commercial	MAD All Built	MAD No Sparse–All Built	MAD Compact–Strong Built	MAD Open–Light–Medium	MAD Commercial
Wuhan	0.91	0.91	0.63	0.73	0.48	9.6	9.6	4.9	6.8	5.8
Wuxi	0.89	0.89	0.47	0.50	0.54	13.5	13.5	18.2	16.9	5.6
Xi'an	0.78	0.78	0.77	0.22	0.54	26.6	26.6	6.3	21.3	5.0
Xiamen	0.86	0.86	0.62	0.62	0.49	10.9	11.1	9.8	12.5	2.8

Note: "none" means correlation is not significant.

APPENDIX C

Berlin

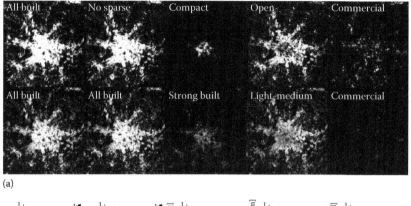

(a)

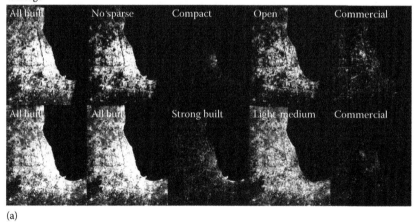

(b)

Chicago

(a)

(b)

(*Continued*)

APPENDIX C (CONTINUED)

Paris

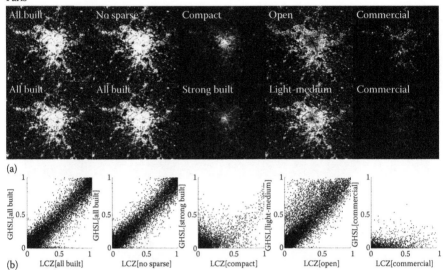

(a)

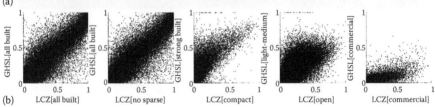

(b)

Pearl River Delta

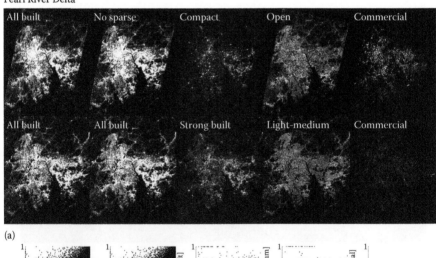

(a)

(b)

(Continued)

APPENDIX C (CONTINUED)

Sydney

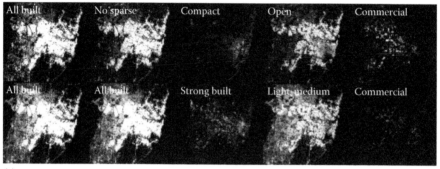

(a)

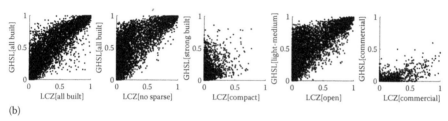

(b)

Vancouver

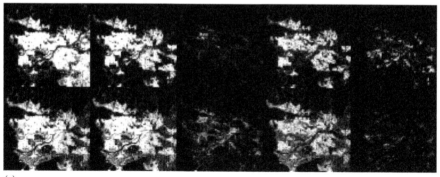

(a)

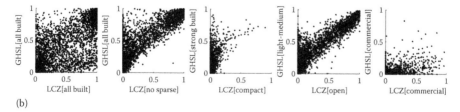

(b)

(Continued)

APPENDIX C (CONTINUED)

Washington

(a)

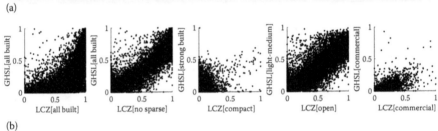

(b)

6 Urban Change Detection Utilizing High-Resolution Optical Images Taken from Different Viewing Angles and Different Platforms

Shabnam Jabari and Yun Zhang

CONTENTS

6.1 INTRODUCTION

Urban area is one of the most dynamically changing land covers on earth. Detection of urban changes using images has become such an important task in remote sensing and there are numerous papers published in this field. From the literature, the conventional methods of change detection can be divided into two categories: image-based methods and digital surface model (DSM)–based methods.

In image-based change detection methods, most research papers used images of a flat surface for the study to minimize the co-registration errors. Then, polynomial or spline methods were used to co-register the bi-temporal images (Al-Khudhairy et al. 2005; Bouziani et al. 2007; Gueguen et al. 2011; Im and Jensen 2005; Im et al. 2008; Zhou et al. 2008). Some studies used ortho-rectified images for change detection to reduce misregistration caused by relief displacement (Doxani et al. 2012; Niemeyer et al. 2008). Although these methods are relatively successful using moderate- to low-resolution images, existence of distortions is inevitable in the borders of the elevated objects (e.g., buildings) that reduce change detection accuracy using very high resolution (VHR) images.

In DSM-based change detection methods, the DSMs of before and after the change are generated either through stereo photogrammetry or through LiDAR (light detection and ranging). The comparison of the DSMs highlights the change probabilities. Then, the final changes are identified by the comparison of the associated imagery (Jung 2004; Martha et al. 2010; Murakami et al. 1999; Pang et al. 2014; Tian et al. 2014; Waser et al. 2008). Qin (2014) also used stereo space-borne images to generate a DSM and detect changes between the DSM and a city 3D model. In the DSM-based studies, the co-registration of the DSMs and the images is not error free, which introduces false alarms in the change detection results. In addition, in these methods, two sets of DSMs are required, which increase the costs for change detection. It is also possible to use bi-temporal true orthophotos for change detection (Braun 2003). However, true orthophoto generation generally requires using multi-view angle images and high-quality DSMs. Thus, the production costs are high (Braun 2003; Rau et al. 2002). Furthermore, if the ortho-rectification process generates undulated object (e.g., building) edges, which is a very common problem in this approach, the change detection results will be negatively affected.

Although airborne and satellite sensors provide abundant sources of urban imagery, using the conventional methods, finding proper data for urban change detection is still a challenge. This is because these methods require certain specifications for bi-temporal imagery to be used for urban change detection. The specifications that make limitations in data selection are as follows:

- Time of image acquisition: the images are to be selected from similar time of the year to minimize seasonal effects and solar illumination differences. This sort of images is called close-to-anniversary images.
- Sensor spectral characteristics: the spectral bandwidths of the sensors used for change detection should be similar to prevent false alarms.
- Sensor type: the type of the sensors, for example, parallel scanners and point perspective scanners, generating bi-temporal images should be similar unless ortho-rectification is done.

- Geometric distortions: to minimize geometric distortion in images, ortho-rectified aerial photos or close-to-nadir satellite images are preferred.

If we stride toward mitigation of the aforementioned limitations, we can incorporate a wider range of imagery in urban change detection and accordingly reduce the associated costs. From the above, the sensor type and the geometric distortion differences have been addressed in the recent literature.

Pollard et al. (2010) started a voxel-based approach in urban change detection that is capable of incorporating a wide range of off-nadir and airborne images in the process. This method was followed by other studies (Crispell et al. 2012; Kang et al. 2013) as well. Although, the mentioned methods incorporated a wide range of images in change detection, they still require a large number of images to train the system, and it does not decrease the change detection costs.

The majority of VHR satellite images are taken with an off-nadir angle, because of the sensors' agility to quickly capture a ground image within a 45° off-nadir angle. Besides, aerial photos are available for urban areas of industrialized countries. Combining the images of these sources of data can mitigate the bi-temporal data acquisition problem in change detection and reduce the associated costs. This is not easily possible with the conventional change detection methods and motivates our research in this paper.

Highly tilted (off-nadir) images have inherent geometric distortions mainly caused by the relief displacement. The different relief displacement directions in different images will lead to a severe misregistration problem in urban environments. This misregistration problem results in false change detections. Furthermore, the geometric distortion of the satellite images is different from the airborne images because of their scanning mechanism dissimilarities. Normally, VHR satellite images are acquired using push broom sensors that have line perspective geometry, whereas most airborne images have center perspective geometry. Figure 6.1 shows the direction of relief displacement in a standard center perspective scanner and a standard linear scanner (push broom) image. Depending on the satellite image viewing angle, the parallel arrows in Figure 6.1 might have a different direction. This geometric distortion difference causes problems in co-registration between satellite and airborne images.

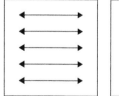

FIGURE 6.1 Examples of relief displacement directions of satellite (left), which is generally a parallel scanner, and airborne (right), which is generally a center-projective scanner, images. Because of the center-projective geometry of airborne images, the relief displacement takes place toward different directions depending on the position of objects. In satellite images, the relief displacement happens in parallel directions.

To utilize off-nadir VHR satellite images for urban change detection, in our previous research, we developed a patch-wise co-registration (PWCR) method to co-register bi-temporal satellite images in a patch-wise manner. In this method, rational polynomial coefficients (RPCs) of satellite images and a DSM of the area are used to perform patch-wise image registration. Then, changes in bi-temporal VHR satellite images were detected by comparing the spectral properties of the co-registered patches (Jabari and Zhang 2016b). We further expanded the PWCR to incorporate aerial images as well, and the initial building change detection results are published in Jabari and Zhang (2016a). In this study, we analyze the application of PWCR for urban change detection while selecting the bi-temporal images from different platforms (i.e., satellite or aerial) and different viewing angles (i.e., off-nadir angles). We assess the accuracy of PWCR by studying error propagation from DSM and exterior orientation (EO) parameters to image spaces. For further accuracy assessment, as suggested by Jabari and Zhang (2016b), area ratios of the segments generated by PWCR to the reference segments generated manually are also calculated. We finally use the multivariate alteration detection (MAD) approach to identify changed patches and assess whether registration errors have negative effects on change detection results. In this study, we incorporated a wider range of bi-temporal image combinations, including satellite images with large off-nadir angles and aerial photos at different spatial resolutions, that is, using images from different platforms and different view angles for urban change detection. This simplifies the data selection step of the change detection and reduces the associated costs.

The DSM used to perform PWCR can be acquired using stereo images or LiDAR. For preventing any confusion, DSM is to be acquired at the same time with one of the images, which is referred to as the *base* image; the other image is the *target* image. The changes of interest in this study are any sort of changes that cause spectral variation such as new constructions and building roof repairs. Plus, considering the occlusion effect, the change detection has to be limited to the objects visible in both images provided that they belong to horizontal surfaces (e.g., building roofs) and not to vertical surfaces (e.g., building façade). This effect is illustrated in Section 6.2.1.3.

6.2   PATCH-WISE MAD APPROACH

The proposed change detection approach is divided into two major steps:

1. Using PWCR to find corresponding patches in the bi-temporal images
2. Comparing the spectral properties of the corresponding patches using MAD Transform

The details of the two steps will be explained in this section.

6.2.1   PATCH-WISE CO-REGISTRATION

The purpose of the PWCR is to transfer segments (patches) from the base image to their corresponding positions in the target image considering the geometric

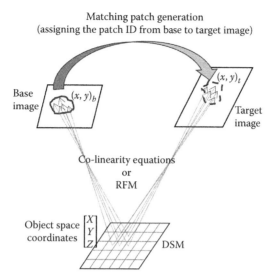

FIGURE 6.2 Schematic representation of the PWCR method. The DSM pixels are transferred to the image spaces showing the hypothetical matching points. The patch IDs of the points in the base image are transferred to their corresponding points in the target image. After this process is done for the whole points in the DSM, those points in the target image that have the same ID are merged to make the corresponding patches. (Modified after Jabari, Shabnam, and Yun Zhang. 2016a. Building change detection using multi-sensor and multi-view-angle imagery. In *IOP Conference Series: Earth and Environmental Science*, 34: 12018.)

differences of the images (Jabari and Zhang 2016b). The reason the co-registration is done in a patch-wise manner is that because of various geometric distortions of the images, the relative positions of the objects in the images are different. Thus, a global co-registration cannot be performed.

In the PWCR, as shown in Figure 6.2, a DSM, which presents the elevations of the top parts of the objects in an orthogonal projection way, is used for relating the corresponding points in the bi-temporal images. Using the co-linearity equations in the airborne images and the rational function model (RFM) in the satellite images, each DSM pixel is transferred to the bi-temporal image spaces. The projected DSM pixels in the image spaces represent the hypothetical matching points. We use the term "hypothetical matching points" to reflect that there might be changes attributed to the time difference between the bi-temporal images. Finally, from the hypothetical matching points, the corresponding patches are generated (Jabari and Zhang 2016b). Figure 6.2 illustrates a schematic representation of the co-registration method used in this study.

In this process, the first step is to segment the base image to generate patches, and the next step is to regenerate the patches in the target image using the object coordinates given by DSM.

6.2.1.1 Segmentation

In order to generate the initial patches, image segmentation is performed. In this study, in order to produce optimized segments, a method developed by Tong et al. (2012)

called the FbSP (Fuzzy-based Segmentation Parameter) tool is used. First, the image is over-segmented using the multi-resolution segmentation method in the eCognition software. Then, the optimized multi-resolution parameters (scale, shape, and compactness) are produced by the FbSP tool. Finally, the segments are merged using the optimized parameters to generate optimized segments.

Since under-segmentation negatively affects the change detection results, we manually checked the segments to prevent this issue. It was ensured that each segment includes only one object or an object is divided into more than one segment. For example, either each building roof is a single segment that does not include any parts of the building façade or the roof is divided into a few segments. This sort of segmentation result is referred to as *patch* in this study. Geographic information system (GIS) layers for roads or building borders can also be used to help produce the patches. In this way, with the help of the GIS layer, the existing road and building segments are generated in the images and then the rest of the images are segmented to look for new developments.

After image segmentation, each pixel in the base image is given a patch ID denoted by S_{k_i}, $1 < k < K$, where K is the total number of segments in the base image (Jabari and Zhang 2016b).

Additionally, in order to assess the accuracy of the PWCR, some test segments are selected in each data set. The test segments are manually edited to fit the actual object borders, for example, building borders. Later on in this study, the level of the fitness of the test segments, which are transferred to the target image, to the actual object borders in the target image is measured using a parameter called area ratio to assess the accuracy of the PWCR.

6.2.1.2   Co-Registered Patch Generation

The corresponding pixels, and accordingly the corresponding patches, in bi-temporal images are detected by transforming the object coordinates, given by DSM, to the image space coordinates using the EO parameters in the airborne images or RPCs in the satellite images. This transformation is referred to as *DSM projection*. In order to clarify the DSM projection, we go through the related formulas for parallel scanners (here, referred to as satellite images) and center-projective scanners (here, referred to as airborne images).

In photogrammetry, image–object coordinate transformation is given by scales, rotations, and shifts in the 3D space in Cartesian systems. The model for this relation is called the co-linearity equations and the coefficients in the model are given by the sensor position parameters, sensor attitude parameters (rotation angles), and the scale, which are called EO parameters. Thus, if the EO parameters are known, image–object transformation can be established.

This relation is different for satellite images. Because of parallel scanning (line scanning) mechanism, each scanned line has a different coordinate system; therefore, instead of the co-linearity equations, a generic model comprising an extended direct linear transform (with second-degree polynomials) is used, which is called RFM. And the coefficients of the polynomial terms, called RPCs, are calculated by imaging vendors and presented to the users. Thus, RFM can be used to relate satellite image coordinates to object coordinates, which, in this case, is a normalized geodetic coordinate system.

From the above, it can be inferred that, if the object coordinates and the parameters relating the object coordinates to the image spaces are known, the establishment of the object-image coordinate relation can lead to indirectly relating the corresponding points in the bi-temporal image spaces.

6.2.1.2.1   DSM Projection into Satellite Images

As explained by Jabari and Zhang (2016b), in PWCR, the projection of DSM pixels to their image spaces is performed by RFM (Grodecki 2001),

$$\tilde{x} = \frac{P_1(X,Y,Z)}{P_2(X,Y,Z)}$$

$$\tilde{y} = \frac{P_3(X,Y,Z)}{P_4(X,Y,Z)}$$

$$P(X,Y,Z) = \sum_{c=0}^{m}\sum_{b=0}^{m}\sum_{a=0}^{m} A_{a,b,c} X^a Y^b Z^c, \tag{6.1}$$

where $\tilde{x}$ and $\tilde{y}$ are normalized image coordinates, and X, Y, and Z are normalized ground coordinates. m is generally set to 3 (Grodecki 2001).

Since the RPCs provided by imaging vendors have inherent uncertainties owing to attitude or ephemeris errors, a projected DSM pixel into the related image space might not represent the exact position of interest. Nevertheless, Fraser and Hanley (2003) demonstrated that the aforementioned uncertainties introduce biases in the images spaces and can be modeled using an affine transformation.

$$\begin{bmatrix} \hat{x} \\ \hat{y} \end{bmatrix}_{ij} = \mathbf{T}_i G_i\left(\begin{bmatrix} X \\ Y \\ Z \end{bmatrix}_j\right), \tag{6.2}$$

where $\hat{x}$ and $\hat{y}$ are bias-compensated image coordinates, $G_i\left(\begin{bmatrix} X \\ Y \\ Z \end{bmatrix}_j\right)$ is the transformation from the object space to the image space based on the RFM equations for image i and pixel j, and $\mathbf{T}_i$ is a 2D affine transformation given in

$$\begin{bmatrix} \hat{x} \\ \hat{y} \end{bmatrix}_{ij} = \mathbf{T}_i\begin{bmatrix} \tilde{x} \\ \tilde{y} \end{bmatrix}_{ij} = \begin{bmatrix} n_{11} & n_{12} & n_{13} \\ n_{21} & n_{22} & n_{23} \\ 0 & 0 & 1 \end{bmatrix}_i\begin{bmatrix} \tilde{x} \\ \tilde{y} \\ 1 \end{bmatrix}_{ij}, \tag{6.3}$$

where n_{kl}, $k \in 1{:}3$, $l \in 1{:}2$, are the unknown coefficients of the affine transformation. To calculate the affine transformation parameters, either +3 image control points can be selected or the RPCs can be bias-compensated using the ground control points. Using the set of equations (Equations 6.1 through 6.3), each object point (from DSM) can be transferred to its corresponding place in a satellite image space.

6.2.1.2.2 DSM Projection into Airborne Image

As mentioned before, image–object relation in the airborne images (with point perspective geometry) is provided by the co-linearity equations as follows (Jabari and Zhang 2016a):

$$
\begin{aligned}
x &= x_0 - f\,\frac{(X-X_0)m_{11}+(Y-Y_0)m_{12}+(Z-Z_0)m_{13}}{(X-X_0)m_{31}+(Y-Y_0)m_{32}+(Z-Z_0)m_{33}} \\
y &= y_0 - f\,\frac{(X-X_0)m_{21}+(Y-Y_0)m_{22}+(Z-Z_0)m_{23}}{(X-X_0)m_{31}+(Y-Y_0)m_{32}+(Z-Z_0)m_{33}},
\end{aligned}
\tag{6.4}
$$

where x_0 and y_0 are principal point image coordinates; X_0, Y_0, and Z_0 are principal point object coordinates, usually measured by onboard GPS; and m_{ij} are the entries of the rotation matrix that are given in Equation 6.5.

$$
M = \begin{bmatrix}
\cos(\varphi)\cos(\kappa) & -\cos(\varphi)\sin(\kappa) & \sin(\varphi) \\
\cos(\omega)\sin(\kappa)+\sin(\omega)\sin(\varphi)\cos(\kappa) & \cos(\omega)\cos(\kappa)-\sin(\omega)\sin(\varphi)\sin(\kappa) & -\sin(\omega)\cos(\varphi) \\
\sin(\omega)\sin(\kappa)-\cos(\omega)\sin(\varphi)\cos(\kappa) & \sin(\omega)\cos(\kappa)+\cos(\omega)\sin(\varphi)\sin(\kappa) & \cos(\omega)\cos(\varphi)
\end{bmatrix}
\tag{6.5}
$$

where M is the rotation matrix; ω, φ, and κ are sensor rotation angles around the X, Y, and Z axes, respectively. The rotation parameters are usually measured by onboard inertial measurement unit (IMU) (Jabari and Zhang 2016a).

6.2.1.2.3 Corresponding Patch Generation

Having found the corresponding points in the bi-temporal images, the patch IDs of the points from the base image are transferred to their corresponding points in the target image (Jabari and Zhang 2016b). After repeating the process for all the DSM pixels, the coordinates of the points in the target image space are used to generate the corresponding patches based on Equation 6.6.

$$
S_{k_2} = \left\{ \text{round}\left(T_2 G_2 \left(\begin{bmatrix} X \\ Y \\ Z \end{bmatrix}_j \right) \right), 0 \,\middle|\, T_1 G_1 \left(\begin{bmatrix} X \\ Y \\ Z \end{bmatrix}_j \right) \in S_{k_1} \right\}
\tag{6.6}
$$

The round function is used to convert point coordinates into integer pixel coordinates in the images space. This converts the cluster of points into the target image pixels that have the same patch ID. These pixels are merged to generate the corresponding patches of the base image (Jabari and Zhang 2016b).

6.2.1.3   Solution to Occlusion

One of the main problems in urban change detection is dealing with occlusion. This effect is known as Z-buffering in computer graphics (Greene et al. 1993), and a photogrammetric solution has been used to detect occluded areas in remote sensing images as well by Ahmar et al. (1998). As illustrated in Figure 6.3, some areas are visible in the left image while blocked (occluded) in the right image. Occlusion makes uncertainties in change detection since occluded areas in one image do not have a corresponding area in the other image to be compared to.

Considering a DSM as a 2D matrix, for each pixel in (X,Y) position, there exists an elevation value (Z). Therefore, only horizontal surfaces are present in the DSM matrix. Vertical surfaces, such as a building wall, do not have any representation in the 2D matrix of DSM. In this process, by projecting the DSM pixels to the image spaces, vertical objects in the images do not get any projections and accordingly are removed from the process. However, the effect of elevated objects occluding other objects (e.g., buildings occlude roads) still exists and should be removed. In PWCR, the points in the image space that get multiple projections are involved in the occlusion effect; therefore, the intensity of those points in the image is associated with the projected DSM pixel that has higher elevation (Jabari and Zhang 2016b).

Figure 6.4 illustrates a schematic form of this sort of occlusion. Points A, B, and C are occluded by point D. Therefore, the intensity registered in point E belongs to point D with the highest elevation. This limits the change detection process in

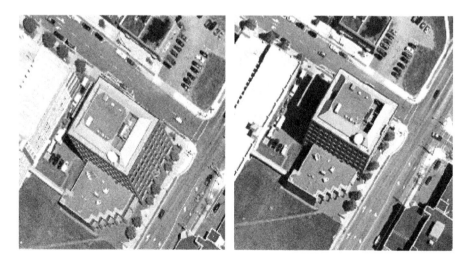

FIGURE 6.3   Comparison of the layout of a building from two view angles. Dissimilarity in the view angles exposes not exactly the same sides of the buildings, while occluding different objects (e.g., a part of the road in the right image).

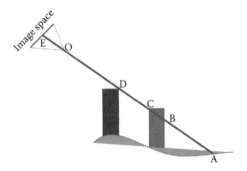

FIGURE 6.4 Occlusion in urban environment. The intensity registered in point E belongs to point D that has the highest elevation and points A and C are occluded.

the urban areas to building roofs and surface objects that are not occluded in both images and prevents false detections.

6.2.1.4 Null Segments

Because of numerical processing and interpolation limitations, there are randomly distributed pixels in the target image that are not projected to by the DSM pixels and thus they do not have a patch ID; this results in pixel-wide null segments in the target image. On the basis of our experiment, the higher the off-nadir angle, the greater the number of null segments. Here, the null segments are removed with neighborhood operations and dissolved in a neighbor patch with a longer common border using Equation 6.7.

$$\forall i \in \{1,..,N\}, \text{ if } S_i = \text{Null, then } \left\{ S_i \subset S_j \middle| \text{Perim}(S_i, S_j) > 0, \right.$$
$$\left. \text{Perim}(S_i, S_j) > \text{Perim}(S_i, S_k) \right\}, \ \forall k, j \in \{1,...,N\}, \ k \neq j \neq i\}$$ (6.7)

where N is the total number of the segments in the target image and Perim() is the length of the common border of two segments.

6.2.2 Comparison of the Spectral Properties Using MAD Transform

After PWCR, the associated intensities of the corresponding patches are compared in order to detect changes. There exist different change criteria such as cosine angle, correlation, principal component analysis, and MAD Transform (Canty 2014). From the above, MAD Transform is capable of compensating for linear radiometric differences between the bi-temporal images (Nielsen et al. 1998). Since the focus of this study is on using images from different sensors, MAD Transform is selected because of its ability to compensate for slight differences between the spectral or radiometric differences of the multi-platform bi-temporal images (Jabari and Zhang 2016b).

This method is based on a linear transformation of the spectral vector of the pixels (in our study: patches) and uses canonical coefficients to maximize the disparity between the brightness values in bi-temporal images. It transfers the bi-temporal spectral vectors $\mathbf{X} = [X_1,...,X_k]^T$ and $\mathbf{Y} = [Y_1,...,Y_k]^T$, k is the number of the spectral bands, into the space D (Nielsen 2011),

$$D = a^T X - b^T Y \tag{6.8}$$

in which a and b are the coefficients of a linear transformation of the spectral bands calculated in such a way that Var $\{a^T X - b^T Y\}$ is maximized considering the constraints of Var $\{a^T X\}$ = Var $\{b^T Y\}$ = 1. Nielsen (2011) uses canonical correlation analysis to solve this problem. Using this approach, the k-dimensional bi-temporal spectral vectors are transferred to another k-dimensional spectral vector in which the first component shows the lowest variation between the spectral bands $\left(a_1^T X - b_1^T Y\right)$ and the k^{st} set shows the highest variation $\left(a_k^T X - b_k^T Y\right)$.

$$\begin{bmatrix} X \\ Y \end{bmatrix} \rightarrow \begin{bmatrix} a_k^T X - b_k^T Y \\ . \\ . \\ . \\ a_k^T X - b_k^T Y \end{bmatrix}$$

$$D\{a^T X - b^T Y\} = 2(\mathbf{I} - \mathbf{R}) \tag{6.9}$$

where $\mathbf{I}$ is the $k \times k$ unit matrix and $\mathbf{R}$ is a $k \times k$ diagonal matrix containing the sorted canonical correlations on the diagonal.

6.3 EXPERIMENTS AND DISCUSSION

6.3.1 STUDY DATA SETS

In this study, different combinations of off-nadir satellite and airborne images are examined. Table 6.1 shows the major specifications of the imagery, including the off-nadir angles of the satellite images used, and Table 6.2 specifies the bi-temporal combinations used in this study.

Data set DS1 is composed of two images from different satellites with four multi-spectral bands red, green, blue (RGB), and near-infrared (NIR) bands. Data set DS2 is composed of off-nadir images. This data set has a rich spectral resolution with eight bands of WV2 satellite imagery. Data set DS3 is a combination of an airborne image and an off-nadir satellite image in which the sensor type, radiometric resolution, and imaging geometries are different.

TABLE 6.1

General Specifications of the Images Used for Change Detection in This Study

Image Name	IKONOS	GeoEye	WV2-2011	WV2-2013	Airborne
Satellite/Airborne	IKONOS	GeoEye	Wordview2	Wordview2	Airborne
Location	Hobart	Hobart	Fredericton	Fredericton	Fredericton
	Australia	Australia	Canada	Canada	Canada
Date	2/22/2003	2/5/2009	7/20/2011	8/18/2013	2005
Approximate GSD[a] (m)	0.9	0.5	0.495	0.582	0.5
Mean off-nadir angle (°)	15	20	15	27.1	–
Number of spectral bands	4(NIR-RGB)	4(NIR-RGB)	8	8	3 (RGB)

[a] Ground sampling distance.

TABLE 6.2

Bi-Temporal Combination of the Satellite Images Used for Change Detection in This Study and the Unique Specifications of the Bi-Temporal Combination

Data Set ID	Base Image	Target Image	Number of Bands Used	Specification	DSM Source
DS1	GeoEye	IKONOS	4	Two images from different satellites	Stereo GeoEye imagery (0.5 m accuracy)
DS2	WV2-2011	WV-2-2013	8	Combination of two off-nadir satellite images	LiDAR (0.5 m accuracy)
DS3	WV2-2011	Airborne	3	Combination of off-nadir satellite image and airborne image	LiDAR (0.5 m accuracy)

6.3.2 Co-Registration Results

Figure 6.5 shows the result of PWCR in the bi-temporal data sets used in this study. Figure 6.5a, c, and e illustrate the scenes of the segmented base images for data sets DS1 to DS3; Figure 6.5b, d, and f show the patches transferred to the related target images using the PWCR method. As illustrated in the images, the borders are properly reproduced in the target images. More detailed images are presented in Figure 6.6.

Figure 6.6 provides an example of how capable the PWCR is in generating the corresponding patches. The borders of two samples of building roofs in the base images (Figure 6.6a and d) are identified as base image patches to the algorithm. Then, the corresponding patches (borders of the roofs) are generated in the target images with the PWCR method. The corresponding patches in the target images are displayed in

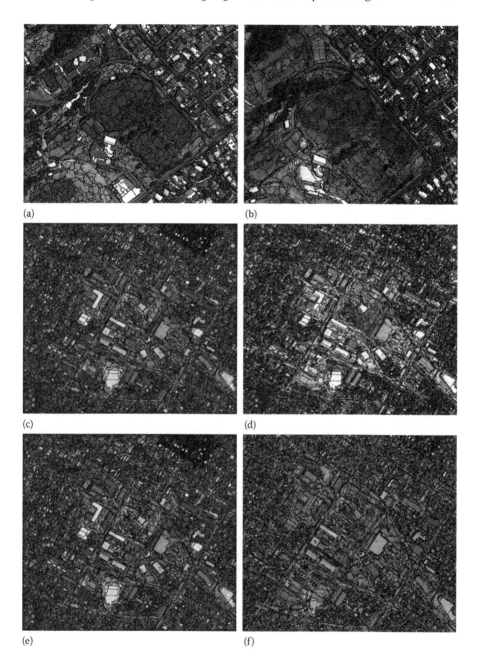

FIGURE 6.5   PWCR results in the study areas. (a, c, and e) Base image segments in the data set DS1, DS2, and DS3, accordingly. (b, d, and f) Target image in DS1, DS2, and DS3, accordingly, with patches generated using the PWCR method.

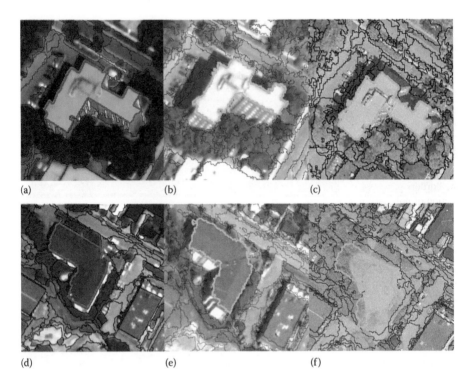

(a)   (b)   (c)

(d)   (e)   (f)

FIGURE 6.6   Borders of building roofs re-generated using PWCR. (a) and (d) are two building roof patches in the WV2-2011 image, which is the base image in data sets DS2 and DS3. The patches are edited manually to properly fit the building borders in the base image. Using PWCR, the patches are transferred to the target images. (b) and (e) are from the WV2-2013 image (target image in data set DS2); (c) and (f) are from the AB image (target image in data set DS3).

Figure 6.6b and e for data set DS2 and also in Figure 6.6c and f for data set DS3. The figures depict that leaning of the buildings are toward different directions, exposing different parts of the building façades. However, with the PWCR method, the borders of the roofs are detected properly. In Figure 6.6f, since the target image is taken 6 years before the base image, the building does not exist; therefore, the borders are transferred to the position where the building roof would be found in the image, if it existed.

6.3.3   CO-REGISTRATION ACCURACY ASSESSMENT AND DISCUSSION

In order to assess the accuracy of the PWCR results, a parameter called *area ratio* is used, which is explained in Section 6.3.3.1. Moreover, the propagation of the existing errors to the image spaces is studied in Section 6.3.3.2.

6.3.3.1   Area Ratio

As presented by Jabari and Zhang (2016b), the overlap percentage of the patches generated by PWCR and the patches produced manually in the target images is used for co-registration accuracy assessment. Figure 6.7 depicts the overlay of a sample patch

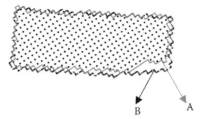

FIGURE 6.7 Example of overlapping polygons in the target image for accuracy assessment. Polygon A represents the borderlines of an object generated manually; polygon B represents the borders generated using the PWCR method. The marker filled area represents the intersection of polygons A and B.

generated manually (*polygon B*) and a patch generated using the co-registration method (*polygon A*) in the target image. In order to specify the accuracy of the co-registration, the ratio of the intersection of the polygons and union of the polygons is calculated using Equation 6.10.

$$\text{Area ratio} = \frac{A \cap B}{A \cup B}, \tag{6.10}$$

where $\cup$ and $\cap$ represent the union and intersection of two polygons A and B, respectively. In the ideal case where the patch generated using the PWCR method and the manually generated polygon are identical, the area ratio is equal to 1. In general, the higher the area ratio, the better the co-registration results. In the work of Jabari and Zhang (2016b), it has been proven that the area ratios of the patches registered by the PWCR method are far higher than the ones generated using a conventional co-registration method that includes ortho-rectification followed by a polynomial co-registration. Thus, here, just the area ratios for the PWCR are generated. Figure 6.8 presents the area ratio for around 40 test building patches with similar areas and various heights in each of the three study data sets. As explained in Section 6.2.1.1, in order to facilitate the comparison, the borders of the test patches are manually edited to fit the actual object borders in the base images.

As shown in Figure 6.8, the average area ratio for data sets DS2 and DS3 (target images are WV2-2013, Airborne of Fredericton city) is 89% and 92%, respectively. And the average for data set DS1 (target image is IKONOS image of the city of Hobart) is 84%. Considering that the spatial resolutions of Fredericton images are approximately similar and around 0.5 m, while the spatial resolution of the bitemporal images of the Hobart data set (DS1) is different (0.5 for the GeoEye image and 0.9 for the IKONOS image), the lower accuracy in the co-registration in the DS1 looks reasonable. DS1 is composed of two images from different satellites, DS2 is composed of two off-nadir satellite images, and DS3 contains off-nadir and airborne imagery. Thus, it can be stated that the off-nadir angle or sensor geometry differences did not result in difficulties in the PWCR.

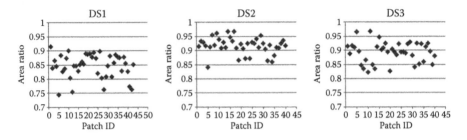

FIGURE 6.8 Area ratio calculated for the test objects in the target images of the three study data sets to assess the accuracy of PWCR. The average values for DS1, DS2, and DS3 are 0.84, 0.92, and 0.89, respectively.

As illustrated by Figures 6.5 and 6.6 and also proven by area ratios shown in Figure 6.8, images acquired by different sensors (airborne and satellite images) under different viewing angles can be co-registered with high accuracies using the PWCR method. Overall, regardless of the sensor types used in the different data sets, the generated patches, using the PWCR method, correspond closely to the manually generated ones.

6.3.3.2 Error Propagation

Considering Equations 6.2 and 6.4, the accuracy of the achieved co-registration results depends on the DSMs and orientation parameters (i.e., RPCs in satellite and the EO parameters in airborne images). Here, the effect of such errors in the final co-registration accuracy is assessed.

In VHR satellite images, as shown by Fraser and Hanley (2005), the RPC errors manifest themselves as biases in the image spaces, which are expected to be removed by a conformal transformation. After the bias compensation, an accuracy of better than one pixel is expected in the projection of DSM pixels to the image spaces. Nevertheless, in airborne images, the exact accuracy of mapping the DSM pixels into the image space is not known. There are three different sources of errors that can propagate into image space and negatively affect the co-registration accuracy: DSM error, the error of principal point position measured by the onboard GPS, and the error of sensor rotation angles measured by the onboard IMU. To validate the co-registration process, the error propagation of the three main sources of errors is calculated in the airborne imagery. To do so, the co-linearity equations, presented in Equation 6.4, are rewritten as

$$
\begin{aligned}
x &= f(X,Y,Z) \\
y &= g(X,Y,Z)
\end{aligned}
\tag{6.11}
$$

where x and y represent image space coordinates and X, Y, Z represent object coordinates (from DSM).

Error propagation based on DSM accuracy is given by

$$\mathbf{C_{x,y}} = \mathbf{JC_{A,B,C}J^T}$$

(6.12)

where $\mathbf{C_{x,y}}$ is the covariance matrix of image coordinates, $\mathbf{C_{A,B,C}}$ is the covariance matrix where (A,B,C) represent the object space coordinates (X, Y, Z), rotation angles $(\omega, \varphi, \kappa)$, and positioning of the principal point (X_o, Y_o, Z_o). J is the Jacobean matrix of co-linearity equations with partial derivatives represented by

$$
\mathbf{J} =
\begin{bmatrix}
\dfrac{\partial f}{\partial A} & \dfrac{\partial f}{\partial B} & \dfrac{\partial f}{\partial C} \\[2mm]
\dfrac{\partial g}{\partial A} & \dfrac{\partial g}{\partial B} & \dfrac{\partial g}{\partial C}
\end{bmatrix}.
$$

(6.13)

Assuming that $C_{A,B,C}$ is a diagonal matrix, that means A, B, and C are independent; Equation 6.13 yields

$$
\mathbf{C_{x,y}} =
\begin{bmatrix}
\sigma_x^2 & \sigma_{xy} \\
\sigma_{yx} & \sigma_y^2
\end{bmatrix}
=
\begin{bmatrix}
\dfrac{\partial f}{\partial A} & \dfrac{\partial f}{\partial B} & \dfrac{\partial f}{\partial C} \\[2mm]
\dfrac{\partial g}{\partial A} & \dfrac{\partial g}{\partial B} & \dfrac{\partial g}{\partial C}
\end{bmatrix}
\begin{bmatrix}
\sigma_A^2 & 0 & 0 \\
0 & \sigma_B^2 & 0 \\
0 & 0 & \sigma_C^2
\end{bmatrix}
\begin{bmatrix}
\dfrac{\partial f}{\partial A} & \dfrac{\partial g}{\partial A} \\[2mm]
\dfrac{\partial f}{\partial B} & \dfrac{\partial g}{\partial B} \\[2mm]
\dfrac{\partial f}{\partial C} & \dfrac{\partial g}{\partial C}
\end{bmatrix}
$$

$$
\sigma_x^2 = \left(\dfrac{\partial f}{\partial A}\right)^2 \sigma_A^2 + \left(\dfrac{\partial f}{\partial B}\right)^2 \sigma_B^2 + \left(\dfrac{\partial f}{\partial C}\right)^2 \sigma_C^2
$$

$$
\sigma_y^2 = \left(\dfrac{\partial g}{\partial A}\right)^2 \sigma_A^2 + \left(\dfrac{\partial g}{\partial B}\right)^2 \sigma_B^2 + \left(\dfrac{\partial g}{\partial C}\right)^2 \sigma_C^2.
$$

(6.14)

In order to achieve a numeric value to make an approximation for the accuracy of the co-registration, we calculate the EO parameters by the values generated by the onboard GPS and IMU given that $f = 153.328$ mm, flight height ≈ 1657 m, image size is 230 mm × 230 mm, and pixel size is 50 μm × 50 μm. The coordinates of image principal point and ground points are in the UTM coordinate system. However, here, the origin of the coordinate system is shifted to (0,0) in order to avoid adding weight parameters to the equations. Table 6.3 presents the accuracies achieved in projecting the DSM pixels into the image space. As can be seen, the accuracies are slightly higher than image pixel size, which means that the co-registered borders of the objects in the target image might have around one pixel

TABLE 6.3

Standard Deviations of the Errors Propagated from the DSM Space to the Image Space in the Airborne Images Used in This Study

Measurement Accuracy	Constants	σ_x, σ_y (µm)
$\sigma_X, \sigma_Y, \sigma_Z = 0.5$ m	$f = 153.328$ mm	58
	$x_0 = 0, y_0 = 0$	
	$\omega = 0.01, \varphi = -0.17, \kappa = -358.19$	
	$X_o = 0, Y_o = 0, Z_o = 1656.958$	
$\sigma_\omega, \sigma_\varphi, \sigma_\kappa = 0.003°$	$f = 153.328$ mm	13
	$x_0 = 0, y_0 = 0$	
	$X = 1100, Y = 1100, Z = 40$	
	$X_o = 0, Y_o = 0, Z_o = 1656.958$	
$\sigma_{X_o}, \sigma_{Y_o}, \sigma_{Z_o} = 2$ cm	$f = 153.328$ mm	1.8
	$x_0 = 0, y_0 = 0$	
	$X = 1100, Y = 1100, Z = 40$	
	$\omega = 0.01, \varphi = -0.17, \kappa = -358.19$	
Total error in image space	$\sqrt{\sigma_{DSM}^2 + \sigma_{IMU}^2 + \sigma_{GPS}^2}$	59.5

displacement with respect to their original places. As an example, in a building with a size of 40 pixels × 20 pixels, this displacement in x and y directions gives an overlap of 740 and a union of 860 pixels, resulting in an area ratio of around 86%. The area ratio is close to the numbers generated by the co-registration accuracy test in the airborne image presented in Figure 6.8, which confirms the PWCR accuracies generated by the area ratio as well.

In addition, internal orientation parameters f, x_0, and y_0 can also cause errors in the accuracy of co-registration in this study that are neglected here.

6.3.3.2.1 MAD Change Detection Results

In PWCR, since there is no one-to-one relation between the corresponding pixels in the bi-temporal images, the mean of the pixel values of each patch is used for change detection. Other parameters related to the distribution of the pixels, such as standard deviation, can also be used in this step. However, for simplicity, only the mean value is used in this study. Thus, for each patch, a multi-spectral mean vector is calculated. Given that the patches are generated in such a way that they only include one object (e.g., one building) or, due to over-segmentation, a part of an object, the mean value is a proper representative of each patch. Using Equation 6.8, the bi-temporal multi-spectral space is transferred to the MAD space (Nielsen 2011). As suggested by Canty (2014), a threshold of ±2σ is used to separate changes from non-changes, where σ is the standard deviation of each MAD band. Figure 6.9 shows the changes highlighted in the target images across the three data sets, whose PWCR results are shown in Figure 6.5.

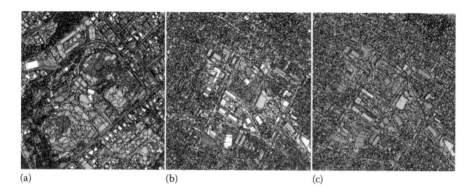

(a)   (b)   (c)

FIGURE 6.9 Snapshots of the results of MAD change detection across the three data sets with changes highlighted. Highlighted areas in (a), (b), and (c) depict MAD change detection result for data sets DS1, DS2, and DS3, respectively. The results of the PWCR of the same images are shown in Figure 6.5.

6.3.3.2.2 MAD Change Detection Accuracy Assessment and Discussion

In order to assess the accuracy of the change detection, a group of test patches, whose change labels were identified manually, were selected in each data set. Then, the change labels of the test patches were checked against the ones generated by the MAD Transform and accordingly the related confusion matrices were generated. Table 6.4 presents the confusion matrices of the change detection results across all three data sets. Since the classification has been divided into two classes, changed or unchanged, there are only two rows and two columns in the confusion matrices.

The objects of interest in this study are the urban structures. Therefore, in the check patch selection step, we retrained using the patches whose change status purely depended on vegetation changes. This is because vegetation status highly depends on seasonal effects and one of the best bands for vegetation change detection is the NIR band that was not available in the airborne image used in this study.

Having generated the confusion matrices of the study data sets, using equations of Table 6.5, ROC (receiver operating characteristic) curves are generated. ROC curves, which plot sensitivity versus fallout for a fine range of thresholds, specify the ability of the change detection method to identify changes from unchanged patches. The closer the ROC curve to the point (0,1), the higher the change detection accuracy.

TABLE 6.4
Confusion Matrices of Change Detection across All Three Data Sets

Data Set ID	Status	DS1		DS2		DS3	
Confusion Matrix	Changed	45	8	32	0	47	1
	Unchanged	7	58	8	38	2	45
Overall Accuracy		0.87		0.9		0.97	

TABLE 6.5
Formulas Related to Making ROC Curves

Parameter Name	Formula	Explanation of Abbreviations
Sensitivity	$\dfrac{tp}{tp+fn}$	**tp** (true positive): patches that are truly identified as changed
		tn (true negative): patches that are truly identified as unchanged
Fallout		**fp** (false positive): patches that are falsely identified as changed
	$\dfrac{fp}{fp+tn}$	**fn** (false negative): patches that are falsely identified as unchanged

As illustrated in Figure 6.10, the ROC curves across all three data sets are close to the point (0,1), which means that the MAD Transform can fulfill identifying the changed patches from the unchanged ones properly. This concept is confirmed with the overall accuracies presented in Table 6.4. All the three data sets have more than 87% accuracies in classifying changed and unchanged patches.

However, within the data sets, DS2 produced the closest ROC curve to point (0,1) and DS1 produced the furthest one. Data set DS2 contains eight multi-spectral bands, and the higher spectral resolution of the images leads to higher change detection accuracies compared to the other data sets. Data set DS3, with the same spatial resolution as DS2, produced the second best curve. The lower spectral resolutions of the images in this data set (the airborne image used has only three bands) is the reason for the slightly lower accuracy of change detection results. Finally, the lower spatial resolution of the IKONOS image compared to the GeoEye image in data set DS1 led to a lower accuracy in the co-registration, and accordingly, the change

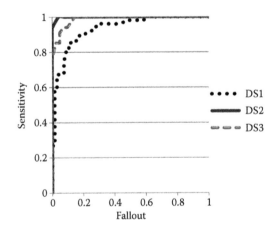

FIGURE 6.10 ROC curves of the MAD change detection across the four data sets. The closer the ROC curve to point (0,1), the better the discrimination capability. All the data sets have good ROC curves. However, DS2 has the best one.

detection results in this data set. The trend of the DS3 ROC curve also demonstrates that the sensor type difference did not severely affect the change detection accuracy.

6.4   FINAL REMARKS

To improve urban change detection accuracies using the presented approach, one needs to consider two separate concepts: co-registration quality improvement and change detection quality improvement.

As far as co-registration accuracy is concerned, as shown in Table 6.3, the accuracy of DSM highly affects the co-registration accuracy. Therefore, a DSM with accuracy better than or equal to the spatial resolution of the images can significantly improve the co-registration results. A detailed LiDAR-generated DSM can be a reliable source for this purpose. A DSM produced by stereo imagery can also be used (e.g., data set DS1 in this study) provided that a proper matching technique is used, which results in high precision elevation model generation. Unlike other 3D-based change detection methods, in this study, only one DSM is required. To prevent any confusion, it is necessary that the DSM is acquired at the same time as one of the bi-temporal images (the base image) or one has to make sure that there is no difference between the objects in the DSM and the base image. Additionally, the methodology presented in this work is capable of being expanded to include oblique airborne imagery such as pictometry images.

To produce patches in the base images, as explained in Section 6.2.1.1, image segmentation is required. This step is required in any other object-based change detection method as well. However, to highlight changes, it is not necessary to achieve a perfect segmentation result in which each image object is identified by a segment. As soon as the image is not under-segmented, the proposed framework can detect changes. Therefore, if the segmentation leans toward over-segmentation rather than toward under-segmentation, proper results can be expected.

In terms of the quality of change detection, bandwidths of bi-temporal images should be as close as possible. Although MAD Transform can compensate for slight linear differences in the spectral values, closer bands can certainly result in better change detection and removal of false alarms. MAD Transform can also compensate for radiometric resolution differences in the images (Nielsen et al. 1998).

One limitation in the airborne image used in this study was the lack on the NIR band that resulted in using only RGB bands of the satellite image for change detection. However, new generations of airborne images have the NIR bands as well. Thus, even better performance in the change detection process can be expected.

On the other hand, although off-nadir imaging does not significantly affect the co-registration accuracy, the spectral properties of objects might vary because of different solar angles that can reduce the accuracy of the change detection. In that case, topographic correction methods can help improve the results (Jabari and Zhang 2017). Besides, the existence of shadows of elevated objects in the images is still a challenge that needs to be compensated.

We limited the change detection study into identifying changed versus unchanged patches detection. Accordingly, if the classification information of the urban object is available, from-to information can also be generated.

6.5 CONCLUSION

In this study, we aimed to increase the data sources for change detection data selection step by utilizing images taken by different platforms and from different view angles as bi-temporal images for urban change detection. We used images with different geometries, including close-to-nadir, off-nadir, and airborne images. However, since the registration of such images is challenging using the conventional co-registration methods, the PWCR method is used to compensate for the difference in the geometric properties of the images. In PWCR, a DSM was used for relating the matching points and thus generating the corresponding patches. We achieved more than 84% accuracy in co-registering the images in a patch-wise manner.

Next, to detect changed patches, the spectral properties of the corresponding patches were compared using MAD Transform. MAD Transform is capable of compensating for slight radiometric differences between bi-temporal images and is preferred in this study since the sources of the bi-temporal images are different.

From the achieved co-registration results, it is concluded that the accurate PWCR is possible regardless of sensor type, provided that accurate orientation parameters (in center-projective sensors) or RPCs (in satellite images) as well as a DSM with the same ground resolution accuracy as the resolution of the images are provided.

From the achieved change detection results, it can be identified that image change detection using MAD Transform is possible; however, the higher the number of independent image bands, the higher the accuracy of the change detection. Thus, we proved that with a PWCR followed by a conventional change criterion (MAD Transform), highly accurate change detection results can be achieved regardless of the sensor type or view angles of the images.

ACKNOWLEDGMENT

This research has been funded by the National Sciences and Engineering Research Council of Canada. We acknowledge the City of Fredericton and the Department of Public Safety of the Province of New Brunswick for providing the airborne and LiDAR data of Fredericton. We also acknowledge Space Imaging LLC and DigitalGlobe for providing the IKONOS and GeoEye satellite images.

REFERENCES

Ahmar, F., J. Jansa, and C. Ries. 1998. The generation of true orthophotos using a 3D building model in conjunction with a conventional DTM. *IAPRS* 32: 16–22.

Al-Khudhairy, D. H. A., I. Caravaggi, and S. Glada. 2005. Structural damage assessments from Ikonos data using change detection, object-oriented segmentation, and classification techniques. *Photogrammetric Engineering and Remote Sensing* 71 (7): 825–37.

Bouziani, M., K. Goïta, and D. C. He. 2007. Change detection of buildings in urban environment from high spatial resolution satellite images using existing cartographic data and prior knowledge. In *IEEE International Geoscience and Remote Sensing Symposium*, 2581–84.

Braun, J. 2003. Aspects on true-orthophoto production. In *Proceedings of 49th Photogrammetric Week*. Vol. 214.

Canty, M. J. 2014. *Image Analysis, Classification and Change Detection in Remote Sensing: With Algorithms for ENVI/IDL and Python.* CRC Press.

Crispell, D., J. Mundy, and G. Taubin. 2012. A variable-resolution probabilistic three-dimensional model for change detection. *IEEE Transactions on Geoscience and Remote Sensing* 50 (2): 489–500.

Doxani, G., K. Karantzalos, and M. Tsakiri Strati. 2012. Monitoring urban changes based on scale-space filtering and object-oriented classification. *International Journal of Applied Earth Observation and Geoinformation* 15: 38–48.

Fraser, C. S., and H. B. Hanley. 2003. Bias compensation in rational functions for Ikonos satellite imagery. *Photogrammetric Engineering & Remote Sensing* 69 (1): 53–57.

Fraser, C. S., and H. B. Hanley. 2005. Bias-compensated RPCs for sensor orientation of high-resolution satellite imagery. *Photogrammetric Engineering and Remote Sensing* 71 (8): 909–15.

Greene, N., M. Kass, and G. Miller. 1993. Hierarchical Z-buffer visibility. In *Proceedings of the 20th Annual Conference on Computer Graphics and Interactive Techniques*, 231–38. ACM.

Grodecki, J. 2001. Ikonos stereo feature extraction-RPC approach. In *Annual Conference of the ASPRS 2001*, 23–27.

Gueguen, L., P. Soille, and M. Pesaresi. 2011. Change detection based on information measure. *IEEE Transactions on Geoscience and Remote Sensing* 49 (11): 4503–15.

Im, J., J. R. Jensen, and J. A. Tullis. 2008. Object-based change detection using correlation image analysis and image segmentation. *International Journal of Remote Sensing* 29 (2): 399–423.

Im, J., and J. R. Jensen. 2005. A change detection model based on neighborhood correlation image analysis and decision tree classification. *Remote Sensing of Environment* 99 (3): 326–40.

Jabari, S., and Y. Zhang. 2016a. Building change detection using multi-sensor and multi-view-angle imagery. In *IOP Conference Series: Earth and Environmental Science*, 34: 12018.

Jabari, S., and Y. Zhang. 2016b. RPC-based coregistration of VHR imagery for urban change detection. *Photogrammetric Engineering & Remote Sensing* 82 (7): 521–34.

Jabari, S., and Y. Zhang. 2017. Building change detection improvement using topographic correction models. *Advances in Remote Sensing* 6 (1).

Jung, F. 2004. Detecting building changes from multitemporal aerial stereopairs. *ISPRS Journal of Photogrammetry and Remote Sensing* 58 (3): 187–201.

Kang, Z., L. Zhang, H. Yue, and R. Lindenbergh. 2013. Range image techniques for fast detection and quantification of changes in repeatedly scanned buildings. *Photogrammetric Engineering & Remote Sensing* 79 (8): 695–707.

Martha, T. R., N. Kerle, V. Jetten, C. J. van Westen, and K. V. Kumar. 2010. Landslide volumetric analysis using Cartosat-1-derived DEMs. *Geoscience and Remote Sensing Letters, IEEE* 7 (3): 582–86.

Murakami, H., K. Nakagawa, H. Hasegawa, T. Shibata, and E. Iwanami. 1999. Change detection of buildings using an airborne laser scanner. *ISPRS Journal of Photogrammetry and Remote Sensing* 54 (2): 148–52.

Nielsen, A. Aasbjerg. 2011. Kernel maximum autocorrelation factor and minimum noise fraction transformations. *Image Processing, IEEE Transactions on* 20 (3): 612–24.

Nielsen, A. A., K. Conradsen, and J. J. Simpson. 1998. Multivariate alteration detection (MAD) and MAF postprocessing in multispectral, bitemporal image data: New approaches to change detection studies. *Remote Sensing of Environment* 64 (1): 1–19.

Niemeyer, I., P. R. Marpu, and S. Nussbaum. 2008. Change detection using object features. In *Object-Based Image Analysis*, 185–201. Springer.

Pang, S., X. Hu, Z. Wang, and Y. Lu. 2014. Object-based analysis of airborne LiDAR data for building change detection. *Remote Sensing* 6 (11): 10733–49.

Pollard, T. B., I. Eden, J. L. Mundy, and D. B. Cooper. 2010. A volumetric approach to change detection in satellite images. *Photogrammetric Engineering & Remote Sensing* 76 (7): 817–31.

Qin, R. 2014. Change detection on LOD 2 building models with very high resolution space-borne stereo imagery. *ISPRS Journal of Photogrammetry and Remote Sensing* 96: 179–92.

Rau, J., N.-Y. Chen, L.-C. Chen, and others. 2002. True orthophoto generation of built-up areas using multi-view images. *Photogrammetric Engineering and Remote Sensing* 68 (6). ASPRS AMERICAN SOCIETY FOR PHOTOGRAMMETRY AND: 581–88.

Tian, J., S. Cui, and P. Reinartz. 2014. Building change detection based on satellite stereo imagery and digital surface models. *IEEE Transactions on Geoscience and Remote Sensing* 52 (1): 406–17.

Tong, H., T. Maxwell, Y. Zhang, and V. Dey. 2012. A supervised and fuzzy-based approach to determine optimal multi-resolution image segmentation parameters. *PE&RS* 78 (10): 1029–44.

Waser, L. T., E. Baltsavias, K. Ecker, H. Eisenbeiss, E. Feldmeyer-Christe, C. Ginzler, M. Küchler, and L. Zhang. 2008. Assessing changes of forest area and shrub encroachment in a mire ecosystem using digital surface models and CIR aerial images. *Remote Sensing of Environment* 112 (5): 1956–68.

Zhou, W., A. Troy, and M. Grove. 2008. Object-based land cover classification and change analysis in the Baltimore metropolitan area using multitemporal high resolution remote sensing data. *Sensors* 8 (3): 1613–36.

Section III

Monitoring, Analyzing, and Modeling Urban Growth

7 Urbanization in India: Patterns, Visualization of Cities, and Greenhouse Gas Inventory for Developing an Urban Observatory

Bharath Haridas Aithal,
Mysore Chandrashekar Chandan,
Shivamurthy Vinay, and T.V. Ramachandra

CONTENTS

7.1 INTRODUCTION

Urbanization is highly dynamic and now ubiquitous, with almost half the population of the world living in cities. This proportion is expected to reach approximately 72% by 2050 (United Nations 2012). Mega cities now represent the most powerful economic poles of growth and the hinterlands would represent a region with loss of biodiversity and degradation of environment (Czamanski et al. 2008) with increasing human environment and human bioproblems that may range from a micro-region to a local region and can affect global issues. This may include increased air and water pollution (Liu and Diamond 2005), local climate alteration, and increased energy demands (Gonzalez et al. 2005; Ramachandra et al. 2012a). It has been argued that the next phase of urban growth will be concentrated in the developing countries, which includes one of the fastest-growing economies—India (Ramachandra et al. 2012b; Taubenböck et al. 2009, 2012; Van Ginkel 2008).

India has been witnessing rapid urban growth, initially leading to concentrated growth with a steep increase in land prices and later to a fragmented outgrowth from the city boundaries toward the urban–rural side. This fragmented outgrowth often results in the phenomenon called urban sprawl (Bharath et al. 2012; Ramachandra et al. 2015a). Sprawl may be defined as regions and settlements that are located in the vicinity of a city and/or outskirts of the growing urban area with very low to low-density development that would evolve into high-density development in a few years depending on the economy and other major factors. These are connected to the urban core without proper recognition from the city development authorities or planners. These areas are mostly devoid of any basic amenities such as water, transport connectivity, sanitation, and electricity (Ramachandra et al. 2014d). Such urban expansions are normally and extensively associated with growth, which is uncoordinated and could result in a serious crisis in terms of sustainable city development and future planning of city growth (Bharath et al. 2013; Ramachandra et al. 2014a,b; Taubenböck et al. 2009).

There is a growing need for landscape monitoring since city managers of mega cities in India face unprecedented challenges with regard to urban planning and land use management owing to the high dynamic growth and change in spatial patterns over time and failures in identifying driving forces and pockets for landscape changes. This necessitates understanding the temporal changes in land use and growth rate, which would help provide vital insights into the decision-making process through understanding the impact of landscape changes, biodiversity, complexity, and fragmentation of the landscape (Bharath et al. 2014; Patino & Duque 2013; Ridd 1995; Zeng & Wu 2005). Therefore, the quantification of landscape changes must consider both modifications in spatial arrangements and their consequences. Simulation-based modeling can provide basic and valuable site-based insights into possible future developments; this includes understanding the paths or pockets of current growth, understanding development corridors attributed to various improving infrastructural facilities, and developmental activities as a result of policy decisions.

Modeling temporal land use dynamics would help in visualizing development scenarios with inputs to sustainable city development aimed at optimizing available resources and decision making (Burgess & Jenks 2002). Urban models based on the principles of cellular automata (CA) are developing rapidly. CA-based urban models usually pay more attention to simulating the dynamic process of urban development and defining the factors or rules driving the development (Batty 2005). Different CA models have been developed to simulate urban growth and urban land use/cover change over time. The differences among various models exist in modifying the five basic elements of CA, that is, the spatial tessellation of cells, states of cells, neighborhood, transition rules, and time (Liu 2009). CA models have been shown to be effective platforms for simulating dynamic spatial interactions among bio-physical and socioeconomic factors associated with land use and land cover change (Ramachandra et al. 2012a). Various other approaches, such as regression modeling, neural networks, artificial intelligence, and so on, have been used effectively (Arsanjani et al. 2013; Mozumder & Tripathi 2014; Puertas et al. 2014).

While new urban models have provided insights into urban dynamics, a deeper understanding of the physical and socioeconomic patterns and processes associated with urbanization is still limited in India. Although emerging geospatial techniques have bridged the spatial data gap recently, there remain very few empirical case studies (Thapa & Murayama 2009). The urban population in India is growing at approximately 2.3% per annum, with the global urban population increasing from 13% (220 million in 1900) to 49% (3.2 billion in 2005) and is projected to escalate to 60% (4.9 billion) by 2030 (Ramachandra & Kumar 2008). The increase in urban population in response to the growth in urban areas is mainly attributed to migration. There are 48 urban agglomerations/cities having a population of more than 1 million in India (in 2011). However, unplanned urbanization coupled with the lack of holistic approaches leads to the lack of infrastructure and basic amenities. Hence, proper urban planning with operational, developmental, and restorative strategies is required to ensure the sustainable management of natural resources. Unplanned growth would involve radical land use conversion of forests, surface water bodies, and so on with the irretrievable loss of ground prospects (Basawaraja et al. 2011). The process of urbanization could be either in the form of townships or unplanned or organic. Many organic towns in India are now influencing large-scale infrastructure development and so on as a result of the impetus from the national government through development schemes such as JNNURM (Jawaharlal Nehru National Urban Renewal Mission). The focus is on the fast-track development through efficient infrastructure and delivery mechanisms, community participation, and so on. Large-scale land use and land cover changes, such as the loss of forests to meet the urban demands of fuel and land (Ramachandra & Kumar 2009), have led to changes in ecosystem structure, affecting its functioning and thereby threatening sustainable development. Cities are often attributed as the engines of economic growth. This compels cities to become smarter in handling large-scale urbanization and in finding new ways to manage complexity, increase efficiency, and improve quality of life. The design of smart cities requires an understanding of spatial patterns of urbanization to implement appropriate mitigation measures. This necessitates spatial information for the city administrators to visualize the patterns of urbanization and also predict the likely changes with the implementation of decisions (such as "what if" scenarios). Smart cities would be self-reliant and self-sufficient systems, providing basic necessities (such as water, sanitation, reliable utility services, and health care), transparent government transactions, and various citizen-centric services. This leads to the first objective of this chapter, that is, to understand the land use changes in five major cities in India.

7.2   CARBON FOOTPRINT ANALYSIS

India has been transitioning from a dependent economy to a more self-reliant manufacturing economy. This essentially involves the manufacturing process, which would entail more greenhouse gas (GHG) emissions, and a large workforce, which would increase the amount of carbon in the atmosphere. Such a transition to suit a very low carbon economy entails the adoption of modern technologies and adaptation strategies that involve improved regulatory mechanisms, better infrastructures, best business practices, low consumption rates, and improved green lifestyles. It should be noted that in the past two decades, an increased concern about global warming has been raised because of the increased emission of GHGs that form a blanket in the atmosphere, with all countries focusing on efforts to minimize these emissions. There has been a large-scale change in industrial output through the transformation of heavy industries in developed countries to a more knowledge-based and service-based economy that is comparatively cleaner (Shafik and Bandyopadhyay 1992). It must also be noted that there have been active citizen participation and awareness campaigns worldwide that have led to improved environmental regulations, which, in turn, decreased the rate of environmental degradation. Given these, international governing bodies should set targets to accomplish tasks and solve global environmental problems.

Also, organizations that govern environmental regulations are trying to establish strategies that would help mitigate GHGs of anthropogenic origin, thus reducing the threat of global warming (Kennedy et al. 2010; Wiedmann & Minx 2008). This resulted in many metropolitan cities across the globe showing interest in estimating GHG emissions and developing strategies to reduce emissions. Thus, carbon footprint is a measure of the impact of human activities on the environment in terms of the amount of GHGs produced (Finkbeiner 2009; Ramachandra et al. 2017). Carbon footprint, also called GHG footprint through estimation of GHG emission, is expressed in terms of carbon dioxide equivalent, indicating the amount of carbon in the atmosphere of a particular region. Carbon dioxide equivalent (CO_2e) is a unit for comparing the radiative forcing of a GHG (measure of influence of a climatic factor in changing the balance of energy radiation in the atmosphere) to that of carbon dioxide (ISO 2006a,b, Ramachandra et al. 2017). It is the amount of carbon dioxide by weight that is emitted into the atmosphere that would produce the same estimated radiative forcing as a given weight of another radiatively active gas (Wiedmann and Minx 2008).

7.3   NECESSITY OF UNDERSTANDING CARBON FOOTPRINT OF A REGION THROUGH QUANTIFICATION

Climate change is now one of most concerning areas for human existence in the next century. Decreasing our carbon footprint has the potential to reduce its impact on climate change by increasing human comfort. It would also add to the valuable information required as a component for sustainable urban planning for planners, policy makers, and local municipalities (Courchene and Allan 2008).

It has been widely reported that the last four decades have witnessed an abrupt increase in concentration of GHGs. All human activities such as industry, agriculture, deforestation, waste disposal, and specifically the burning of fossil fuels for various applications are considered the main factors that contribute to the increase in fossil fuel content in the atmosphere. The concentrations of atmospheric CO_2 increased from approximately 80 ppm considered by volume (ppmv) in the 1950s to 372 ppmv in 2001 and has been increasing approximately 20 ppmv every year as per the 2007 IPCC report. Similarly, CH_4 emissions have increased by 0.02 ppmv. This rapid increase in GHG concentrations in the atmosphere, mainly due to anthropogenic activities of humans, has affected the local and global climate and is said to have far-reaching consequences. The most significant contributors to global warming are the six Kyoto gases and, to a lesser extent, the chlorofluorocarbons (highlighted during the Montreal protocol; IPCC 2006, 2007). CO_2 is the primary GHG (contributing ~60%) followed by methane and nitrous oxide, with the largest GHG-producing sector being the energy supply sector (~26%) (IPCC 1996, 2007).

The effect of these gases is measured in terms of global warming potential (GWP) and depends on radiative forcing and time frame (usually 100 years). The GWP factor for CO_2 is 1 (considered as reference), that for CH_4 is 21, and that for N_2O is 298 (IPCC 2007). Today, it is important to understand and quantify the GHG emissions through a sectorial approach with specific methods. A sectorial approach would help in planning sectorial interventions in reducing carbon footprint in the atmosphere as well as necessary technological interventions based on sectorial needs.

Given these requirements, this chapter addresses three basic objectives based on understanding the steps in building and transforming sustainable urban systems. The first objective is to understand temporal land use change in the last two decades in five major cities in India and to model it based on agents of change. The second objective is to quantify GHG emissions. The third objective is to understand policy interventions required to overcome both land use change effects and effects of GHG emissions.

7.4 STUDY AREA AND DATA

This chapter aims to study five mega cities in India: Mumbai, Delhi, Kolkata, Chennai, and Bangalore, as shown in Figure 7.1. Table 7.1 shows latitude and longitude, population as per the year 2011 census, and metropolitan area spread of these mega cities.

Mumbai, named after the goddess Mumbadevi, has a rich history, and it has been referred to as Bombay province until 1955. The city is located along the west coastline of India. This region has undergone continuous land use change, mainly by reshaping under the Hornby Vellard project, in which the key objective was to merge several islands from the sea. Mumbai has a deep natural harbor, which serves as a significant access point to the Indian subcontinent through many European and Middle Eastern countries and hence getting the nickname "Gateway of India." Mumbai is the most populous city in India and is the ninth most populous urban agglomeration in the world. The Mumbai metropolitan region is composed of three complete districts, Mumbai city, Mumbai suburban, and Thane, and two partial districts, Palghar and Raigad. Because of its unique air, road, rail, and water transportation network,

FIGURE 7.1   Location details of mega cities.

TABLE 7.1

Location, Demographic, and Area Details of Mega Cities

Mega City	Latitude	Longitude	Population (Census 2011)	Area (km²)
Mumbai	18.97	72.82	18,414,288	4355
Delhi	28.61	77.23	16,314,838	1484
Kolkata	22.56	88.36	14,112,536	1886
Chennai	13.08	80.26	8,696,010	1189
Bangalore	12.96	77.56	8,499,399	741

Mumbai ranked first in commerce and industries by contributing 6.16% of India's gross domestic product. Mumbai enjoys a tropical dry and wet climate with an average annual precipitation of 2167 mm.

Delhi is the capital city and also a union territory of India, located at the northern part of the country, and stands on the west bank of river Yamuna. Delhi is surrounded by two states, namely, Haryana and Uttar Pradesh. With a population of 16,314,838 (Census 2011), Delhi is the second most populous urban agglomeration in India. Delhi and its adjoining urban region have been given the special status of National Capital Region under the Constitution of India's 69th Amendment Act of 1991. The Delhi metropolitan area consists of five municipal corporations: North Delhi Municipal Corporation, South Delhi Municipal Corporation, East Delhi Municipal Corporation, New Delhi Municipal Corporation, and Delhi Cantonment Board. Various historical monuments, such as Qutub Minar, Humayun's tomb, Iron

Pillar, and Red Fort, and political places of interest, such as Rashtrapati Bhavan, the Parliament House, the Supreme Court of India, the Cabinet Secretariat, and the Reserve Bank of India, are located in the central business district of New Delhi. The Delhi region has excellent railway, metro, and road connectivity such as the 28-km Delhi–Gurgaon Expressway, National Highway-1 connecting Delhi and Attari (near the Pakistan border), and National Highway-2 connecting Delhi and Kolkata (one of the busiest routes covering four states in between). Culture, history, and tourism, on one hand, and politics, economy, utility services, and transport, on the other hand, have made Delhi the second highest populated mega city next to Mumbai.

Kolkata is situated in the eastern part of India, and it is the capital city of West Bengal state. Much of Kolkata's development happened during the British rule during the late 1870s to 1910s. The region also served as the capital for British-held territories in India until 1911. With the increase in population from 9,194,000 in 1981 to 14,112,536 in 2011, Kolkata is the third most populous metropolitan city in India after Mumbai and Delhi. Kolkata is situated on the delta of River Ganga along the east bank of the Hoogly River, near Bay of Bengal. Majority of the landscape within the region was originally wetland, which has been consistently decreasing because of the unprecedented increase in population. The Kolkata Metropolitan Area (KMA) consists of four municipal corporations and 36 municipalities. The city has a rich culture and, unlike other cities in India, shows extreme passion for sports, especially football, which tends to attract a large number of tourists throughout the year. Kolkata is also known for its education sector, IT sector, banking sector, and heavy-scale industries. The city has completely taken advantage of the Port of Kolkata, which is one of the oldest operating ports in India, to export various consignments to other parts of the world.

Chennai is the capital city of the Tamil Nadu state. It is located at the eastern coast—Coromandel Coast, popularly known as the "Gateway to South India." It is one of the major metropolitan cities in India and has been one of the favorite destinations for tourism, industries, education, culture, and commerce. Chennai, which has a wide range of automobile industries, is famously called "Detroit of India" and is located in between two major rivers (i.e., Cooum and Adyar). Chennai has a tropical wet and dry climate with temperatures ranging from 15°C to 40°C. The jurisdiction of the Chennai (city) Corporation was expanded from 174 to 426 km^2 in 2011. The Chennai Metropolitan Area (CMA) has an area of 1189 km^2 comprising the Chennai city district and partially extending to two districts, Kancheepuram and Tiruvallur. Chennai is presently the fourth most populous city in India. The population of Chennai City has increased steeply from 4.34 million (2001) to 4.68 million (2011), whereas the CMA population shows an increase of 1.86 million based on the 2001 and 2011 census.

Bangalore is located in the southern part of India with an elevation of 900 m from the mean sea level. The city has been the capital of the Karnataka state since 1956. In 2006, the city administrative jurisdiction was expanded (known as Greater Bangalore) by merging existing city limits, eight urban local bodies, and 111 villages of the Bangalore urban district (Ramachandra and Kumar 2008). The city is located on a ridge and has numerous planned lakes (for storing water) that are interconnected and drain out to rivers from the three watersheds: Hebbal, Kormangala-Chellaghatta, and Vrishabhavathi (Ramachandra et al. 2015a). Bangalore was once called "Garden City" because of its lush green vegetation and large number of parks such as Lalbagh and Cubbon Park. After 2000, Bangalore experienced a huge population increase because of the establishment of a large number of information technology companies (the city has the nickname "Silicon Valley"). The city also houses public sector industries such as Bharat Electronics Limited, Hindustan Aeronautics Limited, National Aerospace Laboratories, Bharat Heavy Electricals Limited, and Bharat Earth Movers Limited. Because of the city's excellent transportation network and other infrastructure facilities, Bangalore is ranked third most productive metropolitan area of India.

Administrative boundaries of all mega cities were obtained from topographic maps or toposheets provided by Survey of India. The toposheets were geo-referenced and projected to respective Universal Transverse Mercator (UTM) zones. Bhuvan was used to supplement along with toposheets to ensure and delineate base layers such as administrative boundary, road and rail network, and drainage network. Each study area includes a 10-km buffer (from the administrative boundary centroid) to analyze and account for the land use changes in the future. Data collection involved obtaining ground control points (GCPs) using Global Positioning System (GPS). Bhuvan and Google Earth interfaces were used to collect data from remote areas and restricted areas where manual GPS data collection was not possible. They were also used to collect data in the form of points, lines, and polygons of industries, IT companies, educational institutes, healthcare units, road network, railway network, natural drainage network, restricted area, ecologically sensitive areas, coastal zones, and so on.

Satellite data starting from early 2000 to 2015 depending on availability were obtained from the US Geological Survey public domain website (https://earthexplorer .usgs.gov), which is available for free. Landsat 4, 5 (TM), 7 (ETM+), and 8 (OLI-TIRS) data were used for the analysis with a spatial resolution of 30 m. IRS LISS-III data were obtained from NRSC in case of missing Landsat data, with a spatial resolution of 23.5 m. GCPs were used to geo-register the satellite imageries to reduce any kind of discrepancy of temporal data.

7.5   METHOD

Figure 7.2 depicts the method adopted to assess urban growth patterns, which included three significant stages: data creation, land use and land cover analysis, and integrated model generation and validation.

7.5.1   DATA CREATION

Data were obtained from a survey of India topographic sheets. The sheets with scales of 1:50,000 and 1:250,000 were scanned with high resolution. City administrative maps were obtained from the respective metropolitan development authority database. Ground truth data of various locations within the study region were taken using handheld GPS. Places that were inaccessible and remote were visualized using Google Earth and Bhuvan.

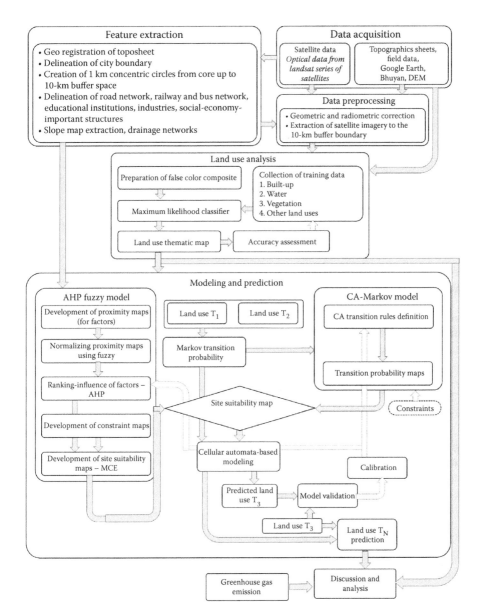

FIGURE 7.2 An integrated model approach.

All these data sets corresponding to different study regions were geo-referenced and re-projected to common datum world geodetic system 1984 (WGS84) and respective UTM zones to ensure uniformity in mapping. Hence, region-specific administrative boundary maps were obtained. Further, a 10-km buffer was drawn from the centroid of the central business district to better understand the urbanization process beyond city administrative boundaries. Concentric circles of 1-km radius were also delineated from the centroid along with four major directional grids (i.e., NE, NW, SE, and SW) to analyze minute changes within these fragments.

7.5.2 Land Use Analysis

Satellite data for various time periods were geo-referenced (to WGS84 and respective UTM zones), geo-corrected, rectified, and cropped, pertaining to different study regions. To maintain similar resolution and for better comparison, satellite data were resampled to 30 m using nearest-neighborhood function. Land use analysis was performed to understand change in landscape pattern throughout the study regions temporally. It involved the following process: (a) generation of a false color composite (FCC) image, (b) digitizing training polygons using FCC as base layer to distinguish heterogeneous features, (c) collection of training polygons via Google Earth (https://www .google.com/earth) (used as ancillary data for classification), (d) classification using Gaussian Maximum Likelihood (GML) classifier, and (e) validation and accuracy assessment. Land use analysis was performed using Geographical Analysis Support System open source software with four different categories as shown in Table 7.2.

A classification process is complete only when its accuracy is tested. Accuracy can be obtained by preparation of an error matrix or a confusion matrix (Congalton and Green 2008). User accuracy, producer accuracy, overall accuracy, and kappa statistics were calculated from the error matrix. Overall accuracy considers only diagonal elements of reference map and classified map, whereas kappa statistic takes into account off-diagonal elements as well (Lillesand et al. 2014).

7.5.3 Integrated Model Generation and Validation

Urbanizing agents and constraints were delineated in the form of points, lines, and polygons using Google Earth interface. Proximity maps were generated using minimum and maximum distance functions from each agent layer. Data values were normalized using fuzzy functions, and the entire range was between 0 and 255 (0 indicating no changes and 255 indicating maximum probability of change in land use types). The analytical hierarchical process was employed to estimate principal eigenvectors, and therefore, priority maps were created. Classified land use maps for initial stages, say, T_0, T_1, and T_2, were considered along with slope, and drainage layers analyzed using digital elevation model imageries were given as inputs to generate land use transitions

---

TABLE 7.2

List of Different Categories Used for Land Use Analysis and Corresponding Features

Category	Features Involved
Built-up	Houses, buildings, road features, paved surfaces, etc.
Vegetation	Trees, gardens, and forest
Water body	Seas, lakes, tanks, rivers, and estuaries
Others	Fallow/barren land, open fields, quarry site, dry river/lake basin, etc.

---

from T_0 to T_1 and from T_1 to T_2. Markov chain analysis was used to calculate the transition probability matrix based on two different time periods of classified images (Guan et al. 2008), where the probability of each land use type changing to the other remaining categories within the region considered is recorded after a specific number of iterations. Cellular automation with Markov chain was used to predict future land use for a specified year. Model validation was done by generating an error matrix for time period T_2 of both classified image versus simulated image.

7.5.4   ESTIMATION OF GHG FOOTPRINT

The method involved for calculating GHG can be divided into two significant steps. In the first step, a sector-wise approach was adopted to quantify GHG emissions. Various sectors and their description are given in Table 7.3. GHG

TABLE 7.3
Sector-Wise GHG Emission and Description

Sl. no.	Sector	Description
1	Energy sector: electricity consumption and fugitive emissions	Combustion of fossil fuels in thermal power plants; gases occurring during extraction, production, processing, and transportation of fossil fuels. Gas emissions include CO_2, SO_x, NO_x, SPM, and CH_4.
2	Domestic sector	Major contribution from activities like cooking (using LPG, firewood, and kerosene), lighting, and heating and from household appliances. CO_2, SO_x, NO_x and SPM.
3	Transportation sector	Emissions calculated by considering fuel consumed or distance traveled by various types of vehicles. Emissions from number of vehicles using compressed natural gas (CNG) were also accounted for. CO_2, SO_x, N_2O and CH_4.
4	Industrial sector	Major industries considered to emit CO_2 are iron and steel, ammonia manufacturing units, chemical products from fossil fuels, cement industry, petrochemical plants, fertilizer plants, power plants, glass industry, etc.
5	Agriculture sector	Includes paddy cultivation—the main phenomenon of CH_4 emissions from rice fields to atmosphere, agricultural soils, and burning of crop residues. Gases quantified under this sector are CH_4, N_2O (direct and indirect emissions), NO_x, and CO_2.
6	Livestock sector	Chief activities listed under animal husbandry include enteric fermentation and manure management. CH_4 and N_2O gases are calculated in this sector.
7	Waste sector	Methane is the major GHG emitted under the waste sector. The broad classes under waste can be listed as municipal solid waste disposal, domestic wastewater and industrial wastewater. Nitrous oxide (N_2O) emissions can occur both directly (from wastewater treatment plants) or indirectly (from wastewater after disposal of effluent to water bodies).

footprint was calculated using equation factors and various attributes as per Ramachandra et al. 2015b.

The major GHGs that were taken into consideration were carbon dioxide (CO_2), methane (CH_4), and nitrous oxide (N_2O). The second step included the conversion of non-CO_2 gases into units of CO_2 equivalent (CO_2e), and the aggregation of CO_2e was conducted as a region-specific analysis covering all the five study regions.

7.6   RESULT AND OUTCOMES

7.6.1   LAND USE ANALYSIS

Land use for different time periods was assessed using GML classifier. Dramatic changes in the landscape of various parts of India for over three decades are shown in Figure 7.3. Mumbai has shown a significant increase in built-up areas from 3.32% in 1973 to 14.26% in 2009. An extensive decrease of 74% can be seen in vegetation

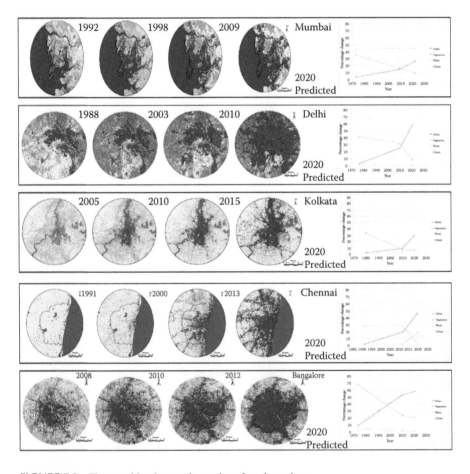

FIGURE 7.3   Temporal landscape dynamics of study regions.

category from 1973 to 2009. In terms of urbanization and urban sprawl, Delhi is no different from Mumbai, since it also shows an increase in impervious urban areas from 3.6% to 25.06% over a period of 33 years. This tremendous change in landscape makes surrounding places more vulnerable to various hazards and issues such as ecological imbalance (Riley et al. 2005; Xian et al. 2007), change in rainfall pattern leading to urban floods (Bronstert et al. 2002; McCuen and Beighley 2003), formation of urban heat island, traffic density congestion and therefore increases in air pollution, stress on surface as well as subsurface water resource, and so on. Further, density gradient analysis showed agents that fueled rapid urban growth in and around the Indira Gandhi International Airport (IGI) and the Indian Air Force base (Hindon). Classification results for Kolkata indicate a steep increase in urban category, which occurred at the cost of wetland and agriculture land degradation in various places (Ramachandra et al. 2014c).

Results for the Chennai region revealed a surge in built-up areas of 72% (1991–2000) and 646% (2000–2013). It is necessary to note that the growth that occurred especially in 2000 and 2010 is attributed to industrial and information technology policy measures. Industries play a key role in the urban growth process by attracting a huge number of populations from core city areas to urban fringe or transition zones; these zones later become part of a greater city region. Land use results for the Bangalore region shows a twofold increase in urban area from 24.86% to 48.39% within a span of 4 years (2008–2012). During this period, Bangalore lost a huge amount of trees especially due to pressure from construction of buildings, expansion of roadways, introduction of the metro-rail system, and so on. The vegetation cover decreased from 38.34% to 26.40%, showing unplanned urban growth.

7.6.2 VALIDATION

Accuracy assessment for land use classification was performed based on kappa statistics and overall accuracy as shown in Table 7.4. The results of overall accuracy indicate that classified maps along with ground truth have a lot of resemblance and are therefore accurate.

7.6.3 MODELING AND PREDICTION

Land use transitions were calculated to determine land use change for the year 2020 for Mumbai, Delhi, Kolkata, Chennai, and Bangalore. Markov chain was used to derive probability of change between two time periods. Knowing the land use of T_0 and T_1, the land use of time period T_2 was predicted. Prediction was made considering important urbanizing agents such as industries, educational institutes, road network, rail network, hospitals, and other service facilities. Constraints such as water bodies, wetland areas, protected areas, and so on were assumed to remain constant over all periods of time. Predicted results for time T_2 were compared with classified maps for time period T_2 to validate the data sets. Predicted maps were observed to be in good agreement, obtaining high accuracy and kappa values. Therefore, the process was repeated to predict land use for the period T_n. Cities like Delhi and Kolkata have shown a twofold increase in paved areas from time period T_2 to T_n.

TABLE 7.4
Overall Accuracy and Kappa Statistics

City	Mumbai				Delhi			Chennai		Kolkata			Bangalore		
Year	1992	1998	2009	1988	2003	2010	1991	2010	2013	2005	2010	2015	2008	2010	2012
Overall accuracy (%)	73	99	97	99	88	91	92	91	97	88	93	91	94	98	97
Kappa	0.81	0.82	0.81	0.99	0.84	0.96	0.92	0.9	0.9	0.9	0.99	0.96	0.98	0.96	0.95

Density and gradient analysis for the predicted areas confirmed that core urban areas are clumped as one large built-up patch, thereby not giving any other category the chance to develop.

7.6.4 TRANSITION PROBABILITY

Based on land use during T_1 and T_2, Markov transition potentials were computed, and the probability matrices of each land use type are given in Table 7.5. Diagonal elements indicative of land use classes being the same and other elements indicate the probability of land use to transition into land use classes. The results for most of the cities and built-up areas remain almost constant; non-urban land use classes have a higher

TABLE 7.5
Markov Transitional Probability for Study Regions

		Built-Up Areas	Water Bodies	Vegetation	Others
	Markov Transition Probabilities for Mumbai for the Period 1998–2009				
1998–2009	Built-up areas	0.990	0.000	0.000	0.010
	Water bodies	0.332	0.634	0.030	0.004
	Vegetation	0.486	0.003	0.475	0.036
	Others	0.623	0.001	0.002	0.374
	Markov Transition Probabilities for Delhi for the Period 2003–2010				
2003–2010	Built-up areas	0.970	0.000	0.002	0.028
	Water bodies	0.38	0.604	0.008	0.008
	Vegetation	0.528	0.001	0.431	0.04
	Others	0.613	0.001	0.018	0.368
	Markov Transition Probabilities for Kolkata for the Period 2010–2015				
2010–2015	Built-up areas	0.992	0.000	0.000	0.008
	Water bodies	0.18	0.754	0.005	0.061
	Vegetation	0.612	0.001	0.386	0.001
	Others	0.693	0.000	0.006	0.301
	Markov Transition Probabilities for Bangalore for the Period 2010–2015				
2008–2012	Built-up areas	0.998	0.000	0.000	0.002
	Water bodies	0.604	0.386	0.004	0.006
	Vegetation	0.793	0.001	0.177	0.029
	Others	0.486	0.002	0.006	0.506
	Markov Transition Probabilities for Chennai for the Period 2010–2015				
2008–2012	Built-up areas	0.948	0.000	0.030	0.022
	Water bodies	0.326	0.645	0.008	0.021
	Vegetation	0.516	0.000	0.467	0.017
	Others	0.294	0.004	0.013	0.689

probability of transitioning into an urban land use class. Bangalore showed substantial loss of water bodies and vegetation (as also reported by Ramachandra et al. 2017; Ramachandra and Bharath 2016). Kolkata and Chennai also follow the same trend.

7.6.5 GHG FOOTPRINT

Results from the energy sector revealed that Delhi has the highest commercial CO_2e emission. The Delhi region consumed 5339.63 MU of electricity during 2009–2010, releasing 5428.55 Gg of CO_2. The Kolkata region, which released approximately 1746.34 Gg of CO_2, contributed the least among the five cities. Delhi also showed maximum CO_2e for the domestic sector, with 11,690.43 Gg, followed by Chennai (8617.29 Gg), Greater Mumbai (8474.32 Gg), Kolkata (6337.11 Gg), and Bangalore (4273.81 Gg). Transportation sector emissions were recorded for vehicles using fuel CNG and non-CNG separately. The lowest value under this sector was for the Kolkata region, with 1886.60 Gg of emission from the non-CNG category only, whereas the peak values were observed for the Delhi region with 10,867.51 Gg under the non-CNG category and 1527.03 under CNG vehicles. The industry sector emission results revealed that ammonia production and fertilizer industries in Greater Mumbai and Chennai yielded 654.5 and 223.28 Gg of CO_2e, respectively. The Delhi region had a large amount of CO_2e emission from the agriculture sector (17.05 Gg from paddy cultivation, 248.26 Gg from soils, and 2.68 Gg from crop residue burning), followed by Bangalore (5.10 Gg from paddy cultivation and 113.86 Gg from soils). The livestock management and waste sector also had similar results, where Delhi was at the peak in terms of CO_2e emission followed by other regions. The combination of all these sectors clearly indicates that the range varies from 38,633.20 Gg/year (Delhi region) to 14,812.10 Gg/year (Kolkata region). Furthermore, sector-wise analysis shows that, of all the considered sectors, the transportation sector (Delhi, 32.08%; Bangalore, 43.48%) and the domestic sector (Mumbai, 37.20%; Chennai, 39.01%; Kolkata, 42.78%) have the highest emissions. The huge emissions from the transportation sector can be attributed to unplanned urbanization as well as to a lack of public transportation system, as a result of which people tend to use private vehicles. The results sector-wise are presented in Figure 7.4.

7.7 DISCUSSION AND CONCLUSION

Land use assessment that showcased urban growth has been phenomenal and would prosper with the help of initiatives that a growing economy like India would be providing. Accuracy assessment provided insights into accurate land use maps and their usability. Further GHG emission quantification would help toward the development of an indicator that would govern all aspects of sustainable development goals (SDGs) as described by the United Nations. In June 2015, the government of India has envisioned the development of smart cities through physical, institutional, and social infrastructures under its smart cities mission. That would help in understanding city growth and in improving quality of life. More possibly to improve socioeconomic cultural visibility of these smart cities to attract foreign investments. However, with this mission, there would be rapid urbanization as predicted, which means inadequacy of resources to match the rate of urbanization. The smart cities

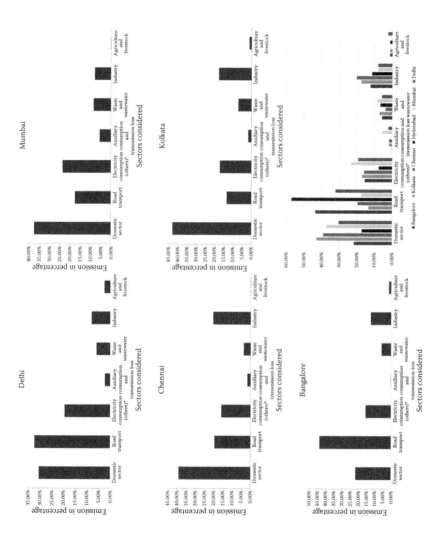

FIGURE 7.4 GHG emissions of study regions.

mission should develop an urban observatory to fill in the gaps in current land use and to monitor the sustainable goals of the urbanization process and all its dimensions and then create plans using these digital observations to improve strategies.

Strategies here are in the form of major components such as the following: (i) Green field development through smart townships. A holistic land management approach should be used here. (ii) Adoption of smart applications like transport, reuse, and recycle of wastewater; smart metering; recovering energy from solid waste; and so on. This would reduce pressure on nonrenewable sources of energy. (iii) Retrofitting current cities to match the comfort and security offered by existing climate-resilient infrastructures by reducing GHG footprint. (iv) Development of existing built-up areas through the creation of new combinations of land uses and improving infrastructure and amenities considering the location and needs of the citizens. (v) Maintaining the carrying capacity of the city, considering the available resources for citizens' use; otherwise, there would be increased GHG emissions.

This necessitates very efficient and citizen-friendly decision making through (i) understanding integrated land use planning based on the carrying capacity of a city, (ii) improving public transit system and making it more citizen friendly (especially because of the large amount of GHG emissions from the transport sector), (iii) improving and developing mass rapid-transport systems, and (iv) improving the application of ICTs and forming an urban data observatory as enabling technologies.

As shown in analyses, most cities are facing a crisis in terms of senseless unplanned rapid urbanization. Environmentally friendly urban centers with basic amenities and advanced infrastructure (such as sensors, electronic devices, and networks) would stimulate sustainable economic growth and improvements in citizen-centric services. This improvement in citizen-centric services can be realized if the data collected by the government and other stakeholders can be accessed and used by citizens. This could be possible by building different tiers of data repositories. Each tier would represent a specific need and a specific objective. This also includes developing a spatial database infrastructure that is connected and more accessible, forming an interconnected database network system or Internet of things, which, in turn, leads to huge databases that can be adopted for decision making with data security aspects in mind. This will eventually develop into an urban observatory. This would allow piling up of sufficient useful data in databases that would contribute to effective and coordinated governance, which would then support urban growth through improved economy and active participation of citizens. Although these cities are undergoing smart technological innovations through connectedness of data, this is limited to a certain network of users and should be used to focus on increased living comfort through providing adequate infrastructure, green spaces, and basic amenities to every citizen.

ACKNOWLEDGMENT

We are grateful to SERB, India, the Ministry of Science and Technology, Government of India, and Ranbir and Chitra Gupta School of Infrastructure Design and Management, Indian Institute of Technology Kharagpur, for financial and infrastructure support.

REFERENCES

Arsanjani, J. J., H. Marco, W. Kainz, and A. D. Boloorani. 2013. Integration of logistic regression, Markov chain and cellular automata models to simulate urban expansion. *International Journal of Applied Earth Observation and Geoinformation* 21 (April): 265–275. doi:10.1016/j.jag.2011.12.014.

Basawaraja, R., K. B. Chari, S. R. Mise, and S. B. Chetti. 2011. Analysis of the impact of urban sprawl in altering the land-use, land-cover pattern of Raichur City, India, using geospatial technologies. *Journal of Geography and Regional Planning* 4: 455–486.

Batty, M. 2005. *Cities and Complexity: Understanding Cities through Cellular Automata, Agent-Based Models and Fractals.* MIT Press.

Bharath, H. A., and T. V. Ramachandra. 2016. Visualization of urban growth pattern in Chennai using geoinformatics and spatial metrics. *Journal of the Indian Society of Remote Sensing* 44(4): 617–633. doi:10.1007/s12524-015-0482-0.

Bharath, H. A., D. Sannadurgappa, and Ramachandra, T. V. 2012. Status of wetlands in urbanising Tier II cities of Karnataka, In online proceedings Lake 2012: Wetlands, National Conference on Conservation and Management of Wetland Ecosystems, 06-08, Nov 2012, School of Environmental Sciences, M. G. University, Kerala, India.

Bharath, H. A., S. Vinay, S. Durgappa, and T. V. Ramachandra. 2013. Modeling and simulation of urbanisation in Greater Bangalore. In *National Spatial Data Infrastructure (NSDI)*. November. 34–50

Bharath, H. A., S. Vinay, and T. V. Ramachandra. 2014. Landscape dynamics modeling through integrated Markov, fuzzy-AHP and cellular automata. In *International Geoscience and Remote Sensing Symposium (IGARSS)*, 3160–3163. doi:10.1109/IGARSS.2014.6947148.

Bhuvan. A virtual land repository. Available at: http://bhuvan.nrsc.gov.in/

Bronstert, A., Niehoff, D. and Bürger, G. 2002. Effects of climate and land-use change on storm runoff generation: Present knowledge and modelling capabilities. *Hydrological Processes* 16(2): 509–529. doi:10.1002/hyp.326.

Burgess, R., and Jenks, M. eds. 2002. *Compact Cities: Sustainable Urban Forms for Developing Countries.* Routledge.

Congalton, R. G., and Green, K. 2008. *Assessing the Accuracy of Remotely Sensed Data: Principles and Practices.* CRC Press.

Courchene, T. J., and Allan, J. R. 2008. Climate change: The case for a carbon tariff/tax. *Policy Options—Montreal* 29(3): 59.

Czamanski, D., Benenson, I., Malkinson, D., Marinov, M., Roth, R., and Wittenberg, L. 2008. Urban sprawl and ecosystems—Can nature survive? *International Review of Environmental and Resource Economics* 2(4): 321–366. doi:10.1561/101.00000019.

Finkbeiner, M. 2009. Carbon footprinting—Opportunities and threats (editorial). *The International Journal of Life Cycle Assessment* 14(2): 91.

Gonzalez, J. E., Luvall, J. C., Rickman, D., Comarazamy, D., Picon, A., Harmsen, E., Parsiani, H., Ramirez, N., Vásquez, R. et. al. 2005. Urban heat islands developing in coastal tropical cities. *Eos, Transactions American Geophysical Union* 86(42): 397. doi:10.1029/2005EO420001.

Guan, D., Gao, W., Watari, K., and Fukahori, H. 2008. Land use change of Kitakyushu based on landscape ecology and Markov model. *Journal of Geographical Sciences* 18(4): 455–468. doi:10.1007/s11442-008-0455-0.

IPCC. Report of the twelfth season of the intergovernmental panel on climate change. Mexico City; 11–13 September 1996.

IPCC. 2006. National greenhouse gas inventories: Land use, land use change and forestry. Hayama, Japan. http://www.ipcc-nggip.iges.or.jp/public/2006gl/.

IPCC. 2007. Climate change 2007: Synthesis report: Contribution of working groups I, II and III to the fourth assessment report. https://www.ipcc.ch/report/ar4/syr/

ISO. 2006a. ISO 14064-1:2006. Greenhouse gases part 1: Specification with guidance at the organization level for quantification and reporting of greenhouse gas emissions and removals. http://www.ipcc-nggip.iges.or.jp/public/2006gl

ISO. 2006b. ISO 14064-2:2006. Greenhouse gases part 2: Specification with guidance at the project level for quantification, monitoring and reporting of greenhouse gas emission reductions or removal enhancements. http://www.ipcc-nggip.iges.or.jp/public/2006gl

Kennedy, C., Steinberger, J., Gasson, B., Hansen, Y., Hillman, T., Havránek, M., Pataki, D., Phdungsilp, A., Ramaswami, A., and Mendez, G. V. 2010. Methodology for inventory-ing greenhouse gas emissions from global cities. *Energy Policy* 38(9), 4828–4837.

Landsat data. USGS Earth Explorer. Available at: https://earthexplorer.usgs.gov/

Lillesand, T., Kiefer, R. W., and Chipman, J. 2014. *Remote Sensing and Image Interpretation*. John Wiley & Sons.

Liu, J., and Diamond, J., 2005. China's environment in a globalizing world. *Nature* 435(7046): 1179–1186. doi:10.1038/4351179a.

Liu, Y. 2009. *Modelling Urban Development with Geographical Information Systems and Cellular Automata*. Taylor and Francis Group.

McCuen, R. H., and Beighley, R. E. 2003. Seasonal flow frequency analysis. *Journal of Hydrology* 279(1–4): 43–56. doi:10.1016/S0022-1694(03)00154-9.

Mozumder, C., and Tripathi, N. K. 2014. Geospatial scenario based modelling of urban and agri-cultural intrusions in Ramsar Wetland Deepor Beel in Northeast India using a multi-layer perceptron neural network. *International Journal of Applied Earth Observation and Geoinformation* 32(October): 92–104. doi:10.1016/j.jag.2014.03.002.

Patino, J. E., and Duque, J. C. 2013. A review of regional science applications of satellite remote sensing in urban settings. *Computers, Environment and Urban Systems* 37: 117–128. doi:10.1016/j.compenvurbsys.2012.06.003.

Puertas, O. L., Henríquez, C., and Meza, F. J. 2014. Assessing spatial dynamics of urban growth using an integrated land use model. Application in Santiago metropolitan area, 2010–2045. *Land Use Policy* 38: 415–425. doi:10.1016/j.landusepol.2013.11.024.

Ramachandra, T. V., & Kumar, U. 2008. Wetlands of greater Bangalore, India: Automatic delineation through pattern classifiers. *Electronic Green Journal* 1(26).

Ramachandra, T. V., & Kumar, U. 2009. Land surface temperature with land cover dynamics: Multi-resolution, spatio-temporal data analysis of Greater Bangalore. International *Journal of Geoinformatics* 5(3).

Ramachandra, T. V., Aithal, B. H., and Beas, B. 2014a. Urbanisation pattern of incipient mega region in India. *Tema. Journal of Land Use, Mobility and Environment* 7: 83–100. doi:10.6092/1970-9870/2202.

Ramachandra, T. V., Aithal, B. H., and Sanna, D. D. 2012a. Insights to urban dynamics through landscape spatial pattern analysis. *International Journal of Applied Earth Observation and Geoinformation* 18(August): 329–343. doi:10.1016/j.jag.2012.03.005.

Ramachandra, T. V., Aithal, B. H., and Sowmyashree, M. V. 2014b. Monitoring spatial pat-terns of urban dynamics in Ahmedabad City, textile hub of India. *Spatium* 31: 85–91.

Ramachandra, T. V., Aithal, B. H., and Sowmyashree, M. V. 2014c. Urban structure in Kolkata: Metrics and modelling through geo-informatics. *Applied Geomatics* 6(4): 229–244. doi:10.1007/s12518-014-0135-y.

Ramachandra, T. V., Asulabha, K. S., Sincy, V., Sudarshan, B., and Bharath H. A. 2015a. Wetlands: Treasure of Bangalore. ENVIS Technical Report 101, Energy and Wetlands Research Group, CES, IISc, Bangalore, India.

Ramachandra, T. V., Bajpai, V., Kulkarni, G., Aithal, B. H., and Han, S. S. 2017. Economic disparity and CO_2 emissions: The domestic energy sector in Greater Bangalore, India. *Renewable and Sustainable Energy Reviews* 67: 1331–1344. doi:10.1016/j.rser.2016.09.038.

Ramachandra, T. V., Bharath, H. A., 2016. Bengaluru's reality: Towards unliveable status with unplanned urban trajectory, Current Science Editorial, 110.

Ramachandra, T. V., Bharath, H. A., and Sanna, D. 2012b. Land surface temperature analysis in an urbanising landscape through multi-resolution data. *Research & Reviews: Journal of Space Science & Technology* 1(1): 1–10.

Ramachandra, T. V., Bharath H. A., and Shreejith, K. 2015b. GHG footprint of major cities in India. *Renewable and Sustainable Energy Reviews* 44: 473–495. doi:10.1016/j .rser.2014.12.036.

Ramachandra, T. V., Bharath, A. H., and Sowmyashree, M. V. 2015c. Monitoring urbanization and its implications in a mega city from space: Spatiotemporal patterns and its indicators. *Journal of Environmental Management* 148: 67–81. doi:10.1016/j .jenvman.2014.02.015.

Ramachandra, T. V., Bharath, H. A., and Sowmyashree, M. V. 2014d. Urban footprint of Mumbai—The commercial capital of India. *Journal of Urban and Regional Analysis* 6(1): 71–94.

Registrar General, India. 2011. Census of India 2011: Provisional population totals—India data sheet. *Office of the Registrar General Census Commissioner, India. Indian Census Bureau.*

Ridd, M. K. 1995. Exploring a V–I–S (vegetation–impervious surface–soil) model for urban ecosystem analysis through remote-sensing—Comparative anatomy for cities. *International Journal of Remote Sensing* 16(12): 2165–2185.

Riley, S. P., Busteed, G. T., Kats, L. B., Vandergon, T. L., Lee, L. F., Dagit, R. G., Kerby, J. L., Fisher, R. N., and Sauvajot, R. M. 2005. Effects of urbanization on the distribution and abundance of amphibians and invasive species in southern California streams. *Conservation Biology* 19(6): 1894–1907. doi:10.1111/j.1523-1739.2005.00295.x.

Shafik, N., and Bandyopadhyay, S. 1992. *Economic Growth and Environmental Quality: Time-Series and Cross-Country Evidence* (Vol. 904). World Bank Publications.

Taubenböck, H., Esch, T., Felbier, A., Wiesner, M., Roth, A., and Dech, S. 2012. Monitoring urbanization in mega cities from space. *Remote Sensing of Environment* 117: 162–176. doi:10.1016/j.rse.2011.09.015.

Taubenböck, H., Wegmann, M., Roth, A., Mehl, H., and Dech, S. 2009. Urbanization in India—Spatiotemporal analysis using remote sensing data. *Computers, Environment and Urban Systems* 33(3): 179–188. doi:10.1016/j.compenvurbsys.2008.09.003.

Thapa, R. B., and Murayama, Y. 2009. Examining spatiotemporal urbanization patterns in Kathmandu Valley, Nepal: Remote sensing and spatial metrics approaches. *Remote Sensing* 1(3): 534–556. doi:10.3390/rs1030534.

United Nations. 2012. *World urbanization prospects: The 2011 revision.* New York: United Nations Department of Economic and Social Affairs/Population Division. UN Proceedings.

Van Ginkel, H. 2008. Urban future. *Nature* 456: 32–33.

Wiedmann, T., and Minx, J. 2008. A definition of 'carbon footprint'. *Ecological Economics Research Trends* 1: 1–11.

Xian, G., Crane, M., and Su, J. 2007. An analysis of urban development and its environmental impact on the Tampa Bay Watershed. *Journal of Environmental Management* 85(4): 965–976. doi:10.1016/j.jenvman.2006.11.012.

Zeng, H., and Wu, X. B. 2005. Utilities of edge-based metrics for studying landscape fragmentation. *Computers, Environment and Urban Systems* 29(2): 159–178. doi:10.1016/j .compenvurbsys.2003.09.002.

8 Mapping Impervious Surfaces in the Greater Hanoi Area, Vietnam, from Time Series Landsat Image 1988–2015

Hung Q. Ha and Qihao Weng

CONTENTS

8.1 INTRODUCTION

Urban sprawl is a global phenomenon, which comes along with economic development, expansion of settlements, and industrial development, but it also brings environmental problems such as loss of agricultural land (Burchell et al. 1998) and deterioration of water quality (Civico and Hurd 1997). As urbanization occurs, large proportions of the land surface will be covered by impervious materials, leading to higher rates of surface runoff, and putting stress on urban infrastructure (Hoang et al. 2007; Kim et al. 2016). Impervious surfaces, including rooftops, roads, pavement, and concretized surfaces, do not allow water to infiltrate into the soil beneath and can indicate the degree of urbanization.

Urbanization is increasingly becoming a critical issue in Vietnam, a developing country in Southeast Asia, because of ineffective land use planning and management in the country since the early 1950s. Although land administration in Vietnam has advanced to some degree in terms of the methodology and technology employed over recent years, many tasks are still performed using field-based mapping and surveys. Currently, international organizations such as World Bank, Japan International Cooperation Agency, and the Government of Vietnam have invested in improving the efficiency of land administration in Vietnam. One of the priorities is to develop a cost-effective and sustainable nationwide land monitoring system based on remotely sensed data (General Department of Land Administration of Vietnam 2016; World Bank 2008).

In Vietnam, specialized organizations such as the National Department of Remote Sensing (Ministry of Natural Resources and Environment) and Department of Survey and Mapping of Vietnam have studied land use and land cover (LULC) from remotely sensed data for decades, using either visual interpretation or automatic classification. However, the use of remote sensing in LULC mapping, especially in impervious surface mapping, was not fully recognized or employed at a nationwide scale. Furthermore, the time series analysis of urbanization in Vietnam is still limited in both the quantity and quality of research projects.

Arnold and Gibbons (1996) pointed out the need for obtaining relevant information about the impervious surfaces in urban studies because of its efficiency in mapping urbanization. In particular, in the context of developing countries where annual monitoring and observation systems are not put in place, the annual mapping of impervious surfaces is of crucial relevance.

The growth of residential lawn and the renovation of existing urban infrastructures are the other dynamics of urban LULC change besides impervious surfaces. In mapping urbanization, however, the focus of study should be on impervious surface mapping (Song et al. 2016). In the scope of this study, efforts are made for deriving the magnitude, timing, and duration of changes in impervious surfaces in a 27-year time span from a Landsat time series using normalized difference vegetation index (NDVI) and land surface temperature (LST) parameters.

8.2 TIME SERIES ANALYSIS OF IMPERVIOUS SURFACES

8.2.1 MAPPING IMPERVIOUS SURFACES

The effectiveness of remotely sensed data in mapping urbanization, which is indicated by the impervious surfaces, has been demonstrated in many research studies. Techniques and methods used for mapping impervious surface vary, including spectral mixture analysis (Deng and Wu 2013; Lu and Weng 2004, 2006; Wu 2004; Wu and Murray 2003) or decision tree (Dougherty et al. 2004; Im et al. 2012; Jantz et al. 2005).

Different approaches have been employed to study changes in impervious surfaces over time. Temporal analysis of impervious surface change can be achieved using normalized multitemporal remotely sensed data (Lu et al. 2011), time series analysis of different parameters (Zhang and Weng 2016), or NDVI time series using regression tree models (Mantas et al. 2016). These approaches aimed to provide techniques for mapping urbanization by means of impervious surface from remotely sensed data with higher accuracy.

In order to measure impervious area, Stocker (1998) highlighted ground surveys, global positioning systems, aerial photogrammetry, existing maps or digital data, or satellite remote sensing as the major approaches. Shahtahmassebi et al. (2016) suggested that hybrid techniques (visual interpretation, image subtraction, rule-based technique with pixel-by-pixel comparison, change vector analysis, post mapping analysis, and geographic information system), time series approaches (multidate classification using temporal rules, write memory function, per-pixel level comparison), modeling approaches (conjunction of vegetation–impervious surface–soil [VIS] technique, VIS model, spatial variance, logistic function and classification), and a spatial metrics approach (developing metrics) are the most effective techniques for change detection of impervious surfaces. These approaches performed different change detection tasks with various requirements for a number of impervious fraction inputs and provided different final results ranging from a change difference map (multidate classification) to urban change maps: simulation of urban growth (per pixel level comparison).

Using the VIS model proposed by Ridd (1995), a series of studies have been made for urban analysis using spectral mixture analysis (Deng and Wu 2013; Lu and Weng 2004, 2006; Wu 2004; Wu and Murray 2003) or decision tree classification (Dougherty et al. 2004; Im et al. 2012; Jantz et al. 2005). Changes in impervious surfaces are studied using different methods: conservative change thresholds for change detection (Xian and Homer 2010), normalized multitemporal remotely sensed data (Lu et al. 2011), time series analysis of different parameters (such as LST, biophysical composition index [BCI], and NDVI) (Zhang and Weng 2016), and NDVI time series with regression tree model (Mantas et al. 2016). Furthermore, the changes in time of imperviousness can be optimally studied with time series analysis, which helps to avoid phenological or long-term changes of vegetation (Gao et al. 2012; Lu et al. 2011; Sexton et al. 2013; Zhang and Weng 2016).

8.2.2 TIME SERIES IMAGE ANALYSIS

Using time series analysis with observations or variable set, their trends, periods, or cycles, unusual observations, or a combination of patterns can be detected (Montgomery et al. 2008). Time series analysis of impervious surfaces turned out to be an optimal approach for avoiding phenological and long-term changes of vegetation that might largely influence the accuracy of the image classification (Gao et al. 2012; Lu et al. 2011; Sexton et al. 2013; Song et al. 2016; Zhang and Weng 2016). The magnitude, timing, and duration of impervious surface changes can be derived from the time series of remotely sensed data (Song et al. 2016).

Changes in LULC or impervious surfaces in a time series image can be detected by change detection on a per-pixel basis between adjacent images. Using time series analysis for impervious surface mapping, different types of time series images can be generated including NDVI and LST for characterizing the surface objects and accounting for variation in these parameters. This step helps define pixels with less variation (stable pixels) in the time series data, which are considered stable and can be used as a consistent collection of training data for image classification. The training data shall consist of impervious, pervious, and water samples for each image.

The time series analysis takes all the available images into account rather than specific ranges of images, which shall provide better capture of the phenomena. Since all the available images are employed for image classification, a more comprehensive picture of the phenomena can be acquired.

In the time series analysis of urbanization, the urban transition process might be mapped inaccurately becuase of the phenological changes caused by highly varied vegetation cover such as rice. Given that, a temporal filtering is required to eliminate pseudo changes, where false changes are detected for areas on the ground with no actual land cover change. Actual changes can represent either real urban transitions or pseudo ones. Hence, a model for temporal filtering shall help eliminate pseudo changes in urban transitions shown by the increase in impervious surfaces.

8.2.3 FRAMEWORK FOR TIME SERIES ANALYSIS OF IMPERVIOUS SURFACES

Figure 8.1 illustrates the workflow for mapping impervious surfaces using Landsat time series.

8.2.3.1 Stacking Time Series of Landsat Data for Annual Impervious Surface Estimation

A stack of time series images of the Hanoi area, Vietnam, was made consisting of 27 years of data from 1988 to 2015 (method adapted from Song et al. 2016). Image stacking helps remove the inconsistency in atmospheric correction between different types of Landsat data and it ignores the matter of various units of the remotely sensed data (Song et al. 2001).

8.2.3.2 Estimation of NDVI and LST for Time Series Analysis

The strong correlations between impervious surfaces and other parameters such as NDVI, LST, and other biophysical factors were demonstrated throughout many

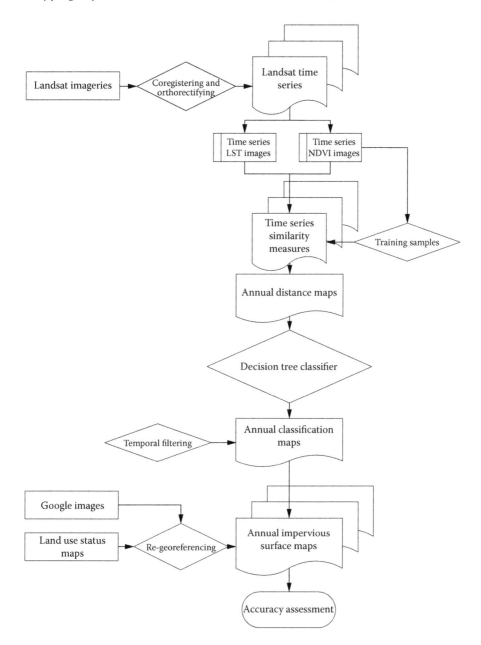

FIGURE 8.1   Workflow to map impervious surfaces from time series Landsat images.

research studies (Mathew et al. 2016; Wang et al. 2016; Weng and Lu 2008; Xiao and Ouyang 2007; Yuan and Bauer 2007; Zhang et al. 2009). The inverse relationship between vegetation cover (NDVI in this case) and impervious surfaces can be used effectively in mapping the spatial extent of impervious surfaces since it helps in delineating the abundance of vegetation (Weng 2012).

The NDVI, developed by Deering (1978), reflected the important vegetation characteristics by the ratio of two spectral bands consisting of infrared and red. NDVI is calculated as below:

$$NDVI = \frac{IR - R}{IR + R},$$
(8.1)

where IR is the infrared band and R is the red band.

Before calculating land surface emissivity (ε) using the model of Qiuji and Chuting (2015) for LST estimation, the land surface was divided into three groups of cover types including natural surface (bare land, grassland, and arable land), town or built-up (road, industrial, city, and rural residential area), and water body.

Qiuji and Chuting (2015) proposed the use of NASA Atmospheric Correction Parameter Calculator for vegetation cover and ε based on NDVI images:

$$P_v = \frac{(NDVI - NDVI_s)}{(NDVI_v - NDVI_s)}$$
$$\varepsilon_{surface} = 0.9625 + 0.0614 * P_v - 0.0461 * P_v^2$$
$$\varepsilon_{built-up} = 0.9589 + 0.086 * P_v - 0.0671 * P_v^2$$
$$\varepsilon_{water} = 0.995,$$
(8.2)

where P_v is vegetation cover, and $\varepsilon_{surface}$ and $\varepsilon_{built-up}$ are ground emissivity for vegetation and towns, respectively. $NDVI_v$ is the NDVI value of pure vegetation area or agricultural land, and $NDVI_s$ is the NDVI value of bare soil or little vegetation; $NDVI_v = 0.70$; $NDVI_s = 0.50$. If NDVI > 0.7, pixel is considered to have full vegetation cover; hence, $P_v = 1$; if NDVI < 0.05, $P_v = 0$. Finally, atmospherically corrected radiance was converted to at-sensor (at satellite) brightness temperature (referenced to blackbody temperature, T_B) as below:

$$T_B = \frac{K_2}{\ln\left(\frac{K_1}{B_T} + 1\right)},$$
(8.3)

where T_B is the effective at-sensor brightness temperature in Kelvin, and K_1 and K_2 are the prelaunch calibration constants for various Landsat sensors (Barsi et al. 2014; Qiuji and Chuting 2015).

With the level 2 data provided by USGS, the LST estimation can be done using the following approach: From at-satellite brightness temperature images (blackbody temperature, T_B), the LST (S_t) values were computed after correcting spectral emissivity (ε) (Artis and Carnahan 1982) according to the nature of land covers (Snyder et al. 1998) and NDVI (Qiuji and Chuting 2015) using the following equation:

$$S_t = \frac{T_B}{1 + \left(\dfrac{\lambda \times T_B}{\rho}\right) \times \ln \varepsilon}, \qquad (8.4)$$

where λ is the wavelength of emitted radiance ($\lambda_1 = 11.5$ μm [for Landsat 5TM and 7ETM+], $\lambda_2 = 10.9$ μm [for Landsat 8 TIRS]), $\rho = h*c/\sigma$ (1.4388×10^{-2} m K), σ is the Boltzmann constant (1.38×10^{-23} J/K), h is Planck's constant (6.626×10^{-34} J s), and c is the velocity of light (2.998×10^8 m/s).

8.2.3.3 Training Sample Selection

Because of the seasonal changes of crop production in the study area, yield status may be the source of improper image classification. Hence, different characterestics in the spectral signature of strongly grown crops and harvested ones in the same areas might be mapped differently in various seasons. In order to have an optimal collection of training data, nature of objects used should be persistent in terms of imperviousness and perviousness throughout the time span in the time series data using the approach illustrated in Figure 8.2.

First, two time series NDVI images were generated for growing season (January to March and July to September) and harversting season (April to June and October to December). By grouping NDVI images this way, NDVI values of the similar yield status in time series images should have comparable ranges of values.

Second, image differencing was applied for both NDVI time series images from m_1 and m_2 dimensions (number of images in the time series). For a pixel b in the time series, NDVI value difference d could be defined using the equation below:

$$d = \frac{\displaystyle\sum_{i=1}^{m}(N_t - N_{t-1})}{m}, \qquad (8.5)$$

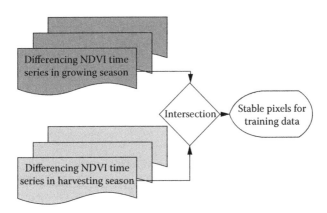

FIGURE 8.2 Training data extraction from image differencing of NDVI time series.

where N_t and N_{t-1} are the NDVI values at specific time t and $t - 1$ in the time series and m is the time series dimension.

After image differencing from two time series NDVI images, two distances d_1 and d_2 were calculated, and potential stable pixels were defined based on the 15% lowest quantiles. The 15% threshold was set where the areas with stable pixels expressed certain level of agreement with pattern of land cover by visual checking and comparison with Google Earth images.

Finally, intersection between d_1 and d_2 was implemented to find pixels satisfying the condition of being stable in both growing and harvesting seasons. The final d_3 image contains stable areas with pixels that can be used for selecting the training data for image classification.

Gao et al. (2012) proposed to select stable pixels in the earlier images for impervious surfaces and to select pixels for pervious ones in the later images in the time series data as training data. In this study, a consistent set of training data was used for the whole time series instead. Urban sprawl represented by the expansion of impervious surfaces is assumed to be an irreversible process, where impervious surfaces should not be converted to pervious ones after being converted from pervious surfaces in a short period. That principle was used later in combining with the 3-year moving window as criteria for temporal filtering of annual impervious surface maps.

8.2.3.4   Time Series Similarity Measures

In time series classification of remotely sensed data, development of similarity measures plays a crucial role. Mahalanobis distance provides the ability to quantify the differences between time series and to account for nonstationarity of variance and temporal cross-correlation (Lhermitte et al. 2011). Despite that, Euclidean distance was proven to be sensitive to noise because of its nonlinear characteristics. Hence, it is not relevant for quantifying temporal correlations (Douzal-Chouakria et al. 2009). Given that, Mahalanobis distance estimation was used to measure the temporal spectral differences between impervious and pervious surfaces in the NDVI and LST time series:

$$D_{\text{Maha}} = \sqrt{(P_{it} - P_{jt})^T \times \sum^{-1} (P_{it} - P_{jt})}, \tag{8.6}$$

where D_{Maha} is the Mahalanobis distance, P_{it} and P_{jt} are the values of pixels i and j at moment t in the time series. Annual distance maps were computed by equally weighting all the available distances for the corresponding years.

8.2.3.5   Decision Tree Classifier

Decision tree classification has been shown to be an effective technique for time series remotely sensed data with continuous nature in many studies (Dougherty et al. 2004; Im et al. 2012; Jantz et al. 2005; Zhang and Weng 2016). According to Im et al. (2012), the binary recurse partitioning process has been used in decision tree classification to classify samples into homogeneous subsets; this has some advantages, including no requirement for assumptions on normal distribution of the data, production of simple and interpretable decisions, and permission of categorical data incorporation. Hence, decision tree classification was used as below:

Training samples used for image classification were of two types: impervious and pervious surfaces (endmembers). Endmembers were the set $P = (p_1, p_2, \ldots, p_n)$, where n is the number of training samples. The class set of traning samples $C = (c_1, c_2, \ldots, c_m)$, where m is the number of classes. A logical test was formulated for attributes if the training samples can find the best splitting attribute. Once the decision tree was built, the classification rules could be generated and used for classification of a testing sample with unknown class labels.

8.2.3.6 Temporal Filtering

Land cover data sets shall have adequate temporal resolution to record the complexity of changes in order to monitor changes in urbanization, where 1- to 5-year basis windows were proposed by Jensen and Cowen (1999), while 3-year frequency was proposed by Lunetta et al. (2004) for monitoring urbanization and change in forests, respectively. Further, class consistency shall be guaranteed using per-pixel temporal filtering within a 3-year window to remove impossible class transitions as proposed by Clark et al. (2010). This 3-year window method shall require verifications of temporally adjacent images.

Figure 8.3 illustrates the principle for a 3-year moving window.

With annual classification maps, $I_1, I_2, \ldots, I_t$, where t specifies the year. If a pixel was classified as an impervious surface in I_1 and I_3, but it was identified as a pervious surface in I_2, then the I_2 is the uncertain pixel. As mentioned earlier, impervious surface is assumed to be irreversible, which is used to set the filter to correct I_2 to impervious surface. By doing it that way, potential errors in land cover transitions can be eliminated with temporal filtering (Zhang and Weng 2016).

8.2.4 Data

Data used in this study include the following:

- 169 Landsat images (level 1 data atmospherically corrected land surface reflectance, level 2 data Top of Atmosphere Reflectance and Brightness Temperature) (Landsat-5 Thematic Mapper TM, Landsat-7 Enhanced Thematic Mapper Plus ETM+, Landsat-8 Operational Land Imager [OLI],

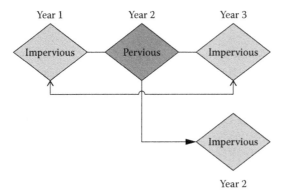

FIGURE 8.3 Temporal filtering with the 3-year moving window.

TABLE 8.1

Characteristics of Landsat Data in the Image Stack

Types of Data	Spatial Resolution (m)	Temporal Resolution (days)	Landsat Path	Landsat Row	Global Notation
Landsat 5-TM, Landsat-7 ETM+, Landsat-8 OLI and TIRS	30 × 30 (60 for thermal band)	16	127	45	World Reference Systems 2

and Thermal Infrared Sensor [TIRS]) in the 1988–2015 period were acquired from EarthExplorer, US Geological Survey, as in Table 8.1 (USGS 2016). No SLC-off data were included in the analysis.

- Land use status maps were collected from the General Department of Land Administration, Ministry of Natural Resources and Environment (Vietnam) for the years 2005 and 2010 (General Department of Land Administration of Vietnam 2016).

8.3 STUDY AREA

The greater Hanoi area (capital of Vietnam, herein called Hanoi area) shown in Figure 8.4 was selected for this study because of its rapid urban growth over the last few decades and its high vulnerability to natural hazards, especially to urban flooding and inundation. Hanoi has other characteristics that make it a good study area, including the fact that its urban population density is among the highest in the world (2.17 thousand people per square kilometer) (General Statistics Office of Viet Nam 2016) and its rapid urban growth over the last 20 years (GlobalSecurity 2011). Rapid urban growth is accompanied by rapid expansion of impervious surfaces. Because of its vulnerability to natural hazards, the Hanoi area was selected as one of the few cities in the Asia-Pacific region involved in a Typhoon Committee project on management and mitigation of floods (Liu et al. 2013).

The Hanoi area covers 3324.7 km² and is located in the Red river delta within the latitude from 20°53′ to 21°23′ N and 105°44′ to 105°02′ E (in the path: 127, row: 45 of the Landsat scene) in northern Vietnam. Since 2008, Hanoi has been extended to include the previous Hanoi city, Ha Tay province, Me Linh district (Vinh Phuc province), and four communes in Luong Son district (Hoa Binh provinces) (People's Committee of Ha Noi 2015). The Hanoi area has a population of 7.216, over the 20.926 million of the Red river delta according to the 2015 census (General Statistics Office of Viet Nam 2016).

Overall, nearly half of the total land area in the Hanoi area is agricultural (46.9%), where rice cultivation is the major agricultural activity. Hence, the vegetation cover is mainly predominated by rice fields, which are characterized by two crop seasons: *spring rice* (from late October or early November to May) and *winter rice* (from late May to mid-November) (People's Committee of Ha Noi 2015). These characteristics

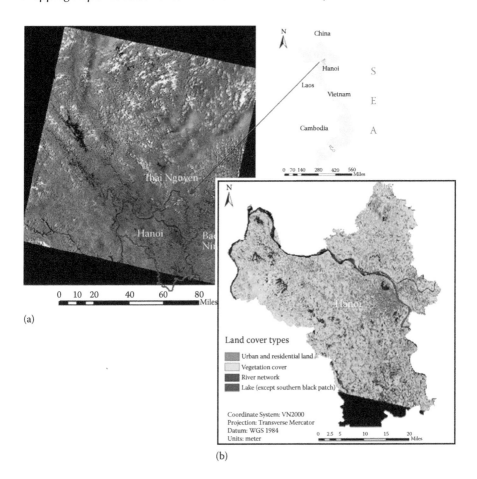

(a)

(b)

FIGURE 8.4 Overview of the study area in the Red river delta. (a) False color composite (FCC) (R:6, G:5, B:4) Landsat 8 OLI image of the Hanoi area acquired on January 6, 2016. (b) Land cover types showing on FCC (6-5-4) of Landsat 8 OLI.

may lead to difficulties in mapping LULC using vegetation index since it may be the source of phenological changes in vegetation indices.

8.4 IMPERVIOUS SURFACE TIME SERIES

8.4.1 ANNUAL DISTANCE MAPS

After calculating the NDVI and LST time series, an initial assessment was made for checking the quality of both NDVI and LST products. Only 114 images with reasonable patterns in both NDVI and LST time series were employed for calculating distance maps.

Figure 8.5 shows the NDVI time series, where pervious surfaces had the highest range of values, impervious surfaces had a moderate range, and water had the smallest range. In Figure 8.6, water had the lowest NDVI values (from −0.6 to 0.22),

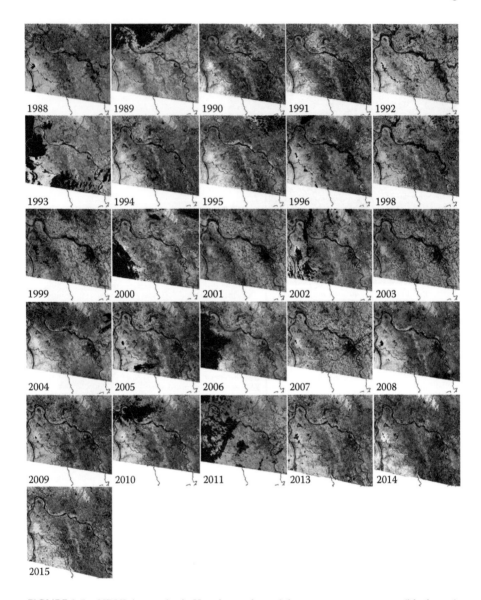

FIGURE 8.5   NDVI time series in Hanoi area showed the contrast among water (black tone), impervious surfaces (dark tone), and pervious ones (bright tone) (except some black areas of cloud masks in the years 1989, 1993, 1995, 1996, 1998, 2000, 2002, 2006, 2010, and 2011).

impervious surfaces had moderate NDVI values (from 0.07 to 0.49), and pervious surfaces had the highest NDVI values (from 0.16 to 0.8).

Figure 8.7 shows the LST time series, where impervious surfaces had the highest range of values, pervious surfaces had a moderate range, and water had the smallest range. In Figure 8.8, water had the lowest LST values ranging from 278 to 293 K (or 5°C to 20°C), pervious surfaces had moderate LST values ranging from 290 to

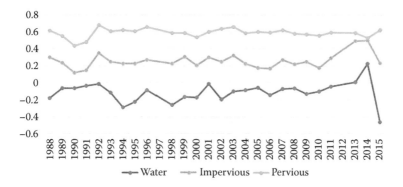

FIGURE 8.6   Separation of water, impervious, and pervious surfaces based on mean NDVI values in stable pixels in the time series from 1988 to 2015.

301 K (or 17°C to 28°C), and impervious surfaces had the highest LST values ranging from 295 to 315 K (or 22°C to 43°C).

8.4.2   ANNUAL IMPERVIOUS SURFACE MAPS

Annual impervious surface maps were generated by taking equally weighted time series similarity measures in each year from all available distance maps for that same year. However, because of the mixed pixel problem or changes in the yield status of paddy cultivation, many unique urban transitions existed in the classified images such as impervious surfaces converted into pervious ones for 1 or 2 years, and then converted back to impervious in the following year. These issues were eliminated using temporal filtering with a 3-year moving window.

As illustrated in Figure 8.9, the impervious surface maps in the years with more than four to five distance maps were considered reliable regarding the sufficiency of data in generating qualified impervious maps. Hence, impervious surface maps in the corresponding years including 1989, 1999, 2001, 2004, 2010, and 2013 can be used as reference data for temporal filtering. An impervious surface map in 2004 was selected as the reference for temporal filtering from 1988 to 2004 and from 2004 to 2015. Because of the low quality and limited number of distance maps for year 1997, and the exclusion of SLC-off data for 2012, no annual impervious surface map was generated for these 2 years.

8.4.3   TREND OF URBANIZATION SHOWN BY THE CHANGES
IN IMPERVIOUS SURFACES

As shown in Figure 8.10, the area of impervious surfaces increased by 305% from 255 to 1034 km² from 1988 to 2015. The increase of impervious surfaces was represented by the two phases of economic development as below:

- Phase 1 (from 1988 to 2000): gentle annual increase of impervious surfaces, totaling 29% overall increase from 255 to 330 km². This pattern was consistent with the economic development during this period, called transitional

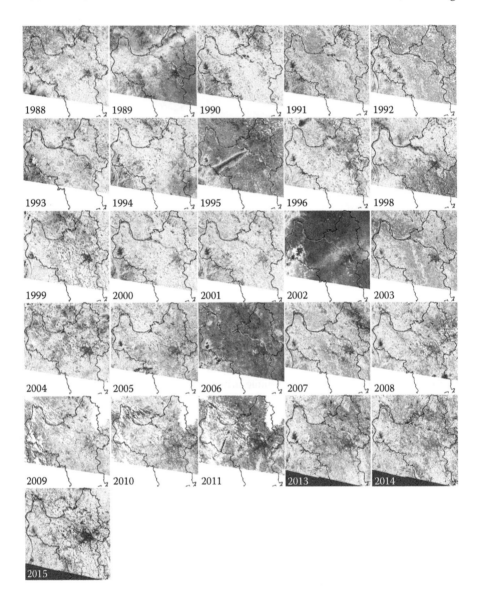

FIGURE 8.7 LST time series (in Kelvin) in Hanoi area showed the high contrast among water (lowest LST values from 278 to 293 K [5°C to 20°C]), pervious surfaces (moderate LST values from 290 to 301 K [17°C to 28°C]), and impervious surfaces (highest LST values from 295 to 315 K [22°C to 43°C]).

economy when the economy started to develop from centrally planned economy to market-based economy under the management and partial control of the government.

- Phase 2 (from 2000 to 2015): steeper and sharper annual increase of impervious surfaces, yielding a total of 213.5% increase from 330 to 1034 km².

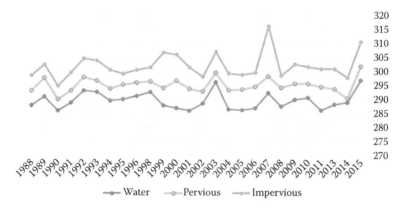

FIGURE 8.8 Separation of impervious, pervious surfaces, and water based on mean LST values (estimated in Kelvin) in stable pixels in the time series from 1988 to 2015.

This pattern was consistent with the economic development in the new century remarked by the explosive establishment of economic corporations and companies.

Because of the differences in the definition and nature of impervious surfaces and urbanization, the area of imperviousness and the urban area might not be the same, but they are considered comparable. The increasing trend of imperviousness, however, reflects the trend in the urbanization process very well.

8.4.4 ACCURACY ASSESSMENT

In order to assess the accuracy of classified images, 150 random points were generated in the impervious class in the area of stable pixels. Visual checking was applied to validate the level of agreement between the classified images, Google images, and land use status maps (for the years 2010 and 2014). Overall accuracy of image classification is further described in Figure 8.11.

If a random point is classified as having impervious surfaces, and it is compared as the identical one in the Google image (or land use status map) for the corresponding year, it is assessed as accurately classified. Vice versa, if a random point is classified as having impervious surfaces, but it has pervious surfaces in the Google image (or land use status map), it was inaccurately classified. By calculating the proportion of accurately classified points and the inaccurately classified ones, the accuracy of the image classification can be assessed.

For 1993, 1999, 2004, 2013, and 2015, image classifications have higher accuracy. This can be inferred as the larger number of accurately classified points between the classified images and the reference data and might have larger number of distance images when compared with other years in the time series.

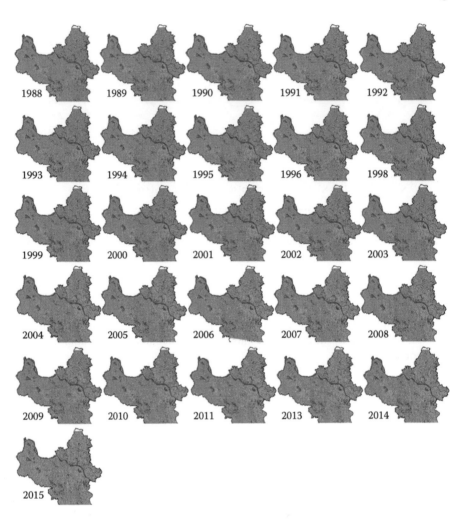

FIGURE 8.9 Annual impervious maps of Hanoi, Vietnam, from 1988 to 2015 generated by equally weighted spectral distances from NDVI and LST time series and decision tree classifier. Years 1997 and 2012 were not included due to the limited number of distance maps and low quality of SLC-off data, respectively.

8.5 DISCUSSION AND CONCLUSIONS

8.5.1 DISCUSSION

This research focused on mapping impervious surfaces of Hanoi, Vietnam, using time series Landsat images. In mapping impervious surfaces, land cover types were categorized into impervious surfaces, pervious surfaces (vegetation), and water. Impervious surface mapping was based on two opposite correlations: the negative correlation between impervious surfaces and vegetation index (NDVI) and the positive correlation between impervious surfaces and LST. More specifically, the higher

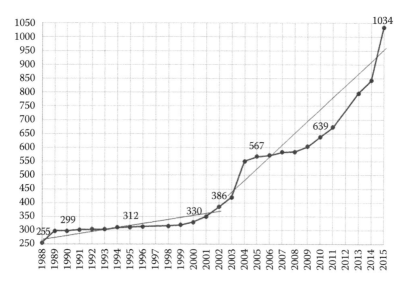

FIGURE 8.10 Area of imperviousness (in square kilometers) in the period of 1988 to 2015 when the increased trend in area was estimated by 305%.

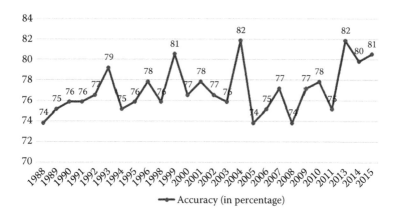

FIGURE 8.11 Accuracy of image classification in the time series from 1988 to 2015.

the NDVI values, the lower the imperviousness, and the higher the LST values, the higher the imperviousness.

One hypothesis was made as a starting point for this research: Imperviousness increased in time from 1988 to 2015, which represented a period of rapid urban sprawl. Annually mapping impervious surfaces provided an assessment of urban expansion, which showed an increase of impervious surfaces. Given that, a time series analysis of impervious surfaces with Landsat data turned out to be an optimal choice.

To effectively monitor impervious surfaces, relevant indicators were required. Hence, NDVI and LST time series were employed to map impervious surfaces because of their strong correlation with imperviousness. Training data extraction from time series, however, was a challenging task since NDVI and LST values largely varied in the time series. Zhang and Weng (2016) effectively used BCI to determine stable areas for time series analysis of impervious surface in the Pearl River delta, China.

Adapting the original research approach by Zhang and Weng (2016) and making further improvements, NDVI, instead of BCI, was used with image differencing in the NDVI time series to determine the stable areas for training data. Characterizing two seasons of rice paddy fields, which represented extensive land cover in Hanoi, was challenging because of the phenological changes in yield status. By categorizing the yield status of paddy cultivation in the study area into two seasons including growing and harvesting, the opposite yield status of paddy can be characterized. Hence, stable pixels with minor changes in the growing and harvesting seasons can be detected from the NDVI time series. As a result, the intersections of stable areas from both seasons were the stable areas to be used for image classification.

In the time series analysis of impervious surfaces from remotely sensed data, annual classified maps in years should be comparable. Hence, the use of time series similarity measures of both NDVI and LST in generating annual spectral distance maps was relevant since the combination of both NDVI and LST in generating annual distance maps helped strengthen the capacity of both NDVI and LST time series in differentiating impervious, pervious, and water body land covers. NDVI time series with higher spatial resolution better differentiated water and impervious surfaces (lower vegetation cover) from pervious surfaces (higher vegetation cover). LST time series better differentiated impervious surfaces from pervious surfaces and water.

In mapping impervious surfaces from time series data, there were cases where urban transition processes were classified inaccurately. A key assumption was the irreversible nature of imperviousness, where impervious surfaces could not be changed to pervious ones after being changed from pervious surfaces for a short period. This concept of surface transformation provided the basis of temporal filtering, and the 3-year moving window was highly effective in this study. Outputs derived from the posttemporal filtering were consistent with the increasing trend of imperviousness, which demonstrated the need to use temporal filtering.

Because of the differences in defining impervious surfaces and urbanization, the area of imperviousness might not be exactly the same as the area of urbanization (including urban built-up, residential area, and so on) since the area of imperviousness might include low infiltrable soils and rocks. In general, the overall trend of urbanization represented by the expansion of urban built-up and residential areas and the trend of changes in impervious surfaces were comparable.

8.5.2 RECOMMENDATIONS

Because of the dense cloud cover in the time series, there was a large amount of data being unused after cloud masking, since the gap-filling approach was not employed in this research. This led to the decrease in available images for generating distance maps. Gap-filling to compensate for missing data after cloud masking has been effective in previous research studies (Garcia 2010; Sirjacobs et al. 2011; Wang et al. 2012; Weiss et al. 2014). In the future, works that employ gap-filling for NDVI and LST time series to compensate for missing data owing to cloud masking would be advised; such efforts would increase the reliability of annual distance maps since they can be generated with more images.

8.5.3 Conclusions

This research employed time series analysis of NDVI and LST in mapping impervious surfaces in Hanoi, Vietnam, where seasonal crops such as rice strongly influence the accuracy of image classification. Temporal filtering helped improve image classifications and increased the agreement between classified images, thus reflecting the increasing trend of imperviousness. However, it also potentially introduced some bias by removing data.

REFERENCES

Arnold, Chester L. Jr., and C. James Gibbons. 1996. Impervious surface coverage: The emergence of a key environmental indicator. *Journal of the American Planning Association* 62 (2):243–258.

Artis, David A., and Walter H. Carnahan. 1982. Survey of emissivity variability in thermography of urban areas. *Remote Sensing of Environment* 12 (4):313–329. doi: http://dx.doi.org/10.1016/0034-4257(82)90043-8.

Barsi, Julia A., John R. Schott, Simon J. Hook, Nina G. Raqueno, Brian L. Markham, and Robert G. Radocinski. 2014. Landsat-* Thermal Infrared Sensor (TIRS) Vicarious Radiometric Calibration. *Remote Sensing* 6:11607–11626. doi: 10.3390/rs61111607.

Burchell, Rorbert W., Naveed A. Shad, David Listokin, Hilary Philips, Anthony Downs, Samuel Seskin, Judy S. Davis, Terry Moore, David Helton, and Michelle Gall. 1998. *The Costs of Sprawl—Revisited: Transportation Research Board Report.* Washington, D.C.: National Academy Press.

Civico, Daniel L., and James D. Hurd. 1997. Impervious surface mapping for the state of Connecticut. *Proceedings of ASPRS/ACSM Annual Convention*, April 7–10, Seattle, Washington.

Clark, Matthew L., T. Mitchell Aide, H. Ricardo Grau, and George Riner. 2010. A scalable approach to mapping annual land cover at 250 m using MODIS time series data: A case study in the Dry Chaco ecoregion of South America. *Remote Sensing of Environment* 114 (11):2816–2832.

Deering, Donald Wayne. 1978. Rangeland reflectance characteristics measured by aircraft and spacecraft sensors. PhD dissertation, Texas A&M University.

Deng, Chengbin, and Changshan Wu. 2013. A spatially adaptive spectral mixture analysis for mapping subpixel urban impervious surface distribution. *Remote Sensing of Environment* 133:62–70. doi: http://dx.doi.org/10.1016/j.rse.2013.02.005.

Dougherty, Mark, Randel L. Dymond, Scott J. Goetz, Claire A. Jantz, and Normand Goulet. 2004. Evaluation of impervious surface estimates in a rapidly urbanizing watershed. *Photogrammetric Engineering & Remote Sensing* 70 (11):1275–1284.

Douzal-Chouakria, Ahlame, Alpha Diallo, and Francoise Giroud. 2009. Adaptive clustering for time series: Application for identifying cell cycle expressed genes. *Computer Statistics and Data Analysis* 53 (4):1414–1426.

Gao, Feng, Eric Brown de Colstoun, Ronghua Ma, Qihao Weng, Jeffrey G. Masek, Jin Chen, Yaozhong Pan, and Conghe Song. 2012. Mapping impervious surface expansion using medium-resolution satellite image time series: A case study in the Yangtze River Delta, China. *International Journal of Remote Sensing* 33 (24):7609–7628. doi: 10.1080/01431161.2012.700424.

Garcia, Damien. 2010. Robust smoothing of gridded data in one and higher dimensions with missing values. *Computer Statistics and Data Analysis* 54 (4):1167–1178.

General Department of Land Administration of Vietnam. 2016. Vietnam Land Administration project. Ministry of Natural Resources and Environment of Vietnam. http://gdla.gov.vn/index.php/vi/redday.html.

General Statistics Office of Viet Nam. 2016. Statistical data of the Red river delta and Hanoi area in year 2015. http://www.gso.gov.vn/default.aspx?tabid=714.

GlobalSecurity. 2011. Red River Delta. Global Security. http://www.globalsecurity.org/military/world/vietnam/red-river-delta.htm.

Hoang, Vinh Hung, Rajib Shaw, and Masimi Kobayshi. 2007. Flood risk management for the RUA of Hanoi. *Disaster Prevention and Management* 16 (2):245–258. doi: 10.1108/09653560710739568.

Im, Jungho, Zhenyu Lu, Jinyoung Rhee, and Lindi J. Quackenbush. 2012. Impervious surface quantification using a synthesis of artificial immune networks and decision/regression trees from multi-sensor data. *Remote Sensing of Environment* 117:102–113. doi: http://dx.doi.org/10.1016/j.rse.2011.06.024.

Jantz, Patrick, Scott J. Goetz, and Claire Jantz. 2005. Urbanization and the loss of resource lands in the Chesapeak Bay Watershed. *Environmental Management* 36 (3):1–19.

Jensen, John R., and Dave C. Cowen. 1999. Remote sensing of urban/suburban infrastructure and socio-economic attributes. *Photogrammetric Engineering & Remote Sensing* 65 (5):611–622.

Kim, Hyomin, Dong-Kun Lee, and Sunyong Sung. 2016. Effect of urban green spaces and flooded area type on flooding probability. *Sustainability* 8 (2):134.

Lhermitte, Stefaan, Jan Verbesselt, Willem W. Verstraeten, and Pol Coppin. 2011. A comparison of time series similarity measures for classification and change detection of ecosystem dynamics. *Remote Sensing of Environment* 115 (12):3129–3152.

Liu, Zhiyu, Xiaotao Cheng, Zuhua Chen, Haotao Wan, Li Zhou, St Lai Edwin, Kunitsugu Masashi, Yang-Su Kim, Tae Sung Cheong, Gunhui Chung, Susan R. Espinueva, and Chang Ning Chen. 2013. Guidelines on urban flood risk management. In *Technical report of Typhoon Committee Cross-Cutting Project on Urban Flood Risk Management in the Typhoon Committee Area*. United Nations—Economic and Social Commission for Asia and the Pacific (ESCAP).

Lu, Dengsheng, Emilio Moran, and Scott Hetrick. 2011. Detection of impervious surface change with multitemporal Landsat images in an urban–rural frontier. *ISPRS Journal of Photogrammetry and Remote Sensing* 66 (3):298–306. doi: http://dx.doi.org/10.1016/j.isprsjprs.2010.10.010.

Lu, Dengsheng, and Qihao Weng. 2004. Spectral mixture analysis of the urban landscape in Indianapolis with Landsat ETM+ imagery. *Photogrammetric Engineering & Remote Sensing* 70 (9):1053–1062.

Lu, Dengsheng, and Qihao Weng. 2006. Use of impervious surface in urban land-use classification. *Remote Sensing of Environment* 102 (1–2):146–160.

Lunetta, Ross S., D.M. Johnson, John G. Lyon, and J. Crotwell. 2004. Impacts of imagery temporal frequency on land-cover change detection monitoring. *Remote Sensing of Environment* 89:444–454.

Mantas, Vasco M., João Carlos Marques, and Alcides J.S.C. Pereira. 2016. A geospatial approach to monitoring impervious surfaces in watersheds using Landsat data (the Mondego Basin, Portugal as a case study). *Ecological Indicators* 71:449–466. doi: http://dx.doi.org/10.1016/j.ecolind.2016.07.013.

Mathew, Aneesh, Sumit Khandelwal, and Nivedita Kaul. 2016. Spatial and temporal variations of urban heat island effect and the effect of percentage impervious surface area and elevation on land surface temperature: Study of Chandigarh city, India. *Sustainable Cities and Society* 26:264–277. doi: http://dx.doi.org/10.1016/j.scs.2016.06.018.

Montgomery, Douglas C., Cheryl L. Jennings, and Murat Kulahci. 2008. *Introduction to Time Series Analysis and Forecasting, Wiley Series in Probability and Statistics*: John Wiley & Sons.

People's Committee of Ha Noi. 2015. Overview of Ha Noi. Accessed November. http://english.hanoi.gov.vn/overview?p_p_id=Cms_WAR_Cmsportlet_INSTANCE_OdEiAOXb1f1R&p_p_lifecycle=0&p_p_state=normal&p_p_mode=view&p_p_col_id=column-1&p_p_col_pos=1&p_p_col_count=2&_Cms_WAR_Cmsportlet_INSTANCE_OdEiAOXb1f1R_jspPage=%2Fhtml%2Fcms%2Fportlet%2Ffrontend%2Fview.jsp&p_p_id=OdEiAOXb1f1R&_Cms_WAR_Cmsportlet_INSTANCE_OdEiAOXb1f1R_categoryId=2002&_Cms_WAR_Cmsportlet_INSTANCE_OdEiAOXb1f1R_articleId=23231&_Cms_WAR_Cmsportlet_INSTANCE_OdEiAOXb1f1R_title=overview-of-ha-noi&_Cms_WAR_Cmsportlet_INSTANCE_OdEiAOXb1f1R_command=details&_Cms_WAR_Cmsportlet_INSTANCE_OdEiAOXb1f1R_counter=1.

Qiuji, Chen, and Li Chuting. 2015. Land surface temperature retrieval based on Landsat ETM/TM: Taking Xi'an city as an example. *Open Cybernet System* 9 (1).

Ridd, Merrill K. 1995. Exploring a V–I–S (vegetation–impervious surface–soil) model for urban ecosystem analysis through remote sensing: Comparative anatomy for cities. *International Journal of Remote Sensing* 16 (12):2165–2185.

Sexton, Joseph O., Xiao-Peng Song, Chengquan Huang, Saurabh Channan, Matthew E. Baker, and John R. Townshend. 2013. Urban growth of the Washington, D.C.–Baltimore, MD metropolitan region from 1984 to 2010 by annual, Landsat-based estimates of impervious cover. *Remote Sensing of Environment* 129:42–53. doi: http://dx.doi.org/10.1016/j.rse.2012.10.025.

Shahtahmassebi, Amir Reza, Jie Song, Qing Zheng, George Alan Blackburn, Ke Wang, Ling Yan Huang, Yi Pan, Nathan Moore, Golnaz Shahtahmassebi, Reza Sadrabadi Haghighi, and Jing Song Deng. 2016. Remote sensing of impervious surface growth: A framework for quantifying urban expansion and re-densification mechanisms. *International Journal of Applied Earth Observation and Geoinformation* 46:94–112. doi: http://dx.doi.org/10.1016/j.jag.2015.11.007.

Sirjacobs, Damien, Aida Alvera-Azcárate, Alexander Barth, Geneviève Lacroix, YoungJe Park, Bouchra Nechad, Kevin Ruddick, and Jean-Marie Beckers. 2011. Cloud filling of ocean colour and sea surface temperature remote sensing products over the Southern North Sea by the Data Interpolating Empirical Orthogonal Functions methodology. *Journal of Sea Research* 65 (1):114–130. doi: http://dx.doi.org/10.1016/j.seares.2010.08.002.

Snyder, William C., Zhengming Wan, Yulin Zhang, and Yue Z. Feng. 1998. Classification-based emissivity for land surface temperature measurement from space. *International Journal of Remote Sensing* 19 (14):2753–2774.

Song, Conghe, Curtis E. Woodcock, Karen C. Seto, Mary Pax-Lenney, and Scott A. Macomber. 2001. Classification and change detection using Landsat TM data: When and how to correct atmospheric effects. *Remote Sensing of Environment* 75:230–244.

Song, Xiao-Peng, Joseph O. Sexton, Chengquan Huang, Saurabh Channan, and John R. Townshend. 2016. Characterizing the magnitude, timing and duration of urban growth from time series of Landsat-based estimates of impervious cover. *Remote Sensing of Environment* 175:1–13. doi: http://dx.doi.org/10.1016/j.rse.2015.12.027.

Stocker, Joel, 1998. Method for measuring and estimating impervious surface coverage. In *NEMO Technical Paper No. 3*.

USGS. 2016. Landsat Missions. http://landsat.usgs.gov/.

Wang, Guojie, Damien Garcia, Yi Liu, Richard De Jeu, and A. Johannes Dolman. 2012. A three-dimensional gap filling method for large geophysical datasets: Application to global satellite soil moisture observations. *Environmental Modelling and Software* 30:139–142. doi: http://dx.doi.org/10.1016/j.envsoft.2011.10.015.

Wang, Jiong, Zhan Qingming, Huagui Guo, and Zhicheng Jin. 2016. Characterizing the spatial dynamics of land surface temperature–impervious surface fraction relationship. *International Journal of Applied Earth Observation and Geoinformation* 45, Part A: 55–65. doi: http://dx.doi.org/10.1016/j.jag.2015.11.006.

Weiss, Daniel J., Peter M. Atkinson, Samir Bhatt, Bonnie Mappin, Simon I. Hay, and Peter W. Gething. 2014. An effective approach for gap-filling continental scale remotely sensed time-series. *ISPRS Journal of Photogrammetry and Remote Sensing* 98:106–118. doi: http://dx.doi.org/10.1016/j.isprsjprs.2014.10.001.

Weng, Qihao. 2012. Remote sensing of impervious surfaces in the urban areas: Requirements, methods, and trends. *Remote Sensing of Environment* 117:34–49. doi: http://dx.doi.org/10.1016/j.rse.2011.02.030.

Weng, Qihao, and Dengsheng Lu. 2008. A sub-pixel analysis of urbanization effect on land surface temperature and its interplay with impervious surface and vegetation coverage in Indianapolis, United States. *International Journal of Applied Earth Observation and Geoinformation* 10 (1):68–83.

World Bank. 2008. Land Administration Project in Vietnam. World Bank. http://www.worldbank.org/projects/P096418/land-administration-project?lang=en&tab=overview.

Wu, Changshan. 2004. Normalized spectral mixture analysis for monitoring urban composition using ETM+ imagery. *Remote Sensing of Environment* 93 (4):480–492.

Wu, Changshan, and Alan T. Murray. 2003. Estimating impervious surface distribution by spectral mixture analysis. *Remote Sensing of Environment* 84 (4):493–505.

Xian, George, and Collin Homer. 2010. Updating the 2001 national land cover database impervious surface products to 2006 using Landsat imagery change detection methods. *Remote Sensing of Environment* 114 (8):1676–1686. doi: http://dx.doi.org/10.1016/j.rse.2010.02.018.

Xiao, Rong B., and Zhi Y. Ouyang. 2007. Spatial pattern of impervious surfaces and their impacts on land surface temperature in Beijing, China. *Journal of Environmental Sciences* 19 (2):93–99.

Yuan, Fei, and Marvin E. Bauer. 2007. Comparison of impervious surface area and normalized difference vegetation index as indicators of surface urban heat island effects in Landsat imagery. *Remote Sensing of Environment* 106 (3):375–386.

Zhang, Lei, and Qihao Weng. 2016. Annual dynamics of impervious surface in the Pearl River Delta, China, from 1988 to 2013, using time series Landsat imagery. *ISPRS Journal of Photogrammetry and Remote Sensing* 113:86–96. doi: http://dx.doi.org/10.1016/j.isprsjprs.2016.01.003.

Zhang, Youshui, Inakwu O.A. Odeh, and Chunfeng Han. 2009. Bi-temporal characterization of land surface temperature in relation to impervious surface area, NDVI and NDBI, using a sub-pixel image analysis. *International Journal of Applied Earth Observation and Geoinformation* 11 (4):256–264. doi: http://dx.doi.org/10.1016/j.jag.2009.03.001.

9 City in Desert: Mapping Subpixel Urban Impervious Surface Area in a Desert Environment Using Spectral Unmixing and Machine Learning Methods

Chengbin Deng and Weiying Lin

CONTENTS

9.1 INTRODUCTION

More than half of the world's population live in cities, and this number will increase from the current 54% to 66%, according to United Nation's prediction (United Nations 2014). The increasing population requires various needs such as water, food, and living space. Urbanization is taking place all over the world, and an apparent consequence is the modification of land cover from natural land to anthropogenic urban land, also known as urban impervious surfaces, such as rooftops, parking lots, driveways, sidewalks, and other types of pavements (Arnold and Gibbons 1996; Weng 2012). Increasing imperviousness leads to ecosystem change because of its impermeable nature. For instance, a growing amount of impervious surface will increase volume of surface runoff and increase the probability and severity of flooding, as well as reduce groundwater contribution to stream flow (Schueler 1994, 2003). Additionally, urban impervious surface areas lead to the increase of surrounding air temperature, known as "urban heat island," decreasing vapor evaporation (Weng 2009). Impervious surface becomes increasingly important for a variety of urban and environmental applications, including metropolitan master planning, stream protecting strategy analysis, and watershed conservation and restoration (Arnold and Gibbons 1996; Schueler 1994, 2003; Schueler et al. 2009; Walsh et al. 2005). In order to provide better urban planning and ecosystem monitoring and protection, there is a great demand for accurate and up-to-date data on the acquisition of urban impervious surface information.

With a synoptic view of large geographic area and frequent revisit cycle, remote sensing images have been extensively used as a cost-effective means for mapping urban impervious surface. Various image processing methods have been developed to derive impervious surface information in different urban areas, which can be categorized as the per-pixel and subpixel mapping approaches. In terms of the per-pixel method, classification was carried out to generate different imperviousness-level classes (Flanagan and Civco 2001; Ji and Jensen 1999). This per-pixel classification method was further incorporated with other remotely sensed data (Zhang et al. 2012, 2014), improved by using artificial immune network (Gong et al. 2011), and used in multi-temporal time series for urban land use and land cover change detection (Jin and Mountrakis 2013). Because of the mixed pixel issue in all scales, per-pixel classification may not be an appropriate method to extract urban impervious surface. In terms of subpixel mapping methods, they can be further divided into two groups, spectral mixture analysis (SMA) and machine learning approaches. For spectral unmixing, normalized SMA (NSMA; Wu 2004) tends to reduce within-class brightness differences by averaging reflectance of each pixel in all bands of a Landsat image. Multiple endmember SMA (MESMA) was developed to test all potential endmember combinations in SMA and select the best-fit model with the lowest error indicator (Roberts et al. 1998). This method has been employed for mapping subpixel distribution of different land covers, including chaparral (Roberts et al. 1998), burn severity (Quintano et al. 2013), plant species (Roth et al. 2015), desert landforms (Balletine et al. 2005), and urban impervious surface (Deng 2016; Powell et al. 2007; Rashed et al. 2003). Also, the endmember bundling method examines all endmembers in the selected bundle to obtain minimum, medium, and maximum fractions of subpixel land covers (Bateson et al. 2000). More recently, spatially adaptive SMA (SASMA; Deng and Wu 2013b) was proposed to generate the "most representative" endmembers by synthesizing the signatures of all adjacent extracted endmembers weighted by their respective distance to the target mixed pixel. In addition to these spectral unmixing techniques, machine learning techniques were also applied to estimate various land covers at the subpixel level. Examples include classification and regression tree (CART; Xian 2007; Yang et al. 2003), artificial network (ANN; Pu et al. 2008; Shao and Lunetta 2011), support vector regression (SVR; Okujeni et al. 2013), and random forest (Deng and Wu 2013b).

In addition to these different image processing methods, different climates, urban settings, and vegetation types and natural landscapes also affect the accuracy of impervious surface fraction estimation. Most of the existing studies place their emphasis on temperate areas in the contiguous United States, Europe, China, and Australia. Moreover, these temperate areas in the literature are dominated by vegetated covers, and the spectral contrast between vegetation and urban impervious surface is high. Previous studies of desert cities only focus on image classification for land use and land cover change detection, particularly for urban expansion studies in Middle East and North Africa (Al-sharif and Pradhan 2014; Alqurashi and Kumar 2014; Haregeweyn et al. 2012; Kashaigili and Majaliwa 2010; Madugundu et al. 2014; Mahmoud and Alazba 2016; Shair and Nasr 1999; Weber and Puissant 2003; Yagoub 2004; Yagoub and Al Bizreh 2014). With complicated desert landforms that are prone to result in spectral confusion among different land covers, impervious surface extraction and quantification in desert environments, however, is not well studied. Sand and bright impervious surface, as well as shadow and dark impervious surface, tend to be spectrally confused with each other, and these land covers are manually separated before performing spectral unmixing (Myint and Okin 2009). Therefore, the major objectives of this study include (1) mapping subpixel urban impervious surface distribution in a desert city using different spectral analytical techniques, and (2) evaluating their performances so as to provide potential guidance for future studies.

9.2 STUDY AREA AND DATA

We selected Dubai, United Arab Emirates (UAE), a typical desert city in West Asia, as our study area. As shown in Figure 9.1, Dubai is located on the south coast of the Arabian Gulf and lies within the Arabian Desert in the Middle East. It has a population of approximately 2.753 million in 2017. Summers in Dubai are extremely hot, windy, and humid, with an average temperature of 106°F, while winters are warm with an average temperature of 75°F with a mean annual precipitation of 94.3 mm. On the basis of its historical temperature and precipitation records, Dubai is categorized as having a tropical desert climate in the Koppen climate classification. Dubai covers a geographical area of 1588 square miles, increasing from the initial 1500 square miles by land reclamation from the Arabian Sea. There are a great number of land cover and land use types in the study area. The medium-/high-density developed areas, mainly located along the Arabian Sea, are dominated by commercial and residential land uses. The flat sandy desert, which provides living environment for wild grasses and date palms, surrounds the entire urban area. There is a natural inlet, Dubai Creek, running northeast–southwest through the city. In particular, imperviousness can be found on a variety of constructions and infrastructures, including parking lots, concrete for building roofs, and asphalt for roads. In addition, Dubai is one of the world's fastest-growing economies, and its gross domestic product per capita reaches US$24,866 in 2014. Today, tourism is the most important source of revenue for Dubai, which was the seventh most visited city in the world according to air traffic in 2013. Examples of famous landmarks include the Palm Islands, the World Islands, Burj Khalifa, and Dubai Marina. Real estate is another major industry in Dubai, and most of the developments are undergoing on the west of the Dubai Creek.

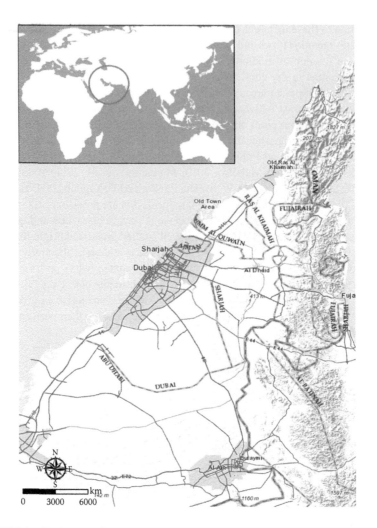

FIGURE 9.1 Study area of Dubai, UAE.

In terms of the involved data, to fully cover the greater Dubai region, two cloud-free scenes of Landsat 8 Operational Land Imager (OLI) images both acquired on October 14, 2016, were downloaded from the website of the US Geological Survey (USGS). These two images were mosaicked into one large scene and then subset using a rectangle shapefile, as the metropolitan Dubai area lies in between. This image subset covers both the urbanized human settlements and its surrounding rural desert areas. Because of the cloud-free atmospheric condition on the acquisition date, no atmospheric correction was carried out for our data. Following the *Landsat 8 Data Users Handbook* (USGS 2015), the digital number of each band was converted into top-of-atmospheric (TOA) reflectance. A water shapefile was also downloaded and used as a water mask to remove any water body in the study area before fractional cover estimation.

9.3   METHODOLOGY

Three popular approaches were performed to map subpixel urban imperviousness cover in urban environments: two spectral unmixing techniques, simple SMA and NSMA, and one machine learning method, random forest.

9.3.1   Spectral Unmixing

Endmember selection is the first step for both SMA approaches. The use of three endmembers based on the vegetation–impervious surface–soil (VIS) model (Ridd 1995) is common in temperate cities. Such a strategy, however, is not suitable in desert environments. This is because sand is the prevalent land cover in desert environments, rather than vegetation and soil. Instead of using the VIS endmembers, we modified it for our study area in desert as a model with four different endmembers: vegetation, bright impervious surface, dark impervious surface, and sand. These endmembers can be obtained directly from the feature space of the Landsat image. Feature space was created as a two-dimensional plot between the first two principal components after executing principal component analysis of all seven Landsat optical bands. The summation of the derived fractions of bright and dark impervious surface from the initial simple unmixing was used as the final urban impervious surface fraction (Wu and Murray 2003). Hereafter, this approach is called simple SMA throughout this paper. Simple SMA can be expressed as the following formulas:

$$R_b = \sum_{i=1}^{n} f_i R_{i,b} + e_b \tag{9.1}$$

subject to

$$\sum_{i=1}^{N} f_i = 1 \text{ and } f_i \geq 0, \tag{9.2}$$

where R_b is the reflectance for each band b of the Landsat OLI image, N is the total number of endmembers (four in this study), f_i is the resultant fraction of endmember i, $R_{i,b}$ is the reflectance of endmember i in band b, and e_b is the model residual.

Comparatively, NSMA was developed in an attempt to address the issue of spectral variability caused by brightness differences (Van de Voorde et al. 2009; Wu 2004; Yang et al. 2010). This method is designed to follow Ridd's VIS model to derive fractional urban land covers by using these three endmembers. Therefore, the endmember candidates of bright and dark impervious surface are combined to construct the general impervious surface endmember. Not only can it reduce brightness difference, it can also preserve the relative spectral shape. This is done by a simple but effective preprocessing step of reflectance normalization. The selected endmembers from the previous step of simple SMA were normalized using the mean reflectance of all seven Landsat OLI optical bands. These normalized endmembers were then employed to unmix the entire image, of which each band was also normalized using the same normalization method. With NSMA, fractional covers can be derived by using Equation 9.1, while the endmember normalization is calculated as follows:

$$\bar{R}_{i,b} = R_{i,b}/\mu_b \qquad (9.3)$$

with

$$\mu_b = \frac{1}{N}\sum_{i=1}^{N} R_{i,b}, \qquad (9.4)$$

where $\bar{R}_{i,b}$ is the normalized reflectance for endmember i of band b and μ_b is the mean reflectance of band b. All other variables are the same as those in Equations 9.1 and 9.2.

9.3.2 MACHINE LEARNING

Random forest was developed by Breiman (2001, 2002), which is employed as the machine learning method for fractional urban impervious surface estimation. This algorithm has been used widely in many remote sensing applications, such as per-pixel image classification and subpixel imperviousness estimation (Belgiu and Drăguţ 2016). Random forest is found to outperform other machine learning techniques, such as Cubist CART (Deng and Wu 2013b). The feature of random forest is that it is composed of a large number of regression models (each model as an "individual tree"), which therefore is compared to a large "forest." The generation of these different regression models (trees) is based on repetitively random selection of training samples. Specifically, four fundamental steps are contained in this nonparametric ensemble learning technique to predict continuous responses. The first is the random selection of bootstrap sample from the original training data set. The second is the random generation of an independent variable subset from all predictor variables. Third, the first two steps are repeated, and then a variety of regression models are yielded. The last stage is to predict the final result through averaging the estimates from all assembled regression model trees. In this research, by randomly selecting a subset of predictors at each decision split, 50 bagged classification trees were generated, and their results were combined as the final output for every estimate.

To build the random forest model in our experiment, 500 training sample pixels were obtained by a stratified random sampling strategy. Among them, one-half was chosen in the urban core region by using the city boundary from OpenStreetMap. The other half was selected in the rest areas outside the urban core region, and these areas are dominated by natural land covers in the desert (e.g., shrubs and sand sheets). The TOA reflectance of each Landsat optical band was used as the input variables. To acquire impervious surface fractions as training samples, we manually digitized imperviousness on high-resolution satellite images on Google Earth.

9.3.3   ACCURACY ASSESSMENT

Similar to the collection of training samples, 500 testing samples were also collected using the stratified random sampling strategy: one-half in the urban core region and the other half in non-urban areas. To reduce possible geometric errors caused by the multiple source images, each testing sample covers a polygon of a 90 m by 90 m geographic area, corresponding to a 3 by 3 Landsat pixel. Three extensively used accuracy assessment measurements, including root-mean-square error (RMSE), mean absolute error (MAE), and systematic error (SE), were applied to evaluate the estimation results. These three metrics can be expressed as follows:

$$\text{RMSE} = \sqrt{\frac{1}{n}\sum_{i=1}^{n}(\hat{f}_i - f_i)^2} \tag{9.5}$$

$$\text{MAE} = \frac{1}{n}\sum_{i=1}^{n}\left|\hat{f}_i - f_i\right| \tag{9.6}$$

$$\text{SE} = \frac{1}{n}\sum_{i=1}^{n}(\hat{f}_i - f_i), \tag{9.7}$$

where $\hat{f}_i$ is the estimated abundance of urban impervious surface of sample i using spectral unmixing or random forest, f_i is the reference impervious surface fraction of sample i from digitizing on the high-resolution satellite images, and n is the total number of all sample polygons. Besides, the modeled urban impervious surface abundance is plotted against the reference fraction. Consequently, their scatterplots are drawn, their regression models are built, and the model coefficients are derived for comparison. A good estimation has a slope and a coefficient of determination close to 1, and an intercept close to 0.

In addition, the accuracy for overall, rural and low-density developed, and medium-/high-density developed area was evaluated. A 30% threshold of percent impervious surface is used to determine the low-density developed area and the medium-/high-density developed area among the testing samples (Deng 2016; Deng and Wu 2013a,b; Wu 2004; Wu and Murray 2003; Zhang et al. 2015).

9.4   RESULTS

After implementing the two SMAs and random forest approaches, three resultant maps of subpixel urban impervious surface distribution in Dubai were generated and are shown in Figure 9.2. The three assessment metrics are calculated and reported in Table 9.1. For convenient visual comparisons among the performances of the three

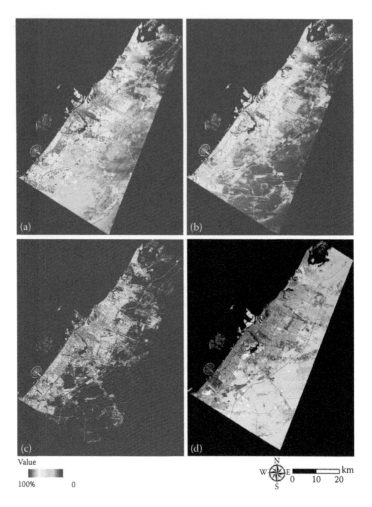

FIGURE 9.2 Resultant subpixel urban impervious surface maps using (a) simple SMA, (b) normalized SMA, and (c) random forest. (d) False-color Landsat composite.

methods, these indicators are further shown as bar charts in Figure 9.3. The scatter figures are also created by plotting the modeled urban impervious surface fraction against digitized reference abundance, as illustrated in Figure 9.4.

For the simple SMA result shown in Figure 9.2a, visual observation found that although major urban impervious surface areas are located along the coast, as expected, evident overestimations can be clearly discerned in the surrounding rural area. This is attributed to the severe spectral confusion between sand and bright impervious surface. Such overestimations spread throughout almost all the rural areas outside the urban core region. This finding is also reflected by the evaluation metrics in Table 9.1 and Figure 9.3a, as well as the scatterplot in Figure 9.4a. For all samples, RMSE, MAE and SE are approximately 30%, 23%, and 19%, respectively. The regression model of the scatterplot (see Figure 9.4a) has a slope of 0.57,

TABLE 9.1

Accuracy Assessment of the Three Approaches Using Three Evaluation Metrics

Method	Area	RMSE (%)	MAE (%)	SE (%)
	Overall	29.832	23.530	18.557
Simple SMA	Low-density developed	32.188	25.107	25.107
	Medium-/high-density developed	26.221	21.345	9.480
	Overall	24.098	17.444	5.590
Normalized SMA	Low-density developed	22.478	15.355	15.355
	Medium-/high-density developed	26.177	20.338	−7.943
	Overall	16.707	9.867	0.757
Random forest	Low-density developed	9.797	4.537	4.537
	Medium-/high-density developed	23.085	17.253	−4.481

an intercept of 27.86%, and an R^2 of 0.5. All these numbers indicate that the estimation accuracy is very poor, and this simple SMA method easily overestimates fractional impervious surface. With further examinations in both medium-/high-density developed and low-density developed areas, overestimations occur in both regions, especially with an RMSE (32%) and MAE (25%) in low-density developed areas. This shows that the most severe overestimations are attributed to the spectral confusions in rural areas dominated by desert landforms (with an SE of 25%). The unsatisfactory result is not unexpected: the issue of spectral variability is not taken into consideration with simple SMA.

For the NSMA result, despite the preprocessing step of reflectance normalization, visual examinations in Figure 9.2b show that the overestimations have not been fully addressed, most of which still occur in rural areas dominated by sand and sediments. Both the evaluation measurements in Table 9.1 and the bar chart in Figure 9.3 (RMSE of 24%, MAE of 17%, and SE of 5%) indicate that overestimations are somewhat improved when compared with their counterparts of the result using simple SMA. A more intuitive observation with NSMA is that the majority of pixels in the urban areas are cyan (only with medium percent impervious surface), and there are many fewer red and orange pixels (with higher impervious surface fractions) than the result of simple SMA in Figure 9.2a. This suggests the underestimations of impervious surface in the central urban area. Apparent examples include the runways in the three airports. As shown in the bar charts in Figure 9.3, the white bar in medium-/high-density developed areas with NSMA is higher than that of low-density developed areas. As indicated in the scatterplot in Figure 9.4b, the slope is 0.46, the intercept is 17%, and the coefficient of determination is 0.49. It shows that, in contrast to simple SMA with which the primary errors are from rural areas, the errors with NSMA originate mainly from medium-/high-density developed areas. It also indicates that while the reflectance normalization step improves the overestimations in

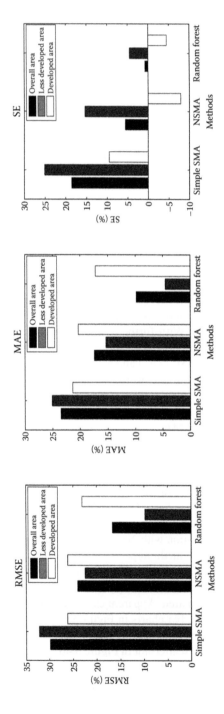

FIGURE 9.3 Assessment metrics of the resulting fractional urban impervious surface using the three methods.

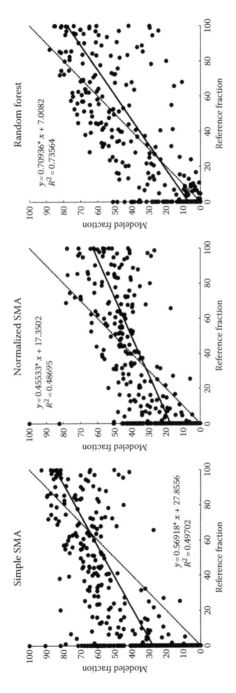

FIGURE 9.4   Scatterplots of the modeled results of impervious surface against reference fractions using the three methods.

rural areas caused by spectral confusion between sand and impervious surface, this feature of NSMA also introduces uncertainty at the same time, shown as the underestimation in urban core regions.

For the random forest result shown in Figure 9.2c, an evident and reasonable distribution pattern of subpixel urban impervious surface can be observed, and its pattern is similar to that of the false-color Landsat composite in Figure 9.2d. This is shown as the majority of blue pixels with low percent impervious surface fraction in areas that are far from the Arabian Sea and the red/orange pixels with high impervious surface fraction cluster along the coast that are surrounded by the yellowish pixels. The former pattern shows the effective separation between sand and impervious surface without severe overestimations, while the latter suggests the relatively satisfactory performance in the built-up area without severe underestimations. Another example of the adequate estimation is that even some highways in the outer ring of the metropolitan area can be identified from visual examination, and they are noticeably discriminated from sand and other desert land covers in rural areas. Compared with the results of the two spectral unmixing techniques, the overall performance has been substantially improved, as supported by the metrics in Table 9.1 and the scatterplot in Figure 9.4c. For the overall area, the RMSE, MAE, and SE are 16.7%, 9.87%, and 0.76%, respectively. For the regression model coefficients, both slope and R^2 are at a level of around 0.7, and the intercept is only 7%. For low-density developed areas, the estimation accuracy is high, with an RMSE of 9.8% and an MAE of 4.5%. For medium-/high-density developed areas, the accuracy is lower than that for low-density developed areas, but still much better than the counterparts of the two SMAs. The different degrees of red of medium-/high-density areas in Figures 9.2 and 9.5 correspond to the metrics in Table 9.1, including overestimation (an SE of 9.5% using simple SMA) and underestimation (an SE of −8% and −4.5 using NSMA and random forest, respectively). All these comparisons suggest the considerable improvements with random forest when compared with the other two SMA methods.

9.5 DISCUSSION

Random forest has the best performance among the three techniques, which is comparable with the performance in the literature. For example, an MAE of 13.8% and an SE of 5.2% are reported in Powell et al. (2007) in which MESMA was used. Similarly, a slope of 1.43, an intercept of 7.7, and an R^2 of 0.789 were reported at 15-m resolution, and a slope of 1.44, an intercept of 11, and an R^2 of 0.718 were reported at 60-m resolution in Roberts et al. (2012). The resultant metrics in our study are all slightly better than those indicators described in the literature. Even with a study area of a desert city, random forest still yields relatively accurate results. This is likely because the performance of random forests is stable and random forests outperform other methods when a large number of samples are readily available (Deng and Wu 2013b). In this research, when implementing random forest, samples were collected not only from endmember pixels but also from mixed pixels. Comparatively with the two spectral unmixing methods, only the endmember spectra were collected and used for the unmixing. As a result, the performances of the other two spectral unmixing methods are relatively poor.

With careful examinations and comparisons among the three results, most of the mapping errors are likely to be attributed to the issue of spectral variability. This issue is pointed out as the primary error source in spectral unmixing (Somers et al. 2011), which is an unsolved technical problem despite a large number of spectral analytical methods. In particular, spectral variability seems to be more severe owing to the presence of a variety of complicated desert landforms with different compositions (such as sand, sediment, and rock). These desert landforms include alluvial fan systems, ergs and dune fields, dry lakebeds, open water bodies, basaltic volcanoes and flows, sedimentary mountain ranges, regs or serirs, stripped bedrock surfaces, sandsheets, and vegetation (Ballantine et al. 2005; Clements 1957; Raisz 1952). In our experiments, NSMA was employed in an attempt to accommodate spectral variability. However, its performance is also unsatisfactory. This may be explained by the principle of NSMA: the designed reflectance normalization aims to minimize the brightness differences of the intraclass variability, but the interclass variability is not taken into account. One solution may be the MESMA (Roberts et al. 1998). By considering both interclass and intraclass spectral variability, MESMA with various endmember candidates in a library may provide better estimation results. In this pilot study, we did not perform this algorithm, since more local knowledge of this study area is needed to collect representative endmembers to build an appropriate spectral library. But our future effort will continue to test different spectral analytical techniques.

Although the best performance is associated with random forest, some common estimation errors can be found in the desert environment with these approaches. This is mainly attributed to spectral confusions in the desert environments with complicated desert landforms. Unlike the high spectral contrast between vegetation and impervious surface in temperate areas where vegetation can be commonly seen, the spectral contrast is very low between urban impervious surface and desert landscapes (e.g., sand, sediment, and rock), and the spectral confusion is more severe in a desert environment. Two types of overestimation errors are illustrated in Figure 9.5, which occur commonly in all three resultant maps. Each figure covers an area about 3 km by 6 km. Figure 9.5a through d illustrate the overestimations caused by spectral variability from dark sand located at the center. In addition to the examples in the locations in Figures 9.5a through d, overestimations can be discerned in construction sites (e.g., on the Deira Islands, but not shown in Figure 9.5). Another example is the vegetated covers shown in Figure 9.5e through h. These covers include shrublands and agricultural lands. The mixtures of sand and shadow shed from shrubs and agricultural or other desert plants make it difficult to separate dark urban from impervious surface, thus resulting in the overestimations. For the spectral unmixing techniques, such inaccurate estimates may be alleviated by using MESMA through exploring more endmembers in the global library. For random forest, since samples of dark sand and vegetation are not prevalent covers, these covers may not be adequately represented and trained by the random samples. As a result, these spectrally similar covers may become the source of error. To alleviate such spectral confusions to achieve better estimates, more training samples are helpful to characterize these lands (Deng and Wu 2013b).

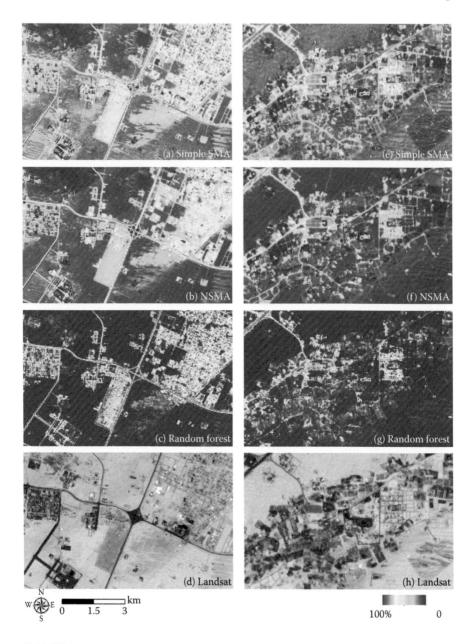

FIGURE 9.5 Overestimations caused by dark sand (a through d) and mixed vegetation (e through h) shown in the resultant maps of the three methods and the false-color Landsat composites.

9.6   CONCLUSIONS

Urban impervious surface information is an essential indicator for planning and scientific research. The derivation of such an important urban morphological parameter is not an easy task owing to spectral confusions from other land covers, particularly in desert environments with complicated desert landforms. In this paper, we used Dubai, UAE, as our study area, and extended topical areas in the literature by exploring three spectral analytical techniques to map the subpixel distribution of urban impervious surface in this desert city. On the basis of the mapping results, we further evaluated the major error sources of these three methods.

Four major findings can be concluded from our experiments in this desert metropolitan area. First, random forest considerably outperforms the other two spectral unmixing approaches in a desert environment, with an RMSE of ~16%, an MAE of ~10%, and an SE of 0.7%. Second, both simple SMA and NSMA substantially overestimate fractional impervious surface in rural desert areas because of the spectral similarity of bright sand, while random forest can effectively separate them. Third, compared with the results in the literature, the performance of SMAs in a desert environment is less satisfactory than that in temperate cities where vegetation is dominant. Finally, commonly found with the three spectral analytical techniques, overestimation errors are likely to come from the spectral confusions from dark sand/sediments and certain types of desert plants (e.g., shrubs, agricultural plants, and their casting shadow). In the future, more efforts are needed to explore the performance of other spectral unmixing and machine learning techniques for different desert cities, such as MESMA, SASMA, endmember bundle, SVR, ANN, and CART.

REFERENCES

Al-sharif, A. A., & Pradhan, B. (2014). Monitoring and predicting land use change in Tripoli Metropolitan City using an integrated Markov chain and cellular automata models in GIS. *Arabian Journal of Geosciences*, 7(10), 4291–4301.

Alqurashi, A. F., & Kumar, L. (2014). Land use and land cover change detection in the Saudi Arabian desert cities of Makkah and Al-Taif using satellite data. *Advances in Remote Sensing*, 3(03), 106.

Arnold, C. L., & Gibbons, C. J. (1996). Impervious surface coverage: The emergence of a key environmental indicator. *Journal of the American Planning Association*, 62(2), 243–258.

Ballantine, J. A. C., Okin, G. S., Prentiss, D. E., & Roberts, D. A. (2005). Mapping North African landforms using continental scale unmixing of MODIS imagery. *Remote Sensing of Environment*, 97(4), 470–483.

Bateson, C. A., Asner, G. P., & Wessman, C. A. (2000). Endmember bundles: A new approach to incorporating endmember variability into spectral mixture analysis. *IEEE Transactions on Geoscience and Remote Sensing*, 38(2), 1083–1094.

Belgiu, M., & Drăguţ, L. (2016). Random forest in remote sensing: A review of applications and future directions. *ISPRS Journal of Photogrammetry and Remote Sensing*, 114, 24–31.

Breiman, L. (2001). Random forests. *Machine Learning*, 45, 5–32.

Breiman, L. (2002). Manual on setting up, using, and understanding random forests v3.1 (http://oz.berkeley.edu/users/breiman/Using_random_forests_V3.1.pdf).

Clements, T. (1957). A study of desert surface conditions. Headquarters Quartermaster General Research and Development Command. Environmental Protection Research Division.

Deng, C. (2016). Automated construction of multiple regional libraries for neighborhood-wise local multiple endmember unmixing. *IEEE Journal of Selected Topics in Applied Earth Observations and Remote Sensing*, 9(9), 4232–4246.

Deng, C., & Wu, C. (2013a). A spatially adaptive spectral mixture analysis for mapping subpixel urban impervious surface distribution. *Remote Sensing of Environment*, 133, 62–70.

Deng, C., & Wu, C. (2013b). The use of single-date MODIS imagery for estimating large-scale urban impervious surface fraction with spectral mixture analysis and machine learning techniques. *ISPRS Journal of Photogrammetry and Remote Sensing*, 86, 100–110.

Flanagan, M., & Civco, D. L. (2001). Subpixel impervious surface mapping. In *Proceedings of the 2001 ASPRS Annual Convention*, St. Louis, MO, April (Vol. 23).

Gong, B., Im, J., & Mountrakis, G. (2011). An artificial immune network approach to multi-sensor land use/land cover classification. *Remote Sensing of Environment*, 115(2), 600–614.

Haregeweyn, N., Fikadu, G., Tsunekawa, A., Tsubo, M., & Meshesha, D. T. (2012). The dynamics of urban expansion and its impacts on land use/land cover change and small-scale farmers living near the urban fringe: A case study of Bahir Dar, Ethiopia. *Landscape and Urban Planning*, 106(2), 149–157.

Ji, M., & Jensen, J. R. (1999). Effectiveness of subpixel analysis in detecting and quantifying urban imperviousness from Landsat Thematic Mapper imagery. *Geocarto International*, 14(4), 33–41.

Jin, H., & Mountrakis, G. (2013). Integration of urban growth modelling products with image-based urban change analysis. *International Journal of Remote Sensing*, 34(15), 5468–5486.

Kashaigili, J. J., & Majaliwa, A. M. (2010). Integrated assessment of land use and cover changes in the Malagarasi river catchment in Tanzania. *Physics and Chemistry of the Earth, Parts A/B/C*, 35(13), 730–741.

Madugundu, R., Al-Gaadi, K. A., Patil, V. C., & Tola, E. (2014). Detection of land use and land cover changes in Dirab region of Saudi Arabia using remotely sensed imageries. *American Journal of Environmental Sciences*, 10(1), 8.

Mahmoud, S. H., & Alazba, A. A. (2016). Land cover change dynamics mapping and predictions using EO data and a GIS-cellular automata model: The case of Al-Baha region, Kingdom of Saudi Arabia. *Arabian Journal of Geosciences*, 9(5), 1–20.

Myint, S. W., & Okin, G. S. (2009). Modelling land-cover types using Multiple Endmember Spectral Mixture Analysis in a desert city. *International Journal of Remote Sensing*, 30(9), 2237–2257.

Okujeni, A., van der Linden, S., Tits, L., Somers, B., & Hostert, P. (2013). Support vector regression and synthetically mixed training data for quantifying urban land cover. *Remote Sensing of Environment*, 137, 184–197.

Powell, R. L., Roberts, D. A., Dennison, P. E., & Hess, L. L. (2007). Sub-pixel mapping of urban land cover using multiple endmember spectral mixture analysis: Manaus, Brazil. *Remote Sensing of Environment*, 106(2), 253–267.

Pu, R., Gong, P., Michishita, R., & Sasagawa, T. (2008). Spectral mixture analysis for mapping abundance of urban surface components from the Terra/ASTER data. *Remote Sensing of Environment*, 112(3), 939–954.

Quintano, C., Fernández-Manso, A., & Roberts, D. A. (2013). Multiple Endmember Spectral Mixture Analysis (MESMA) to map burn severity levels from Landsat images in Mediterranean countries. *Remote Sensing of Environment*, 136, 76–88.

Raisz, E. (1952). Landform Map of North Africa. Environmental Protection Branch, Office of the Quartermaster General.

Ridd, M. K. (1995). Exploring a VIS (vegetation-impervious surface-soil) model for urban ecosystem analysis through remote sensing: Comparative anatomy for cities. *International Journal of Remote Sensing*, 16(12), 2165–2185.

Rashed, T., Weeks, J. R., Roberts, D., Rogan, J., & Powell, R. (2003). Measuring the physical composition of urban morphology using multiple endmember spectral mixture models. *Photogrammetric Engineering & Remote Sensing*, 69(9), 1011–1020.

Roberts, D. A., Gardner, M., Church, R., Ustin, S., Scheer, G., & Green, R. O. (1998). Mapping chaparral in the Santa Monica Mountains using multiple endmember spectral mixture models. *Remote Sensing of Environment*, 65(3), 267–279.

Roberts, D. A., Quattrochi, D. A., Hulley, G. C., Hook, S. J., & Green, R. O. (2012). Synergies between VSWIR and TIR data for the urban environment: An evaluation of the potential for the Hyperspectral Infrared Imager (HyspIRI) Decadal Survey mission. *Remote Sensing of Environment*, 117, 83–101.

Roth, K. L., Roberts, D. A., Dennison, P. E., Alonzo, M., Peterson, S. H., & Beland, M. (2015). Differentiating plant species within and across diverse ecosystems with imaging spectroscopy. *Remote Sensing of Environment*, 167, 135–151.

Schueler, T. R. (1994). The importance of imperviousness. Watershed Protection Techniques 1.

Schueler, T. R. (2003). Impacts of impervious cover on aquatic systems. Watershed protection research Monograph No. 1, Center for Watershed Protection. Silver Spring, MD.

Schueler, T. R., Fraley-McNeal, L., Cappiella, K. (2009). Is impervious cover still important? Review of recent research. *Journal of Hydrologic Engineering*, 14.

Shair, I. M., & Nasr, A. H. (1999). Using satellite images to detect landuse change in Al-Ain City, United Arab Emirates. *The Arab World Geographer*, 2(2), 139–148.

Shao, Y., & Lunetta, R. S. (2011). Sub-pixel mapping of tree canopy, impervious surfaces, and cropland in the Laurentian Great Lakes Basin using MODIS time-series data. *IEEE Journal Selected Topics Applied Earth Observations and Remote Sensing*, 4(2), 336–347.

Somers, B., Asner, G. P., Tits, L., & Coppin, P. (2011). Endmember variability in spectral mixture analysis: A review. *Remote Sensing of Environment*, 115(7), 1603–1616.

United Nations. (2014). World Urbanization Prospects: Final Reports. Department of Economic and Social Affairs, Population Division. Available online: https://esa.un.org/unpd/wup/Publications/Files/WUP2014-Report.pdf

USGS. (2015). *Landsat 8 Data Users Handbook*. Available online: https://landsat.usgs.gov/documents/Landsat8DataUsersHandbook.pdf

Van de Voorde, T., De Roeck, T., & Canters, F. (2009). A comparison of two spectral mixture modelling approaches for impervious surface mapping in urban areas. *International Journal of Remote Sensing*, 30(18), 4785–4806.

Walsh, C. J., Roy, A. H., Feminella, J. W., Cottingham, P. D., Groffman, P. M., & Morgan II, R. P. (2005). The urban stream syndrome: Current knowledge and the search for a cure. *Journal of the North American Benthological Society*, 24(3), 706–723.

Weber, C., & Puissant, A. (2003). Urbanization pressure and modeling of urban growth: Example of the Tunis Metropolitan Area. *Remote Sensing of Environment*, 86(3), 341–352.

Weng, Q. (2009). Thermal infrared remote sensing for urban climate and environmental studies: Methods, applications, and trends. *ISPRS Journal of Photogrammetry and Remote Sensing*, 64(4), 335–344.

Weng, Q. (2012). Remote sensing of impervious surfaces in the urban areas: Requirements, methods, and trends. *Remote Sensing of Environment*, 117, 34–49.

Wu, C. (2004). Normalized spectral mixture analysis for monitoring urban composition using ETM+ imagery. *Remote Sensing of Environment*, 93(4), 480–492.

Wu, C., & Murray, A. T. (2003). Estimating impervious surface distribution by spectral mixture analysis. *Remote sensing of Environment*, 84(4), 493–505.

Xian, G. (2007). Mapping impervious surfaces using classification and regression tree algorithm. *Remote Sensing of Impervious Surfaces*, 39–58.

Yagoub, M. M. (2004). Monitoring of urban growth of a desert city through remote sensing: Al-Ain, UAE, between 1976 and 2000. *International Journal of Remote Sensing*, 25(6), 1063–1076.

Yagoub, M. M., & Al Bizreh, A. A. (2014). Prediction of land cover change using Markov and cellular automata models: Case of Al-Ain, UAE, 1992–2030. *Journal of the Indian Society of Remote Sensing*, 42(3), 665–671.

Yang, F., Matsushita, B., & Fukushima, T. (2010). A pre-screened and normalized multiple endmember spectral mixture analysis for mapping impervious surface area in Lake Kasumigaura Basin, Japan. *ISPRS Journal of Photogrammetry and Remote Sensing*, 65(5), 479–490.

Yang, L., Huang, C., Homer, C. G., Wylie, B. K., & Coan, M. J. (2003). An approach for mapping large-area impervious surfaces: Synergistic use of Landsat-7 ETM+ and high spatial resolution imagery. *Canadian Journal of Remote Sensing*, 29(2), 230–240.

Zhang, H., Zhang, Y., & Lin, H. (2012). A comparison study of impervious surfaces estimation using optical and SAR remote sensing images. *International Journal of Applied Earth Observation and Geoinformation*, 18, 148–156.

Zhang, Y., Zhang, H., & Lin, H. (2014). Improving the impervious surface estimation with combined use of optical and SAR remote sensing images. *Remote Sensing of Environment*, 141, 155–167.

Zhang, Z., Liu, C., Luo, J., Shen, Z., & Shao, Z. (2015). Applying spectral mixture analysis for large-scale sub-pixel impervious cover estimation based on neighbourhood-specific endmember signature generation. *Remote Sensing Letters*, 6(1), 1–10.

10 Application of Remote Sensing and Cellular Automata Model to Analyze and Simulate Urban Density Changes

Santiago Linares and Natasha Picone

CONTENTS

10.1 INTRODUCTION

10.1.1 REMOTE SENSING AND GIS TO STUDY URBAN AREAS

Weng (2001a) stated that

> The integration of remote sensing and geographic information systems (GIS) has been widely applied and been recognized as a powerful and effective tool in detecting urban land use and land cover change (Ehlers et al. 1990, Treitz et al. 1992, Harris and Ventura 1995). Satellite remote sensing collects multispectral, multiresolution and multitemporal data, and turns them into information valuable for understanding and monitoring urban land processes and for building urban land cover datasets. GIS technology provides a flexible environment for entering, analysing and displaying

digital data from various sources necessary for urban feature identification, change detection and database development (Weng 2001a, p. 2000).

Nowadays, satellite images are a great source of exhaustive, accurate, periodic, and easily updated information on urban areas, and one of the most significant research topics in geographical information technologies is the processing of those sources to come up with these data. Advances in this topic were revitalized by commercial sensors of high spatial resolution* (IKONOS, GeoEye-1, Quickbird, worldview-2, EROS B, etc.) and through the increased availability of sources from alternative data, such as SAR (Synthetic Aperture Radar) and LIDAR (Light Detection and Ranging, or Laser Imaging Detection and Ranging).

Apart from new data source availability, which represent an advance in the subject matter, there are a number of methodological concerns of remote sensing applied to urban analysis that are still in their experimental phase. Two central issues are in discussion: (1) the search for classification procedures that infer the diversity of land uses existing in the urban area and (2) the development of replicable methodologies for classification in different urban contexts that allow increasing dissemination and usefulness of geographic information technology.

The first concern arises from deficiencies to discriminate the different land uses on the basis of the spectral information captured by the satellite sensors; if so, surfaces with identical spectral reflectance values may correspond to very different uses and urban functions.[†] In this sense, research currently aims to improve the quality of ratings (in terms of the number of individual classes and precision with which each one of them can be identified) and ensure that the types of land use may represent categorical spectral signatures in accordance with the administrative nomenclature used to plan and manage urban space. Overcoming attempts have advanced in many ways, and we can point out among them the following: the use of probabilities a priori or a posteriori of a classification (Mesev et al. 2001), the application of classification algorithms known as "Soft" as the blurred classification (Lee 2006), procedures of image segmentation (Schöpfer et al. 2010), the use of auxiliary spatial data (Weng 2010), the combination of spectral data with measures of urban textures and shapes (Ackermann and Mering 2007; Hermosilla et al. 2012), technical data fusion (Ranchin and Wald 2010; Robert and Herold 2004), hyperspectral analysis, and the use of artificial neural network techniques (Lourenço et al. 2005).

The second issue considers that the procedure through which a map of land use is obtained, generated by the classification of satellite images, should not be restricted

---

* Although these sources represent new possibilities, they also presage new problems and stumbling blocks to provide fast solutions to urgent needs. It is common to observe that various specialists in remote sensing used most of their effort to seek new sources of information produced in each technological advance, and not so many devoted themselves to seeking practical operational solutions to everyday requirements arising from different areas of society, being then possible to answer using available resources (Donnay et al. 2001). In fact, in the case of urban areas, the advantages of a higher spatial resolution is positive when performing a visual interpretation, while, when performing a digital interpretation, this increase in resolution produces negative effects, excessively increasing the internal heterogeneity in some categories (Chuvieco 2010; Green 2000).

[†] Unlike those non-urban studies, where it is possible to obtain a direct relationship between the spectral response of the two main components of terrestrial ecosystems (vegetation and water) and the corresponding use.

to a particular area, to a given time or a specific program, because it does not consti-
tute a final product itself but is rather considered a generic procedure for the various
processes of research and planning of the cities. Some examples of these theoretical
and methodological contributions are those made by Molina Mora and Chuvieco
(1997), Jensen and Cowen (1999), Recio et al. (2003), Longley (2002), Stewart and
Oke (2009) and Hermosilla and Ruiz (2009).

It is clear that one of the undeniable virtues of remote sensing applied to urban
spaces is the suitability for studying the extent, magnitude, and evolution of the built-
up area of a city. The current challenges aim to go beyond trying to obtain detailed
information on the differentiation of coverage inside, inferring the settlement densities and the characterization of the different land uses.

Studies on urban densities were frequent and important in the development of
modern geography, from the first half of the twentieth century, and they have not lost
validity. On the contrary, they have increased and contribute to detected connections
between this phenomenon and other aspects of reality. It has been transcendental
for the studies upon the ecological balance of a determining area to explain socio-
demographic behaviors, key factor analysis on economic growth, levels of income,
and variable determinant to define patterns of urban growth.

Parallel to the interest aroused by the discovery of such correlations, different
methods of geovisualization of density in urban areas have evolved. Two groups
stand out, methods to generate continuous surfaces of density gradients obtained
by Kernel estimators and methods of discrete representation of the urban densities
based on detecting coverage using multispectral satellite images processing. GIS and
remote sensing have contributed to the advances in this matter.

10.1.2 CELLULAR AUTOMATA APPLIED TO ANALYZE URBAN GROWTH

Complexity theory–based models are based on delegating the simulation at a macro-
scale urban structure to a set of dynamic sub-models at micro-scale derived from the
complexity theory. These approaches try to represent individual actors (or groups)
in a given system that can interact with each other and/or with an environment. The
macro-level behaviors are configured from the aggregation of these interactions.

The complexity theory can be considered as a new systemic approach to study
the relationship between the parts and the whole in a different way, emphasizing
the idea of an emerging structure from a bottom-up process, where local actions
and interactions produce the overall pattern. The intra-urban dynamic processes are
more important than the structure itself, allowing the understanding of such systems
to go beyond the description (in static terms), to capture the inner essence of the
phenomena of change (Batty 2000; Casti 1997; Wu 2002).

Different models of complexity theory have been used to simulate urban dynamics, within which there are statistical models, agent-based models, fractal models,
and models based on cellular automata (CA) models and artificial neural network
models (Batty 2005). In this paper, we present a software application of a model
based on CA rules.

The seminal work where the concept of CA was conceived to address socio-spatial
processes took place in 1979, when the geographer Waldo Tobler produced the

concept of CA to model and predict urban growth in a city (Tobler 1979). In his paper, he proposed the idea of how CA can function as a useful tool for urban planning and how it can get the best transition rules. Tobler defined a law that would be very important in the development of predictive models for urban growth: "In geography, everything is related to everything else, but near things are more related than distant things." This means that the proximity or distance between certain types of processes or activities inhibits or stimulates the emergence and development of other activities nearby.

Many authors followed the line of the model defined by Tobler, such as Couclelis (1985, 1988), Itami (1988), and Phipps (1989), but it was only from the 1990s onward when models allowed an accurate representation of reality. One of the most important methodological proposals is the SLEUTH model developed by Clarke et al. (1997), which uses six variables as an input, defined as follows: *Slope, Land use area, Excluded area, Urban area, Transportation map*, and *Hillshade area*. At the same time, the model uses five factors to control the system behavior: diffusion, reproduction, spread, slope, and distance to routes.

CA theory assumes that the potential of a cell to undergo some transformation of land use depends on the state of its neighboring cells. White and Engelen's CA (1993) consisted of a finite cellular space representing a hypothetical urban area. The different land uses are identified with elements of the set of states of CA. Two types of states are defined: active and fixed. The first one represents conventional land uses such as residential or commercial, which may change over time. The second one represents road infrastructure or natural features such as a river or a canyon.

State transition rules are defined through a function that relates four different factors:

- The intrinsic convenience between different land uses, representing heterogeneous aspects of the geographic space that is being modeled. These conveniences for land use, located at a specific point, transformed into another, or remained unchanged, are related to issues ranging from soil quality to legal restrictions or speculative economic pressures.
- The effect of a specific land use on the surrounding land uses. This type of effect can be attractive or repulsive, as some land uses attract some and repel others. For example, a residential land use attracts business while repelling industries.
- The effect of local accessibility, which represents the ease of access to the transport network.
- The stochastic disturbance that captures the effect of imperfect knowledge and variable behavior of social actors in relation to land use.

White and Engelen's model introduces a set of changes from classic CA models as it differentiates from the use of a monotonous decay of the influence of the neighborhood as distance increases. Instead, it uses $W_{x,d,j}$ weight to represent the balance of the forces of attraction and repulsion that occur in the different types of land use. As the sum of two opposing forces, the $W_{x,d,j}$ weight is not necessarily monotone and may even be negative. It also introduces a strict order of possible transitions between land uses. The model only allows land use change after a predefined sequence. For example, given a model where there are four possible states of a possible land use, the change sequence is as follows: free → dwelling → industrial → commercial.

More recently, this strict order of possible transitions predefined by the user has in fact been superseded by fuzzy rankings of transitions. According to such type of ranking, not always will the cells with the highest potential to undergo changes in land use actually suffer this change, since transitions are subject to some degree of randomness, so as to render the simulated urban environment as close as possible to reality.

In the last decade, a large number of models based on CA applications have integrated GIS and remote sensing to study changes in coverage and uses of urban land (Mitsova et al. 2011; Sang et al. 2011; Thapa and Murayama 2011; Vaz et al. 2012; Yang et al. 2012; Zhang et al. 2011). In general terms, this process begins with the comparison of two coverage or land use maps of previous dates. This comparison makes it possible to estimate the patterns and processes of change (types of transitions and exchange rates) and calibrate the model. The analysis of pass changes in relation to the explanatory variables allows mapping susceptibility to changes in the future, also referred to in the literature as potential, propensity, or probability. We will use the probability term, although in some models, these values are not probabilities in the strict sense. These maps of probability can be considered as a first product of modeling. Finally, an evaluation of the model performance is usually carried out. It is often based on the spatial coincidence between the simulation and an "observed" map usually obtained through the classification of satellite imagery that serves as a reference.

Land Change Models can be useful tools for environmental and geomatics research concerning land use and cover change (LUCC) (Paegelow et al. 2013). The simulation maps obtained from LUCC models help us to understand, forecast and anticipate the future evolution for a variety of applied environmental problems (Camacho Olmedo et al. 2015, p. 2014).

Considering the work made in the middle cities of Chile (Henríquez 2014), we have selected two main variables to analyze the possible environmental impacts of the urban growth: the temperature spatial distribution and the precipitation runoff.

10.2   STUDY AREA

Tandil City is located in the south-center of Buenos Aires Province, Argentina. It takes up the middle basin of the Langueyú River and the central area of the north face of the Tandilia Hill system. These hills surround the city and control its growth from south to northwest. The main water flows have been anthropologically controlled, using hydraulic structures such as a dam and enclosures (Figure 10.1).

Tandil is considered to be a middle city in the Argentinian urban system. This classification is based on the population of each city: small, if it has between 2000 and 50,000 inhabitants; middle, if it has between 50,000 and 1 million inhabitants; and big, if it has more than 1 million inhabitants (Vapnarsky and Gorojovsky 1990). According to the last census, the city has 116.916 inhabitants, which represent 94% of the department population (INDEC 2010).

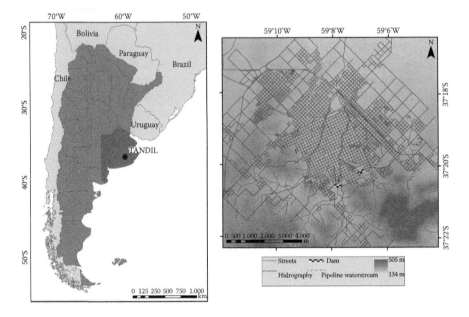

FIGURE 10.1 Localization of Tandil City, Argentina.

As a regional center, its economy is basically concentrated on the services sector, particularly the educational, cultural, and agricultural ones. However, the city has a history of industrial development related to the production of auto parts and the primary production of rock for construction, an activity that was shut down in 2012.

10.3 METHODOLOGY

10.3.1 URBAN DENSITY STUDY

The urban density classes were generated using Landsat 5 TM images. Two images from Path 225 Row 86 were selected (November 03, 2003, and October 24, 2011) and downloaded at the *US Geological Survey* website (http://glovis.usgs.gov/). Bands 1, 2, 3, 4, 5, and 7 were corrected using the DOS model (Chavez 1988).

The generation of the classes consisted of three steps. First, an unsupervised classification of the urban area was made and five urban classes were extracted (Linares and Tisnés 2011). Second, the spectral signature of each class was constructed using samples. For each band, in each class, the mean and the standard deviation were calculated so that they can be used in the supervised classification processes. Finally, spectral mixture analyses were made. Using several spectral libraries (USGS Spectroscopy Lab, Jet Propulsion Laboratory, and Santa Barbara urban spectral libraries). The determination of which *endmember* was part of each class was carried out and their proportion in each class was calculated. Using this information, it was possible to characterize the classes according to the amount of construction and green areas that they had (Picone and Linares 2014).

Table 10.1 shows the results of the image processing with the spectral signature and the characteristics of each class. The proportion of coverage according to classes is included, and the amount of construction and green areas of the class with an image is used as an example.

This methodology was previously applied to analyze the urbanization process of Tandil City in the last 25 years (Picone and Linares 2014). To corroborate the results, the information generated using remote sensing was contrasted with cadastral information, and the confusion matrix was made (Table 10.2). As can be seen, the global accuracy is of 86%; there were no significant differences between the errors observed in each of the classes, showing values around 12% for low and middle density and 15% for high density. The results of the extraction of urban densities were used as inputs in the urban growth modeling.

TABLE 10.1

Spectral Signature and Characteristics of the Density Classes

Classes	Spectral Signature (Minimum and Maximum for Each Band and Class)	Characteristics	Sample Image
Low density	B1: 0.100789–0.144165 B2: 0.100883–0.158503 B3: 0.090548–0.157492 B4: 0.283694–0.403982 B5: 0.199319–0.265179 B7: 0.135501–0.193289	**30% to 45% of constructed area** (20% cement, 15% asphalt, 5% red roofing tile, 5% cement badge) and the rest is **green area** (35% grass and 20% trees).	
Middle density (A)	B1: 0.096573–0.141425 B2: 0.097905–0.150625 B3: 0.100774–0.158934 B4: 0.152680–0.261476 B5: 0.155724–0.232228 B7: 0.091262–0.188210	**55% to 85% of constructed area** (62% asphalt, 7% concrete roads, 4% cement and 5% road of loose materials) and the rest is **green area** (grass).	
Middle density (B)	B1: 0.107656–0.152816 B2: 0.124758–0.174898 B3: 0.131844–0.190040 B4: 0.185933–0.300921 B5: 0.181205–0.282537 B7: 0.143892–0.227124	**55% to 85% of constructed area** (60% cement badge and 25% cement) and the rest is **green area** (grass).	

(Continued)

TABLE 10.1 (CONTINUED)

Spectral Signature and Characteristics of the Density Classes

Classes	Spectral Signature (Minimum and Maximum for Each Band and Class)	Characteristics	Sample Image
Middle density (C)	B1: 0.090806–0.139462￼B2: 0.110516–0.178774￼B3: 0.125206–0.219218￼B4: 0.221674–0.252586￼B5: 0.227136–0.369960￼B7: 0.196043–0.349435	**55% to 85% of constructed area** (45% asphalt, 20% red roofing tile, 20% cement) and the rest is **green area** (tress).	
High density	B1: 0.140074–0.227186￼B2: 0.139103–0.260147￼B3: 0.151912–0.270692￼B4: 0.144505–0.337661￼B5: 0.178452–0.376528￼B7: 0.150452–0.425840	**85% to 100% of constructed area** (40% cement, 50% cement badge, 6% red roofing tile and 4% asphalt)	

Note: The middle class consists of three different spectral mixture signatures, each one representing a particular mix of covers but all having the same amount of construction.

TABLE 10.2

Confusion Matrix: Urban Density Classes Extraction 2011 and Cadastral Information (Pixels)

	Low Density	Middle Density	High Density	Total	User Exactitude	Commission Error
Low density	478	50	16	544	87.9%	12.1%
Middle density	60	882	70	1012	87.2%	12.8%
High density	4	80	468	552	84.8%	15.2%
Total	542	1012	554	2108		
Producer accuracy	88.2%	87.2%	84.5%			
Omission error	11.8%	12.8%	15.5%			

Global accuracy: 86.72%.

10.3.2   Urban Growth Modeling

To generate the urban growth model, the LanduseSIM model and the urban density classes were used. In LanduseSIM, the amount of future changes is calculated from a comparative analysis of land use maps from two historical dates (in this case, 2003 and 2011); this analysis obtains the annual rate of change according to the categories in order to project trends in the future.

To carry out the modeling, it is necessary to determine the urbanization factors. In the case that was analyzed, the ones that were selected were proximity to urbanized areas, communications roads, and central areas. In addition, a slope map was made in order to increase the probability of urbanization in areas with a low slope. When the factors were determined, it was necessary to weigh them according to the importance of each one; the most important factor was distance to communications roads (0.6), then slope (0.2), and finally distance to central areas and proximity to urbanized areas (0.1); the weight of each was derived from a previous research (Linares et al. 2017). The Initial Potential Transition map was generated in two steps: first, the Euclidean distance of each pixel to the factor analyzed was calculated, and then the aforementioned weight was computed (Figure 10.2).

To simulate the neighborhood, a Moore ratio of 1 (3 × 3) was used and the focal operation between the cells was determined to be Sum. After determining the urbanization factor and the kind of neighborhood that was used, the elasticity between each density class had to be determined. This is the probability that each density class has to become another one. In this case, higher values are more likely to change to the class selected. For example, as regards the probabilities to become high density, there is a 0.1 probability for a non-urbanized pixel to change to this density class, a 0.5 for low density to do it and a 1 for middle density to turn into high density (Table 10.3).

| Urbanized areas | Communication roads | Central area | Slope |

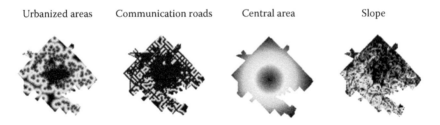

FIGURE 10.2   Euclidian distance to each urbanization factor.

TABLE 10.3
Elasticities between Classes

	Non-Urbanized	Low Density	Middle Density
Low density	1	–	–
Middle density	0.5	1	–
High density	0.1	0.5	1

Finally, the transition rules are generated where all the previous parameters are incorporated. In addition, the land constraints and the land uses where urbanization is not allowed are included (in this case, slopes higher than 30% and areas where public infrastructure is located). Moreover, for each density class, the lower classes are also a constraint because it is not logical, for example, for a high-density pixel to change to a low-density one in normal urbanization patterns. The final input to the model is the number of pixels that each density class will grow in the simulated period (2011–2033). Taking into account the increase in each class between 2003 and

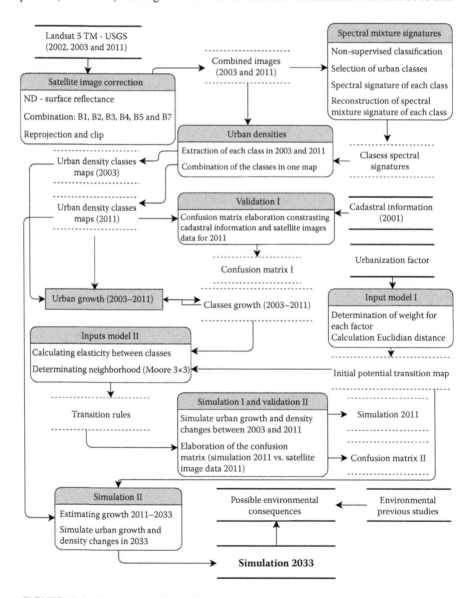

FIGURE 10.3 Flowchart of the methodology.

2011 as a growing parameter, a proportional growth was calculated to add a pixel to each density class to 2033. Low density will grow 5852 pixels, middle density will grow 8800, and high density will grow 6138.

The final simulation was used to analyze the possible environmental consequences that the changes and growth in urban density classes could generate. Considering previous studies that correlate urbanization and urban areas with the behavior of urban climate and runoff in the city for the last decades, exploration of probable future consequences was made. Figure 10.3 presents the flowchart of the methodology.

10.4 RESULTS

10.4.1 URBAN DENSITY 2003–2011 USING SATELLITE IMAGES

Figure 10.4 shows the evolution of urban density classes between 2003 and 2011. Urban densification and the increase of a high-density class are notorious. Besides, the amount of low-density areas that developed during that period is important and their location indicates the spread of urbanization to those areas disconnected from the central urban area. Most of the areas that are involved in this process are family houses and, in some cases, as in the northwest groups, are related to the state urbanization plans.

On the first step of simulation, the 2003 results were used as initial map conditions and the model was run for an 8-year period between the two images. After obtaining the results for 2011, the simulation was contrasted to the remote sensing urban densities to validate the results of the model. The confusion matrix was carried out in order to evaluate the results of the model (Table 10.4). It shows that the global accuracy of the model is 88.72%, encountering some problems distinguishing between the middle- and the high-density class and between the non-urbanized area and the two lower-density classes.

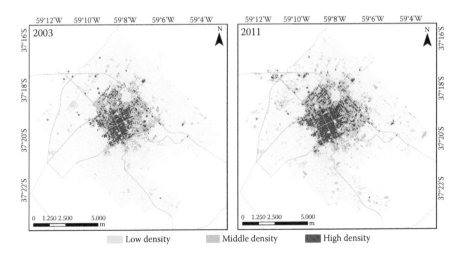

FIGURE 10.4   Urban density evolution 2003–2011. Landsat 5 TM extraction.

TABLE 10.4

Confusion Matrix: Comparison between Simulation 2011 and Satellite Image Data 2011 (Pixels)

	Non-Urbanized	Low Density	Middle Density	High Density	Public Infrastructure	High Slope	User Accuracy	Commission Error
Non-urbanized	82,049	2282	1871	146	0	0	95.0%	5.0%
Low density	1913	1896	816	114	0	0	40.0%	60.0%
Middle density	2193	535	7596	1468	0	0	64.4%	35.6%
High density	230	49	1449	5922	0	0	77.4%	22.6%
Public infrastructure	0	0	0	0	5329	0	100.0%	0.0%
High slope	0	0	0	0	0	21	100.0%	0.0%
Producer accuracy	95.0%	39.8%	64.7%	77.4%	100%	100%	102,813	115,879
Omission error	5.0%	60.2%	35.3%	22.6%	0.0%	0.0%		

Global accuracy: 88.72%.

10.4.2 MODELING SCENERIES OF URBAN GROWTH 2003–2033

After having the global exactitude of the process of modeling, a 22-year simulation was carried out, using the results from the 2011 satellite images as initial condition maps. In this way, a 30-year period (2003–2033) was analyzed in order to study the growth of the urban area in 10-year steps. As can be derived from Figure 10.5, low density is the one with a higher percentage of growth, whereas middle density and high density had a similar increase. In a logical way, the non-urbanized area is the one that loses pixels.

To analyze the spatial distribution of the urbanization process, three maps were generated with LanduseSIM, each one corresponding to a 10-year step. In Figure 10.5, the initial condition map (2003) and the three steps corresponding to 2013, 2023, and 2033 are presented.

The model showed two distinct process. First, it can be said that there is an important compaction process characterized by the densification of the city center and near surroundings, represented by the 150% increase of the high-density class. Second, an important process of suburbanization is shown, indicated by a 300% growth in the low-density class between 2003 and 2033, particularly in the areas near the main roads of the city (Figure 10.6).

According to the results obtained, by 2033, Tandil City will have between 85% and 100% of its surface with impervious areas, drastically reducing the vacant spaces in the compact zone. On the other hand, the dispersion of the lower classes generates new areas of urbanization where the environment is changed.

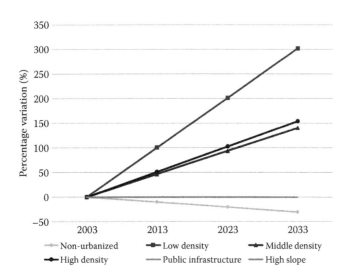

FIGURE 10.5 Percentage variation of each density class 2003–2033.

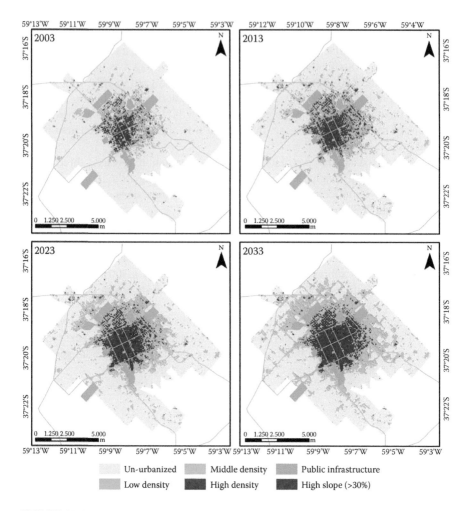

FIGURE 10.6   Simulation of urban growth 2013, 2023, and 2033.

10.4.3   Potential Environmental Consequences
of the Urbanization Process

Although urbanization improves socioeconomic development and life quality, urban sprawl inevitably converts the natural and semi-natural ecosystems into an artificial ecosystem, which results in enormous ecological and environmental challenges, such as heat island effect (Yuan and Bauer 2007), local climate change (Stone et al. 2010), arable land loss (Tan et al. 2005), and water pollution (Yin et al. 2005) (Wang et al. 2016, pp 1073).

The results of urban growth modeling showed that the most important processes that will occur in the city are the densification and compaction of the traditional residential areas. These could have a great environmental impact. Two main aspects will be analyzed in relation to it: the spatial temperature distribution and the surface water runoff.

The historical changes in urban temperature have been analyzed using remote sensing and have shown a great relation with the vegetation. According to Picone (2014), the replacement in the last two decades of the public green areas with impervious areas has led to the development of urban heat sinks during the morning and urban heat islands during the afternoon and night particularly in summer and spring. Taking this into account, if the model shows a more densified urbanization process, this could lead to an increase in the urban temperature differences during the day and higher values of temperature during the afternoon and particularly at night. The direct consequence of this is the generation of more uncomfortable conditions and an increasing possibility of generating conditions for more heat wave effects.

The runoff of cities, as Weng (2001b) stated, should be studied and related to urban land uses because the changes in these last ones have direct impact in the runoff of water, particularly the one from rainfall. Adding more impervious areas would decrease the infiltration of rain; this water is aggregated to the runoff, causing a higher water mass. This could lead to the need of new infrastructure to drain that water to reduce the danger of floods in the city. There is a study of the current condition of the runoff (La Macchia 2015), where it is stated that we are encountering actual problems in the current conditions, particularly in extreme events, which are becoming more frequent. Thus, an increase in runoff could generate more problems to the current condition.

10.5 CONCLUSION

Remote sensing and GIS are key tools to study the urbanization process, particularly in developing countries where there is a lack of information about it. Both technologies can help in the generation of complete and updated databases of urban characteristics and changes, increasing their availability for different uses.

Obtaining discrete densities through satellite image digital processing allows one to obtain more precise information on the actual area occupied by each of the different levels of density, in contrast to the representations obtained through the construction of continuous surfaces. At the same time, one needs to acknowledge a much more real observation of the patterns or distribution models and the changes that have been produced, as well as to discover the trends that may develop in the future.

Using remote sensing, Landsat 5 TM satellite images and the spectral mixture signature methodology, three distinguished density urban classes were detected. An 8-year period of analysis was made to study urban changes. Between 2003 and 2011, Tandil City presented a densification of the city center and at the same time some dispersion of low-density classes. The assessment of the methodology was made with cadastral information from the city hall, showing an 86% global accuracy where errors range from 12% to 15% as maximum.

The results from the previous analysis were used as input for generating suitable sets of the CA model LanduseSIM to model urban density changes for the year 2033. First, the model was run for the same period as the satellite image study to calibrate the model to the current conditions; the results were assessed with the remote sensing data, and an 88% global exactitude was obtained. Second, a predictive model of a 22-year period (2011–2033) was run, resulting in a city with a double spatial patron: a dense city core and a dispersion of the lower densities in the surroundings.

Finally, the potential environmental consequences were inferred from the changes detected by the model. Two main processes could be increased if this pattern of urbanization is accomplished: the development of higher temperature differences and the acceleration of precipitation runoff. The first one could be noticed by the population in the comfort of conditions. Meanwhile, the second one could require the development of new infrastructure to evacuate the water from the city.

REFERENCES

Ackermann, G. and Mering, C. 2007. Extracción de áreas construidas a partir del análisis de imágenes satelitales SPOT. In *Teledetección. Hacia un mayor entendimiento de la dinámica global y regional*, ed. Rivas, R., Grisoto, A., and Sacido M. 369–376. UNCPBA.

Batty, M. 2000. Less is more, more is different: Complexity, cities, and emergence. *Environment and Planning B: Planning and Design.* 27 (2): 167–168.

Batty, M. 2005. Agents, cells, and cities: New representational models for simulating multi-scale urban dynamics. *Environment and Planning A.* 37 (8): 1373–1394.

Camacho Olmedo, M. T., Pontius, R. G., Paegelow, M., and Mas, J. F. 2015. Comparison of simulation models in terms of quantity and allocation of land change. *Environmental Modelling & Software.* 69: 214–221.

Casti, J. L. 1997. *Would Be Worlds: How Simulation Is Changing the Frontiers of Science.* John Wiley and Sons.

Chavez, P. S. 1988. An improved dark-object subtraction technique for atmospheric scattering correction of multispectral data. *Remote Sensing of Environment.* 24: 459–479.

Chuvieco, E. 2010. *Teledetección Ambiental. La observación de la tierra desde el espacio.* Editoral Planeta, S. A. Barcelona.

Clarke, K. C., Hoppen, S., and Gaydos, L. J. 1997. A self modifying cellular automaton of historical urbanization in the San Francisco Bay area. *Environment and Planning B.* 24 (2): 247–261.

Couclelis, H. 1985. Cellular worlds: A framework for modeling micro–macro dynamics. *Environment and Planning B.* 17 (5): 585–596.

Couclelis, H. 1988. Of mice and men—What rodent populations can teach us about complex spatial dynamics. *Environment and Planning A.* 20 (1): 99–109.

Donnay, J. P., Barnsley, M. J. and Longley, P. A. 2001. *Remote Sensing and Urban Analysis.* Taylor & Francis, London.

Ehlers, M., Jadkowski, M. A., Howard, R. R., and Brostuen, D. E. 1990. Application of SPOT data for regional growth analysis and local planning. *Photogrammetric Engineering and Remote Sensing.* 56: 175–180.

Green, K. 2000. Selecting and interpreting high-resolution images. *Journal of Forestry, Society of American Foresters.* 98 (6-1): 37–40.

Harris, P. M. and Ventura, S. J. 1995. The integration of geographic data with remotely sensed imagery to improve classification in an urban area. *Photogrammetric Engineering and Remote Sensing.* 61: 993–998.

Henríquez, C. 2014. *Modelando el crecimiento de ciudades medias: Hacia un desarrollo urbano sustentable.* Ediciones UC.

Hermosilla, T., Ruiz, L. A., Recio, J., and Cambra-López, M. 2012. Assessing contextual descriptive features for plot-based classification of urban areas. *Landscape and Urban Planning*. 106: 124–137.

Hermosilla, T. and Ruiz, L. A. 2009. Detección automática de edificios combinando imágenes de satélite y datos lidar. *Semána Geomática*. Barcelona.

INDEC. 2010. *Censo Nacional de 2010. Instituto Nacional de Estadística y Censos.* Buenos Aires, Argentina.

Itami, R. 1988. Cellular worlds: Models for dynamic conception of landscapes. *Landscape Architecture*. 78 (5): 52–57.

Jensen, J. R. and Cowen, D. C. 1999. Remote sensing of urban/suburban infrastructure socio-economic attributes. *Photogrammetric Engineering & Remote Sensing*. 65 (5): 611–622.

La Macchia, L. 2015. *Modelización y análisis espacial del drenaje urbano de la ciudad de Tandil mediante tecnologías de la información geográfica*. Bachelor Thesis. Facultad de Ciencias Humanas, Universidad Nacional del Centro de la Provincia de Buenos Aires.

Lee, S. 2006. Landsat ETM sub-pixel analysis of urban landscape using fuzzy c-means clustering and differentiated impervious surface classes. *American Society for Photogrammetry and Remote Sensing, Annual Conference*. Bethesda, Maryland, USA. 7 pp.

Linares, S., Etcheverría, S., and Romero, M. C. 2017. Evaluación de factores de expansión urbana aplicando regresión logística. Tandil (2003–2011). VI Congreso Nacional de Geografía de Universidades Públicas. Resistencia, Chaco. 10 pp.

Linares, S. and Tisnés, A. 2011. Extracción y análisis de superficies urbanas construidas empleando imágenes Landsat 5 TM. *I Congreso Nacional de Tecnologías de la Información Geográfica*. Resistencia, Argentina. 180–191.

Longley, P. A. 2002. Geographical Information Systems will developments in urban remote sensing and GIS lead to 'better' urban geography? *Progress in Human Geography*. 26 (2): 231–239.

Lourenço, J., Ramos, L., Ramos, R., Santos, H. and Fernandes, D. 2005. Urban areas identification through clustering trials and neural networks. *14 European Colloquium on Theoretical and Quantitative Geography*. Tomar, Portugal. 11 pp.

Mesev, V., Gorte, B., and Longley, P. A. 2001. Modified maximum-likelihood classification algorithms and their application to urban remote sensing. In *Remote Sensing and Urban Analysis*, eds. Donnay, J.-P., Barnsley, M., and Longley, P. 62–83. Taylor and Francis. London, England.

Mitsova, D., Shuster, W. and Wang, X. 2011. A cellular automata model of land cover change to integrate urban growth with open space conservation. *Landscape and Urban Planning*, 99 (2): 141–153.

Molina Mora, Z. and Chuvieco, E. 1997. Detección del crecimiento urbano en la ciudad de Maracaibo (Venezuela) mediante análisis multitemporal de imágenes MSS y TM. In *Teledetección: Usos y aplicaciones*, eds. Casanova, J. L. and Sanz, J. 361–368. Valladolid, Universidad de Valladolid.

Paegelow, M., Camacho Olmedo, M. T., Mas, J. F., Houet, T., and Pontius Jr., R. G. 2013. Land change modelling: Moving beyond projections. *International Journal of Geographic Information Science*. 27 (9): 1691–1695.

Phipps, M. 1989. Dynamic behavior of cellular automata under the constraint of neighborhood coherence. *Geographical Analysis*. 21 (3): 197–215.

Picone, N. 2014. *Clima urbano de la ciudad de Tandil*. PhD Thesis. Departmento de Geografía y Turismo, Universidad Nacional del Sur.

Picone, N. and Linares, S. 2014. Propuesta metodológica para la extracción y el análisis de densidades urbanas mediante teledetección y SIG. Caso de estudio: Ciudad de Tandil, Argentina. *Revista Universitaria de Geografía.* 23 (2): 77–96.

Ranchin, T. and Wald, L. 2010. Data fusion in remote sensing of urban and suburban areas. In *Remote Sensing of Urban and Suburban Areas*, eds. Rashed, T. and Jürgens, C. 193–218. Series: Remote Sensing and Digital Image Processing, Springer, 10.

Recio, J., Pardo Pascual, J. Ruiz, L. A., Fernández Sarria, A., and Córcoles, P. 2003. Detección y cartografiado de los procesos de expansión urbana mediante técnicas combinadas de teledetección y SIG. *IX Conferencia Iberoamericana de SIG*, Cáceres, Spain. 1–14.

Roberts, D. A. and Herold, M. 2004. Imaging spectrometry of urban materials. In *Infrared Spectroscopy in Geochemistry, Exploration and Remote Sensing*, eds. King, P., Ramsey, M. S., and Swayze, G. 155–181. Mineral Association of Canada, Short Course Series, 33. London, Canada.

Sang, L., Zhang, C., Yang, J., Zhu, D., and Yun, W. 2011. Simulation of land use spatial pattern of towns and villages based on CA-Markov model. *Mathematical and Computer Modelling.* 54 (3–4): 938–943.

Schöpfer, E., Lang, S. and Strobl, J. 2010. Segmentation and object-based image analysis. In *Remote Sensing of Urban and Suburban Areas*, eds. Rashed, T. and Jürgens, C. 181–192. Series: Remote Sensing and Digital Image Processing. Springer, 10.

Stewart, I. and Oke, T. 2009. Classifying urban climate field sites by "local climate zones": The case of Nagano, Japan. *The Seventh International Conference on Urban Climate*, Yokohama, Japan.

Stone, B., Hess, J. J., and Frumkin, H. 2010. Urban form and extreme heat events: Are sprawling cities more vulnerable to climate change than compact cities. *Environmental Health Perspectives.* 118 (10): 1425–1428.

Tan, M., Li, X. and Xie, H. 2005. Urban land expansion and arable land loss in China: A case study of Beijing, Tianjin, Hebei region. *Land Use Policy.* 22 (3): 187–196.

Thapa, R. B. and Murayama, Y. 2011. Urban growth modeling of Kathmandu Metropolitan Region, Nepal. *Computers, Environment and Urban Systems*, 35 (1): 25–34.

Treitz, P. M., Howard, P. J., and Gong, P. 1992. Application of satellite and GIS technologies for land-cover and land-use mapping at the rural–urban fringe: A case study. *Photogrammetric Engineering and Remote Sensing.* 58: 439–448.

Tobler, W. 1979. Cellular geography. In: *Philosophy in Geography*, eds. Gale, S. and Olsson, G. 379–386.

Vapnarsky, C. and Gorojovsky, N. 1990. *El crecimiento urbano en la Argentina.* Grupo Editor Latinoamericano. IIED. América Latina. Buenos Aires.

Vaz, E. D. N., Nijkamp, P., Painho, M., and Caetano, M. 2012. A multi-scenario forecast of urban change: A study on urban growth in the Algarve. *Landscape and Urban Planning*, 104 (2): 201–211.

Wang, H., Ning, X., Zhu, W., and Li, F. 2016. Comprehensive evaluation of urban sprawl on ecological environment using multi-source data: A case study of Beijing. *International Archives of the Photogrammetry, Remote Sensing & Spatial Information Sciences*, 41: 1073–1077.

Weng, Q. 2001a. A remote sensing-GIS evaluation of urban expansion and its impact on surface temperature in the Zhujiang Delta, China. *International Journal of Remote Sensing.* 22 (10): 1999–2014.

Weng, Q. 2001b. Modelling urban growth effects on surface runoff with the integration of remote sensing and GIS. *Environmental Management.* 28 (6): 737–748.

Weng, Q. 2010. *Remote Sensing and GIS Integration: Theory, Methods, and Applications.* The McGraw-Hill Companies, USA.

White, R. and Engelen, G. 1993. Cellular-automata and fractal urban form—A cellular modeling approach to the evolution of urban land-use patterns. *Environment and Planning A.* 25 (8): 1175–1199.

Wu, F. 2002. Calibration of stochastic cellular automata: The application to rural urban land conversions. *International Journal of Geographical Information Science.* 16 (8): 795–818.

Yang, X., Zheng, X. and Lv, L. 2012. A spatiotemporal model of land use change based on ant colony optimization, Markov chain and cellular automata. *Ecological Modelling,* 233: 11–19.

Yin, Z. Y., Walcott, S., and Kaplan, B. 2005. An analysis of the relationship between spatial patterns of water quality and urban development in Shanghai, China. *Computers, Environment and Urban Systems.* 29 (2): 197–221.

Yuan, F. and Bauer M. E. 2007. Comparison of impervious surface area and normalized difference vegetation index as indicators of surface urban heat island effects in Landsat imagery. *Remote Sensing of Environment.* 106 (3): 375–386.

Zhang, Q., Ban, Y., Liu, J., and Hu, Y. 2011. Simulation and analysis of urban growth scenarios for the Greater Shanghai Area, China. *Computer, Environment and Urban Systems.* 35 (2): 126–139.

Section IV

Urban Planning and Socioeconomic Applications

11 Developing Multiscale HEAT Scores from H-Res Airborne Thermal Infrared Imagery to Support Urban Energy Efficiency: Challenges Moving Forward

Bharanidharan Hemachandran, Geoffrey J. Hay,
Mir Mustafiz Rahman, Isabelle Couloigner,
Yilong Zhang, Bilal Karim, Tak S. Fung,
and Christopher D. Kyle

CONTENTS

11.1 INTRODUCTION

Canadian urban energy demand has grown nearly by 20% over the last 5 years and continues to rise with a growing population and increased urbanization. On average, Canadian buildings emit 35% of all atmospheric greenhouse gases (GHGs), generate 10% of airborne particulate matter, consume 50% of Canada's natural resources, and account for more than 30% of all energy used in Canada (CUI 2008), of which the majority is used in space and water heating. Thus, space heating provides one of the best opportunities for energy cost savings through improvements in building design and local alternative energy sources. As part of the Calgary Climate Change Accord (Calgary Climate Change Accord 2009), the Calgary Community GHG Reduction Plan (Calgary Community GHG Reduction Plan 2011), and the City 2020 Sustainability Direction (The 2020 Sustainability Direction 2010), the City of Calgary, Alberta, Canada, is seeking an implementation strategy to reduce GHG emissions and promote low-carbon living that is cost-effective and actionable and reaches a wide city audience. A key implementation strategy identified by the city to achieve these targets is improvements in residential energy efficiency. However, the most cited obstacle to energy efficiency improvements is *a lack of interest* from clients, customers, and consumers (CUI 2008). This comes as little surprise when one considers (i) *what does energy efficiency look like*, (ii) where is *energy efficiency located*, and (iii) *how can a resident know that their home is energy efficient, not the devices inside it?* Darby (2006) and others (Carroll et al. 2009) have shown that effective feedback significantly reduces energy consumption. Additionally, customer energy-use information and behavior programs have explored energy feedback and discovered promising results regarding energy efficiency savings (Briones et al. 2012; Carroll et al. 2009).

The objective of this chapter is to develop and identify the best possible methodology for defining HEAT Scores using thermal imagery as a form of feedback to homeowners. It also describes how this feedback can be provided at different scales to be of interest to different customers such as cities, communities, and house owners. The succeeding sections briefly discuss human behavior and energy efficiency (Section 11.1.1), the Canadian EnerGuide Rating System (ERS) (Section 11.1.2), thermal imaging and home energy efficiency (Section 11.1.3), and the HEAT (Heat Energy Assessment Technologies) project (Section 11.1.4).

11.1.1 Human Behavior and Energy Efficiency

Cherfas (1991) cautions that if human behavior is not analyzed within the energy efficiency context, lifestyle changes could consume all the energy that is saved through other means. Therefore, behavior changes play an important role in today's energy efficiency programs. Taking this as a cue, behavioral change theories have been applied within the energy efficiency context, resulting in a number of insights that are increasingly incorporated within efficiency programs, including (i) social norm, (ii) feedback, (iii) public commitment, and (iv) goal setting (Ashby et al. 2010).

Social norms represent how individuals are influenced by others actions and beliefs, even though they seldom admit to being influenced by others (Cialdini 2007). Research consistently shows that individuals tend to change their behavior closer to the societal *norm* (i.e., the standard, model, or pattern regarded as "typical"). Within an energy efficiency context, *feedback* typically refers to providing individuals with meaningful information on their energy use and related costs. Feedback can be either "indirect" or "direct" (Darby 2006). *Indirect feedback* refers to providing individuals with enhanced billing and estimates of their savings—such as those provided by Opower (www.opower.com). Direct or real-time feedback refers to providing individuals with (near) real-time energy consumption information. Smart meters such as (i) the eMonitor (www.powerhousedynamics.com), (ii) the Ecobee Smart Thermostat (www.ecobee.com), and (iii) the Nest Learning Thermostat (www.nest.com) are direct feedback devices that record detailed consumption information for individual appliances. This detailed information is then made available to consumers in the form of statistical evaluations such as weekly, monthly, and yearly averages, or even second-by-second maximum energy use per monitored circuit. An assessment of several such case studies shows that energy feedback behavior programs conducted in the residential sector prove to be a promising source of energy efficiency savings (Mahone and Haley 2011). These case studies also noted the following critical findings:

i. Depending on the kind of feedback, it is estimated that users will be able to (relatively easily) reduce their total energy consumption somewhere between 1% and 7%.
ii. Comparing and focusing on small social groups (i.e., neighborhoods) to achieve certain energy-use goals can be very successful.
iii. Games and contests are promising areas of further research.
iv. Direct feedback achieves higher savings per participant than indirect feedback; however, indirect feedback has more potential savings opportunity because of its opt-out nature and wider reach.

Similarly, people who commit publicly have higher chances of following through than those who don't. Games and contests could possibly make people commit to reducing their energy consumption and thus increase their chances of achieving it. Similarly, goal setting also makes individuals work toward and reduce their energy consumption (Ashby et al. 2010).

11.1.2 Energy Rating Systems

Worldwide, numerous home and residential energy efficiency rating, monitoring, and feedback system have been implemented to varying degrees of success. In part, this is due to their complexity and costs to implement, interpret, and integrate with systems developed elsewhere, along with challenges related to proprietary formats, public data access, and privacy concerns. For example, some rating systems start at 1 and go to 100 like the Canadian EnerGuide Rating System (ERS), while others like the California Home Energy Rating System (HERS) get as high as 250. For some, a positive outcome is described by a large value, and for others, it is exactly the opposite. Both approaches have merit, but there is no universal agreed upon program.

In Canada, the ERS was created by Natural Resources Canada (NRCan) to provide a standard measure of a home's energy performance. This rating system allows for comparison of the energy efficiency of comparable homes in neighborhoods across Canada (though we note that such comparison details are not provided by NRCan). An ERS shows a home's present level of energy efficiency and provides recommendations for upgrades that would increase the home's energy efficiency. An ERS can be obtained by contacting an *energy advisor*. This advisor will assess a home's structural characteristics (i.e., construction material and building envelope type) and use these attributes to model the home's energy consumption. A home energy efficiency level is rated on a scale from 0 to 100. A rating of 0 means that the home is poorly insulated and has major air leakage and extremely high-energy consumption. A rating of 100 means that the home is well insulated, airtight, and highly energy efficient (Kordjamshidi 2011).

It is important to note that this rating is only an estimate of the home energy consumed each year, not the actual amount (which will vary by number of occupants, consumption habits, and lifestyle). Furthermore, (i) it can only be provided by a certified energy advisor (thus, a typical homeowner cannot conduct it), (ii) it is not easy to calculate and understand, and (iii) it requires the calculation of hundreds of variables and two or more home visits (including a mechanical blower door test), each taking several hours to complete. It is also (iv) expensive to obtain ($300 to $500+), (v) it does not provide the owner with any way to monitor their energy efficiency (i.e., waste heat mapping or energy consumption over time), and (vi) because of privacy restrictions, owners cannot see how their house compares to other houses, unless an owner physically makes their house data available, as the ERS currently has no (web-based) capacity to facilitate comparison.

11.1.3 Thermal Imaging and Home Energy Efficiency

Another form of feedback that can aid in improving a building's energy efficiency is to use thermal infrared (TIR) sensors to identify temperature anomalies in the building envelope where waste heat is leaving the structure and to correct for them (Goodhew et al. 2015). In this research, *waste heat* represents expensive heated air that is leaving a house instead of staying inside and keeping the house warm. Additionally, the waste heat measured leaving a building's envelope may be considered a measure of the energy efficiency of the structure itself, rather than the energy-consuming devices inside it. When appropriately processed, color waste heat maps can (i) be simple and intuitive to understand, (ii) be relatively inexpensive to produce, (iii) show the location of the hottest and coolest areas (i.e., hot and cool spots), (iv) facilitate a meaningful comparison of heat loss between different houses, (vi) validate energy efficiency retrofits, and (vii) provide quantitative evidence of heat loss over time and space through monitoring.

With the advent of airborne TIR imaging, heat loss/waste heat mapping can be multiscale in nature, potentially providing unique information at the house, community, and city scale (Blaschke et al. 2011; Hay et al. 2010, 2011). Over the last 15 years, a number of urban airborne heat loss mapping projects have been conducted around the world. The following list briefly identifies 12 of them:

  i. Winter 2000 and 2007: The London Borough of Haringey conducted airborne TIR heat loss surveys to provide residents with an idea of the energy efficiency of their homes (Haringey 2007).
 ii. 2001 and 2013: The city of Aberdeen, Scotland, conducted thermal heat loss mapping, in order to identify the least thermally efficient areas and house types within the city in order to target home energy efficiency promotions (Aberdeen 2013).
iii. 2009: Worcestershire, England, conducted the *Warmer Worcestershire* project to encourage residents to improve their building's energy efficiency and save their money (Worcestershire 2009).
 iv. 2009: The city of Exeter, England, conducted a heat loss survey using airborne thermal imagery with the goal to increase public awareness of energy efficiency and motivate homeowners to adequately insulate their houses (Exeter 2009).
  v. 2010: Paris, France, processed information from aerial thermography to show heat loss from buildings on a scale of six colors. This was provided as a web-based decision support service to the residents for improving their building's energy efficiency (Paris 2010).

vi. 2011: The inner city of Odense, Municipality of Frederiksberg, and Municipality of Lyngby-Taarbaek in Denmark conducted a similar study. They showed a simple color-coded temperature map showing temperature variations on a building's roof (Denmark 2011).

vii. 2011: The Island City of Jersey, in the United States, States of Jersey, conducted a heat loss map to help residents understand how much heat was being lost from their homes (States of Jersey 2011).

viii. 2012: Portsmouth City Council, England, took a thermal image of the whole city to help residents understand the amount of heat lost from their properties (Portsmouth 2012).

ix. 2013: The EnergyCity project (www.energycity2013.eu) acquired aerial thermography data on seven cities in Central Europe, which was then processed and refined into an online Spatial Decision Support System to visualize and compare the cost-effectiveness and potential of different renewable energy solutions in the project cities.

x. November 2013: The HEAT project was awarded the MIT Climate Co-lab grand prize for web enabling 38,000+ thermal heat loss maps, and hotspots for individual homes and communities in the City of Calgary Alberta Canada (HEAT 2013).

xi. In January 2015, MyHEAT.Inc (a technology start-up based on HEAT intellectual property) released its first commercially available public web-based heat-loss mapping platform for the town of Okotoks, Alberta, Canada. By the end of 2017, it will have web-enabled more than 1.3 million individual homes.

xii. 2016: The German city of Osnabrück conducted a winter aerial survey using TIR imaging to determine which roofs were well insulated versus those that were losing energy. Map results were given a dollar value and communicated to homeowners and businesses (Energy Management in Osnabrück 2016).

In the majority of these heat loss mapping projects, the thermal imagery was published online as very simple temperature class maps, with limited capability for in-depth visual, statistical, or location-aware analysis (Hay et al. 2010). Furthermore, the 2007 survey conducted in London Borough of Haringey had numerous geometry and radiometric normalization problems with the imagery that were not addressed, leading to limited utility (Hay et al. 2011). More recently, the HEAT (Heat Energy Assessment Technologies) project at the University of Calgary, Alberta, Canada, has developed an innovative Geoweb platform to overcome many of these challenges and limitations associated with airborne TIR heat loss surveys (Blaschke et al. 2011; Hay et al. 2011; Rahman et al. 2014), including sophisticated multitemporal, radiometric, and geometric normalization procedures.

11.1.4 HEAT ENERGY ASSESSMENT TECHNOLOGIES

The HEAT project was initially developed as a publicly accessible Geoweb mapping service that built on ideas from behavioral science while improving upon earlier airborne TIR heat loss mapping projects. The HEAT project was designed to allow homeowners to visualize the amount and location of waste heat leaving their homes and communities as easily as clicking on their house in Google Maps (Hay et al. 2010, 2011). The HEAT project provides meaningful visual feedback to homeowners in the form of interactive HEAT Maps, HEAT Scores, hotspots, and results from energy consumption models.

The primary objective of this paper is to describe the evolution of different methodologies used to define HEAT Scores based on high-resolution airborne TIR imagery and the challenges involved. In this work, *HEAT Scores* are ranked numbers that range from 0 to 100 (low heat loss to high heat loss) that represent the amount of waste heat leaving a building. A house with a HEAT Score of 1 represents very low waste heat; consequently, it consumes a small amount of energy for space heating. A house with a HEAT Score of 100 represents very high waste heat, as it consumes a large amount of energy for space heating. Based on the behavioral science concept of *feedback*, HEAT Scores are developed to allow for a meaningful comparison of waste heat of one or more houses with all other houses in their community and city, and ideally for a comparison between participating communities and cities.

11.2   STUDY AREA AND DATA

This study was conducted in the SW quadrant of the City of Calgary, Alberta, Canada, which covers an area of approximately 21 km² (4.19 × 4.93 km) and represents 12 established communities that contains 9278 homes. Within this study area, the Kingsland community* was established in 1957, while the Cedarbrae community[†] was established in 1973. However, houses in this site were constructed as early as 1900 to as recently as 2013. The living area of houses varies from 65 to 500+ m² (i.e., 700 to 5300+ square feet). *Living area* is defined as the living space above and below grade level of a house. The study site also contains a variety of building types including but not limited to (i) bungalow,[‡] (ii) garage + house,[§] and (iii) duplexes.[¶] This study site was chosen because of its diversity of building age, type, and living area.

The TIR data for this study were acquired in the thermal spectrum between 3.7 and 4.8 µm using the TABI-1800 (Thermal Airborne Broadband Imager) sensor, flown by ITRES Research Ltd., Calgary. Imagery was collected on May 14, 2012, at a 50-cm spatial resolution and a 0.05°C thermal resolution in the early morning (between 12:00 a.m. and 4:00 a.m.) under calm skies when the environment was in *thermal equilibrium* (Jensen 2007) (Figure 11.1).

11.2.1   CLIMATIC CONSIDERATIONS FOR THERMAL IMAGING

Predawn collection of thermal data is preferred for a number of reasons, including the following: (i) at predawn, most objects are in thermal equilibrium with the

---

* Kingsland community, Calgary, http://en.wikipedia.org/wiki/Kingsland,_Calgary
[†] Cedarbrae community, Calgary, http://en.wikipedia.org/wiki/Cedarbrae,_Calgary
[‡] Bungalows are single-level wooden structures, typically less than 1000 square feet (93 m²), and normally feature a detached garage.
[§] Garage houses have room for one to several cars, including RVs.
[¶] A duplex is a dwelling having apartments with separate entrances for two households.

environment (i.e., the influence of sun on the thermal characteristics of the objects will be minimal); (ii) since most objects are in equilibrium, winds caused by differential heating of the earth's surface would also have died down (Colcord 1981).

Thermal airborne image acquisitions also pose other challenges including the influence of the atmosphere between the sensor and the surface, the *microclimate* or the local climatic variability (i.e., wind, humidity, etc.), and knowledge of scene objects' emissivity. As a result, houses observed in a thermal image with identical (relative) rooftop temperatures can actually have different kinetic roof temperatures if they are composed of different materials. Ideally, each of these challenges has to be carefully accounted for before performing any analysis that requires highly accurate temperature values. For example, atmospheric influence can be mitigated by applying an atmospheric correction model. Many models have been developed over the years including MODTRAN (Berk et al. 1989) and LOWTRAN (Kneizys et al. 1983). However, the thermal data acquired for this project were processed by the vendor with a digital terrain model (10 m spatial resolution). Thus, nominal elevation related climate errors were already accounted for upon delivery. We note that microclimate normalization was also implemented (see Section 11.2.3).

11.2.2 EFFECTS OF EMISSIVITY IN THERMAL IMAGERY

Important research has been conducted to quantify the effects of emissivity in thermal imagery (Nicole 2009; Stathopoulou et al. 2009; Sugawara and Takamura 2006; Weng 2001). By far, the simplest method is to integrate land-use data with the corresponding thermal image. However, each of these studies has been conducted on ASTER or TM data, which have a coarse spatial resolution (90 and 120 m, respectively). In contrast, this study is conducted using very high spatial resolution imagery (0.5 m), which brings additional challenges owing to high interobject variability (i.e., within a roof) that can confuse traditional classifiers. For example, in this project, rooftops are the objects of interest; however, they will also be composed of other smaller objects such as chimneys, sun lights, and vents. Each of these objects is made of different materials such as metals, glass, and so on, which will have a different emissivity value from that of the dominant roof material. Therefore, correcting for each of these object emissivity values is necessary to calculate the true temperature of the rooftop. However, defining these small objects is a nontrivial process. For simplicity in this project, each rooftop object is assumed to be made of a homogenous material. One way to mitigate this issue would be to resample the TIR imagery from 50 cm to 1 m, thus effectively regularizing the effects of small roof objects within the larger more homogenous background roof signal.

After carefully studying the TIR data set and a corresponding 2012 City of Calgary color infrared (R, G, B, NIR) ortho-mosaic (at a 25-cm spatial resolution), it was concluded that 80% of the rooftops in the study area are composed of asphalt shingles. This has since been verified at 81.2% based on MLS roof data from the Calgary Real Estate Board. The emissivity value of asphalt shingles is 0.91. However, there are other challenges such as the same material (in different locations) having a different surface roughness, moisture content, age, or dominant orientation, which, in turn, will result in different emissivity values (Jensen 2007). We note in general, though, that much of this issue can be mitigated by applying emissivity modulation (Nicole 2009), which effectively integrates appropriate emissivity measures based on predefined land cover classes derived from visible imagery (such as the City of Calgary 25-cm digital ortho-imagery).

11.2.3 Effects of Microclimatic Variability in Thermal Imagery

Microclimate variability also poses a challenge to interpretation of thermal imagery. At the time of this research, there was no simple well-defined procedure to account for its effects in urban settings—especially over fine distances. Since this time, Rahman et al. (2014) have introduced TURN—Thermal Urban Road Normalization—to correct for microclimate variability over an entire city at a 5- to 10-m spatial resolution based on local variability in nighttime road temperature.

To account for microclimate variability in this project, weather data for May 14, 2012, were collected from 21 weather stations across the City of Calgary. These data (ranging from 10°C to 14°C) were used to generate an interpolated surface of the study site's air temperature using *inverse distance weighting* and were used as one of the inputs to develop HEAT Scores (Section 11.3.2).

11.2.4 Cadastral Data and Building Attributes

Cadastral data (i.e., relating to house/parcel ownership boundaries) containing the vector layers of the building outlines (Figure 11.1, red outlines) were obtained from the City of Calgary. Additional attribute data such as the living area, year of construction, and address were also provided by the city.

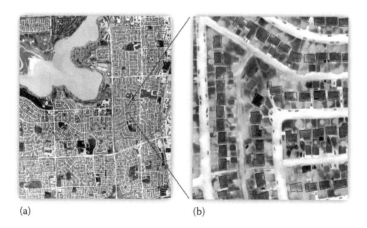

(a)   (b)

FIGURE 11.1 An example (a) of the TABI-1800 illustrating a portion of the study site. Gray-toned temperature variations range from white (e.g., hot roads) to black (e.g., cool houses and parks). City of Calgary GIS building outlines are overlaid the TABI data; which are highlighted in red to give an example of their locational fit (b). A larger reservoir is visible in the top left of (a).

As thermal imagery and cadastral data were collected from different sources at different times and spatial resolution, they do not perfectly geometrically match. Therefore, they must be manually co-registered. This geometric correction step is critical, as the cadastral data will be used in conjunction with thermal imagery (as a mask) to extract thermal building object information. A geometric correction was manually performed using ENVI 5.3 software. The cadastral data were derived from 25-cm RGBNIR data by trained photo-interpreters; thus, they are assumed to be more accurate than the thermal image. As a result, they are used as the base map (i.e., the master) for geometric correction. The thermal image (the slave) was corrected to the cadastral data by selecting recognized Ground Control Points (GCPs)—found in both scenes. Corners of houses that were clear of any obstacles (i.e., trees, sheds, etc.) were selected as GCPs. Approximately 800+ GCPs were manually collected throughout the image. Triangulation (Hosomura 1994) was selected for geometric correction—after a careful visual comparison of results generated using the more conventional least squares method. Nearest-neighborhood interpolation was then used for re-projecting the thermal image in order to preserve their original thermal values, as most spatial errors visually appeared less than 1–2 pixels.

11.3   METHODS—HEAT SCORES

The primary objective of HEAT Scores is to provide building owners with a simple value that they can use to compare the waste heat leaving their house to other houses. Waste heat typically escapes through poorly insulated doors, windows, walls, ceilings, ductwork, and electrical fixtures (i.e., pot lights). This is costly to the homeowner, generates more GHG emissions than necessary, and is invisible to the human eye. Waste heat is calculated from the digital numbers in the thermal image. HEAT Scores are also colored from blue to red—representing low and high heat loss, respectively. For example, a HEAT Score of 95 will be colored *red*, indicating that this house is wasting a relatively large amount of heating energy. Similarly, a HEAT Score of 10 will be colored *blue*, indicating that this house is wasting a relatively small amount of heating energy.

Figure 11.2 summarizes the methodology used in this study. Essentially, the geometrically corrected TIR image was masked using the building GIS vector data obtained from the City of Calgary to produce 9000+ house objects. Statistics such as (i) average temperature, (ii) maximum temperature, (iii) minimum temperature, (iv) standard deviation of temperature, and (v) hotspot locations and their values for each house were calculated after accounting for a general emissivity value of 0.91 for asphalt shingles.

In HEAT, hotspots represent an ordered list of the six hottest locations around the (1 m) roof edge, as well as the six hottest locations covering the remaining rooftop.* Within each of these two zones, hotspots are located at least 1.5 m away from each other, so that they are not clustered around the same location. Their purpose is to guide homeowners to investigate numerous heat loss zones and their potential

---

* We note that hotspots along the roof edge often visually correspond to the doors and windows beneath them, which can be verified in Google Street View.

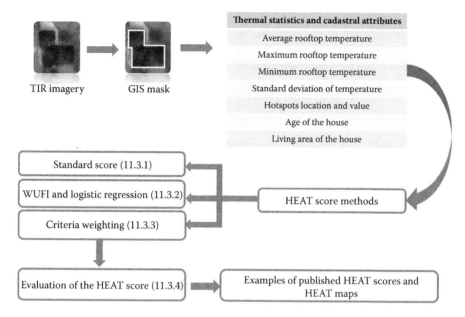

FIGURE 11.2 Flowchart showing the HEAT Score methodology with related section numbers.

causes. This provides a representative visual description of the spatial distribution of hot areas over each of the zones. Along with the GIS house attributes such as age and living area, these statistics were used to calculate HEAT Scores.

As reported in this document, HEAT Score development has been an evolving process based on identifying three key methods and testing their strengths and limitations. Consequently, HEAT Scores have matured from an initial statistical concept based on a standardized score to include more complex WUFI models and relevant weather conditions. The following sections describe each method of the flowchart in detail (and the lessons learned), and then the conceptual validity of HEAT Scores is evaluated using ERS data.

11.3.1 HEAT Score Method 1—The Standardized Score

Given that HEAT Scores are between 0 and 100, a simple standardized score was initially applied to the waste heat of each house. A standard score indicates that the number of standard deviations, an observed value (in this case the waste heat), is above or below the population mean. For HEAT Scores, the population mean is the average of all the waste heat in the "city" or, more specifically, all the houses in the Calgary SW community data set. Conceptually, these same ideas will apply to the full city data set.

Waste heat was calculated as the difference between a house's average rooftop temperature and the minimum of the minimum temperature recorded in the study area multiplied by the living area of the house (Equation 11.1). The assumption made here was that there is a location on some rooftop in the study area that wastes the minimum heat. This was identified using the minimum of the minimum rooftop temperature recorded in the study site and is assumed to be the optimal or ideal rooftop temperature. On the basis of this assumption, waste heat for each house has been defined using the following equation:

$$\text{Waste heat} = \left(\mu_{\text{Temp}} - \text{MIN}\left(\text{MIN}_{\text{Temp}}\right)\right) * \text{Area} \qquad (11.1)$$

where μ_{Temp} is the average rooftop temperature of the house, MIN $(\text{MIN}_{\text{Temp}})$ is the minimum of the minimum rooftop temperature recorded in the whole study site, and Area is the living area of the house. In order to use standard scores, waste heat has to be tested for normality. Analysis showed that waste heat was not normally distributed; therefore, a natural logarithmic transformation was applied to Equation 11.1. Based on the standard score formula (Equation 11.2), where x is the waste heat of the house to

$$z\text{-score} = (x - \mu)/\sigma \qquad (11.2)$$

be standardized, μ is the average of all the waste heat in the study site and σ is the standard deviation of the waste heat in the study site. HEAT Scores are then calculated as shown in Equations 11.3 and 11.4.

$$\text{HEAT Score} = z\text{-score of an item} - (-3.49)/\left(3.49 - (-3.49)\right) * 100 \qquad (11.3)$$

$$\text{HEAT Score} = (z\text{-score} + 3.49)/6.98 * 100 \qquad (11.4)$$

Z-scores follow a standard normal distribution; consequently, 99.98% of these values will lie between −3.49 and 3.49. This z-score is then proportionately converted to a scale from 1 to 100 to represent the HEAT Score for every house. Thus, a HEAT Score of 100 represents a house (or ranked class of houses) with the highest waste heat and a HEAT Score of 1 represents the house (or ranked class of houses) with the lowest waste heat. There is a possibility of outliers that will have a z-score greater than +3.49 or less than −3.49. Therefore, outliers have to be accounted for, which can be achieved using *quartiles*. Quartiles are three points that divide the data set into four equal parts. Outliers are calculated using the following formula:

$$\text{Lower Outer Fence (LOF)} = Q1 - 3 * \text{IQR} \qquad (11.5)$$

$$\text{Upper Outer Fence (UOF)} = Q3 + 3 * \text{IQR}, \qquad (11.6)$$

where $Q1$ is the lower quartile, $Q3$ is the upper quartile, and IQR is the interquartile range ($Q3 - Q1$). Any value below LOF (Equation 11.5) or above UOF (Equation 11.6) is considered as an *extreme outlier* and a HEAT Score of 100 or 1 is assigned, respectively. By using this defined HEAT Score, an individual will be able to compare their house with any other house in any part of the city.

11.3.1.1   Limitations of the Standardized Score Method

Waste heat was calculated using the assumption that there is a location on some rooftop that wastes a minimum amount of heat. Anything above this temperature is considered as waste heat. The problem associated with this assumption is that this minimum temperature might be associated to a metal object on a rooftop. This is important, as metals have a very low emissivity value associated with them and therefore (when uncorrected) appear cold on a thermal image (as they will reflect deep space—which typically records the minimum temperature of the sensor). The result is that waste heat could be miscalculated; thus, the corresponding HEAT Score would be in error.

The second limitation of this method is the use of z-scores of the average rooftop temperature. Unfortunately, these scores cannot be compared between different populations (i.e., different cities) without first aggregating all populations—which then dilutes them. Z-scores are relative; thus, adding or removing a data point from the data set will require recalculation of all other data points. Therefore, every time the study site is extended, or new houses are added to the study, HEAT Scores will have to be recalculated for all other houses potentially changing them. Similarly, as they are relative, they cannot be compared if applied independently to each community or city, which defeats an important reason for defining HEAT Scores in the first place.

Recognition of these conditions resulted in the search for other methods with which to define HEAT Scores. Upon reflection, the new method should (i) allow for a comparison of HEAT Scores between cities, (ii) not change the HEAT Score of a house every time new houses are added to the study area, (iii) take into account the weather conditions of the city, and (iv) take into account the heat transfer through building materials. This last condition will allow for more realistic calculations of waste heat, rather than arbitrarily selecting the optimal or ideal rooftop temperature based on some untested assumption. The following section describes a new method that takes these attributes into consideration.

11.3.2   HEAT Score Method 2—The WUFI Model and Logistic Regression

In this "new" method, waste heat is calculated as the difference between the average roof temperature and the modeled roof surface temperature derived from WUFI software. WUFI (Wärme und Feuchte instationär) is a software family (and modeling environment) that allows realistic calculation of the heat and moisture transport in multi-layer building components exposed to natural weather. Additionally, this software allows the user to "construct" models of building roofs/walls for different climatic conditions.

For this project, a vented roof system with asphalt shingles, typical of the most common rooftop in the City of Calgary, was modeled with the following components:

6	3	4			
7	4	5			
8	2	6			
9	1	7			
10	2	8			
11	3	9			
12	4	10	11	12	13
13	5	11	12	13	14
14	6	12	4	7	3
15	7	13	3	4	6
16	8	14	6	11	12
17	9	15	8	10	8
18	10	16	15	9	14
19	11	17	13	6	7

FIGURE 11.3 Hypothetical rooftop (plan view) and corresponding pixels (rectangles) with numbers representing temperature values. Highlighted pixels (gray) are wasting heat.

(from interior to exterior) 1.25 cm (0.50 in.) interior gypsum wallboard; 19.7 cm (7.75 in.) glass-fiber batt insulation, 19.7 cm (7.75 in.) open-cell polyurethane foam insulation, or 19.7 cm (7.75 in.) closed-cell polyurethane foam insulation between wood rafters; 4 cm (1.50 in.) ventilation space with five air changes per hour; 1.25 cm (0.50 in.) OSB (Oriented Strand Board) sheathing or plywood sheathing; and one layer of felt and asphalt shingles (Nelson and Ananian 2009). This software also allows the user to provide indoor and outdoor climate values as parameters for many cities. However, it allows only two options for outdoor climate, (i) a cold year or (ii) a warm year. These are the averages of weather conditions such as outdoor temperature, humidity, and dew point. Since May 14, 2012, the day of the data collection, was relatively warm compared to the same day for other years, a comparable warm year was selected as input for the outdoor climate. The indoor climate was assumed to be 21°C as this is a commonly set indoor temperature. Running this model for an asphalt vented roof assembly produces the exterior surface temperature of the rooftop. This temperature is considered to be the *ideal* or *optimal* rooftop temperature. Between 00:00 a.m. and 4:00 a.m. on May 14, the modeled temperature varies between 10.5°C and 11°C (highlighted in the orange transparent box—see Figure 11.3).

After correcting for emissivity (0.91) of the roof material, anything above this temperature is considered to be wasting heat. Waste heat for this "new" method is defined as the ratio of pixels that are above the WUFI modeled temperature to the total number of pixels representing the house (Equation 11.7).

$$\text{Waste heat} = \frac{\text{No. of pixels above WUFI modeled temperature}}{\text{Total no. of pixels on the roof}} \qquad (11.7)$$

Consequently, waste heat will have a value between 0 and 1. A waste heat value of 0 means that the house (roof) wastes no heat and 1 means that the entire roof is wasting heat. Figure 11.3 shows a hypothetical roof with each cell in the table depicting the pixels in a rooftop. Assuming 10°C as the optimal rooftop temperature modeled using WUFI, all the pixels that are *above* this temperature are highlighted in gray and considered as waste heat pixels. As shown in Figure 11.3, 32 of the 66 pixels are wasting heat. Therefore, waste heat for this hypothetical house is calculated as the ratio of 32/66, which is equal to 0.49.

As described in Section 11.2.3, the ambient air temperature on the night of acquisition naturally varied throughout the scene (and corresponding TIR data set). Therefore, air temperature (variability) needs to be taken into account while calculating waste heat. To satisfy this condition, a WUFI model was generated for different temperatures, where it was concluded that the variations in air temperature are linearly correlated with rooftop temperature. At least, this was the case for the ambient temperatures that ranged from 10°C to 14°C over May 14, 2012. Consequently, for every 1° rise in air temperature, the modeled rooftop temperature is increased by 1°C.

11.3.2.1   Rationale behind the Selection of Independent Variables

After using Method 2 to calculate the waste heat (adjusted for emissivity and air temperature), it is used as the dependent variable in a logistic regression with the following independent variables: age, living area, average hotspots, and standard deviation of the rooftop temperature. This section describes the rationale behind the selection of these variables.

- *Age*—Assuming that "old" houses are not renovated, older houses are expected to waste more heat through their roofs than new houses—because of earlier limitations in related building codes for older buildings. This is also confirmed from the *Thermal Archetypes** project, which concludes that old houses have very poor insulation compared to the newer ones (Parekh and Kirney 2012).
- *Living area*—As the living area of the house increase, the energy required to heat the house also increases.
- *Hotspots* are geographical representations of unique waste heat locations on the rooftop of a house. When the hotspots are compared with their location in Google Maps Street View, they (typically) visually correspond to heat escaping from specific roof components (i.e., vents, chimneys, etc.) or from the doors, windows, walls, and living envelope situated beneath. The *average* of these six hotspots represents the hottest locations on the rooftop. As the average of these six hotspots increases, it is also expected that the waste heat and related energy consumption will increase.
- *The **standard deviation*** of the rooftop temperature describes the variation of the rooftop temperature. A well-insulated roof should show minimal variation on the thermal image. Therefore, as the variation increases, waste heat is also expected to increase.

---

* The NRCan Thermal Archetypes project provides data of residential energy efficiency archetypes (based on statistically representative data libraries from over 800,000 Canadian homes), including house geometry, thermal characteristics, and operating parameters such as insulation values and furnace efficiency (Parekh and Kirney 2012).

11.3.2.2   Logistic Regression for HEAT Scores

Logistic regression, an inferential statistical method, is used to compute HEAT Scores, as it works well for the analysis and prediction of dichotomous outcomes, and the mean response of such outcomes is a probability (Peng et al. 2002; Walpole et al. 2007); thus, it can vary from 0 to 1. Equation 11.8 shows a general equation of the logistic regression.

$$\text{logit}(Y) = \ln\left(\frac{\pi}{1-\pi}\right) = \alpha + \beta_1 X_1 + \beta_2 X_2 + \ldots + \beta_n X_n, \tag{11.8}$$

where π is the dependent variable or the probability of success, α is the intercept, and β_1, β_2, and β_n are the coefficients of independent variables X_1, X_2, and X_n, respectively. The ratio of pixels that are above optimal rooftop temperature to the total number of roof pixels is used as the dependent variable (Y) - or the binary response in a logistic regression model (Equation 11.8). The predicted value from this model is multiplied by 100 to define the HEAT Score of each house. Equation 11.9 shows the logistic regression model of HEAT Scores computed for all 9000+ homes using R software. R is a free software environment for statistical computation and graphics (http://www.r-project.org/).

$$\begin{aligned} \text{logit (waste heat)} = {} & -3.84 + 0.02 * \text{Age} - 0.001 * \text{Living Area} \\ & + 0.03 * \text{Std. Dev} + 0.44 * \text{Avg Hotspot} \end{aligned} \tag{11.9}$$

11.3.2.3   Limitations of the Logistic Regression Method

The statistical significance of individual regression coefficients (i.e., β in Equation 11.8) is tested using a *t test*. From Table 11.1, it can be concluded that age, living area, and average hotspot temperature are significant predictors of HEAT Score ($p < 0.05$). However, standard deviation is not a significant predictor. In addition, living area is negatively related to HEAT Scores, which contradicts the rationale behind the selection of this variable (Section 11.3.2.1). However, it is *common sense* that as the living area increases, energy consumption and associated waste energy will increase (as there is more space to heat). In order to overcome these problems, the following criteria-weighted method was developed.

11.3.3   HEAT Score Method 3—Criteria Weights

This third method assigns weights to the factors contributing to a HEAT Score. In order to assign weights to each factor, their effects on waste heat and energy consumption need to be assessed. This section describes the logic behind the selection of weights for different factors.

TABLE 11.1

Logistic Regression Model to Calculate HEAT Scores with the *p*-Value for Each Independent Variable

Coefficients:

	Estimate	Std. Error	z value	Pr(>|z|)	
(Intercept)	-3.8370611	0.1851019	-20.729	< 2e-16	***
heatscores_das$buildings$age	0.0220701	0.0002761	9.697	< 2e-16	***
heatscores_das$buildings$livingarea	-0.0006539	0.0002008	-3.257	0.00113	**
heatscores_das$buildings$stdtemp	0.0331122	0.0388426	0.678	0.49781	
heatscores_das$buildings$avghotspottemp	0.4355672	0.0176539	24.673	< 2e-16	***

Signif. codes: 0 '***' 0.001 '**' 0.01 '*' 0.05 '.' 0.1 ' ' 1

(Dispersion parameter for binomial family taken to be 1)

    Null deviance: 2290.9 on 8250 degrees of freedom
Residual deviance: 1589.5 on 8246 degrees of freedom
AIC: 7088.3

Number of Fisher Scoring iterations: 5

Heat leaving a house can be calculated using the heat loss rate* formula shown in Equation 11.10.

$$\text{Heat loss rate} \left(\frac{Q}{t} \right) = \text{Area of the wall} * \frac{T_{\text{inside}} - T_{\text{outside}}}{\text{Thermal resistance of wall}}, \quad (11.10)$$

where T_{inside} is the temperature inside the house in Kelvin (K), T_{outside} is the temperature outside the wall in K; and Thermal resistance of wall (RSI) is in m^2K/W. The total heat loss of a house is calculated as the sum of the heat loss from all its walls and roofs.

According to the Thermal Archetypes project (Parekh and Kirney 2012), houses constructed before 1945 have three times less insulation than the houses constructed after 2005 (Parekh and Kirney 2012). A sample calculation shown in Table 11.2 suggests that, in Calgary, a house constructed before 1945 with a living area of 81 m^2 consisting of four walls and a roof will consume 76 GJ per year at a cost of 460 Canadian dollars (CAD), whereas a similar house constructed after 2005 will consume only 29 GJ per year at a cost of 178 CAD.

To facilitate the calculations shown in Table 11.2, the insulation values of walls, ceilings, and foundation were obtained from the Thermal Archetypes project. Heating degree-days were obtained from http://www.weatherstats.ca/. In simple terms, one heating degree-day is designed to reflect the amount of energy required to heat a house by 1 degree for a day. Based on a fixed rate per GJ, a natural gas price of $5.99 was created by averaging consumption between May 2012 and 2013. (NB: It can also be seen from the calculations in Table 11.2 that as the living area increases, the corresponding heating cost increases.)

From these model calculations, it is evident that the older houses consume 2.6 (460.5/178.9) times more than the new houses for space heating, while the living area of large houses consumes 2.4 (1105.9/460.5) times more than the small houses. Therefore, age affects the cost a little more than the living area; thus, the weight for age needs to be more than the weight for the living area.

If we assume that there is a hole of size 1 m × 1 m in the ceiling where insulation is half the original value, it can be easily calculated that this hole will waste twice the energy than if the ceiling was properly insulated (Equation 11.10). Therefore, it is essential to incorporate the average roof hotspots within the HEAT Score calculation. It is *hypothesized* that the higher the temperature difference between the average roof hotspots temperature and the optimal rooftop temperature calculated from the WUFI model in Section 11.3.2, the higher the chances of wasting heat. The standard deviation of rooftop temperature is not considered as it was not a significant contributor to waste heat according to the logistic regression (Section 11.3.2.2). In addition to this, the average hotspot temperature and the total waste heat already take into account the variation of temperature on the roof. Thus, from these assumptions, there are four critical factors, namely, (i) total waste heat, (ii) average hotspots temperature, (iii) age, and (iv) living area. Consequently, weights for these factors need to be defined in order to calculate HEAT Scores.

---

* Calculating Home Heating Energy, http://hyperphysics.phy-astr.gsu.edu/hbase/thermo/heatloss.html

TABLE 11.2

Example of HEAT Loss and the Cost of Heating Houses in Calgary Based on Different Vintage Types Obtained from the Thermal Archetypes Project

Year of Construction/Living Area	Total Energy Consumed per Year (GJ)	Total Cost to Heat Home per Year(CAD)
<1945/81 m²	76.87	460.5
≥2005/81 m²	29.87	178.9
<1945/225 m²	184.62	1105.9
≥2005/225 m²	68.06	407.7

The absolute *t values* in the logistic regression shows the significance of the factors (Section 11.3.2.2). The higher the absolute *t* value, the higher its significance on the dependent variable. As illustrated in Section 11.3.2.3 (Table 11.1), it can be seen that average hotspot temperature is the most significant contributor to HEAT Score while age and living area are the second and third most significant contributor, respectively. Since this project is primarily focused on thermal energy leaving a house (and as this attribute—waste heat—represents an information set, unique to HEAT), waste heat is assigned a higher weight compared to other factors. Hence, using this logic, waste heat is assigned a weight of 0.35, average hotspot is assigned a weight of 0.3, age is assigned a weight of 0.2, and living area is assigned a weight of 0.15. The following equation is used to calculate the HEAT Score of each house:

$$\text{Heat Score} = \left[(0.35 * \text{Total waste heat}) + (0.3 * \text{Average hotspot}) \right.$$
$$\left. + (0.20 * \text{Age}) + (0.15 * \text{Living area}) \right] * 100 \qquad (11.11)$$

11.3.3.1 Summary of Four Criteria Weights

i. **Waste heat:** Since total waste heat (Equation 11.7) is a ratio, it will always be less than or equal to 1; consequently, it can be simply multiplied by 0.35.

ii. **Average hotspots:** The higher the difference between the hotspot temperature and the WUFI modeled optimal roof temperature, the higher will be the score for the average hotspot temperature. Therefore, the difference between average hotspot temperature and WUFI temperature has to be assessed. Once assessed, this value has to be converted to 0.3 (Equation 11.11). For simplicity, the difference between average hotspot temperature and optimal roof temperature is divided into five classes. If the difference between the optimal roof temperature and the average hotspot temperature is less than or equal to 0, then it is assigned a score of 0; if it is greater than or equal to 5, then it is assigned a score of 0.3; between 0.3 and 0, it is scaled linearly for all other classes (Table 11.3).

TABLE 11.3

Hotspot Classes and Their Weights Used to Define HEAT Scores

Average Hotspot Temperature – Optimal Rooftop Temperature (°C)	Score
>5	0.3
>4 and ≤5	0.25
>3 and ≤4	0.20
>2 and ≤3	0.15
>1 and ≤2	0.10
>0 and ≤1	0.05
≤0	0

iii. ***Age:*** Based on the Thermal Archetypes project (Parekh and Kirney 2012), age is classified into eight vintage groups based on changes to the Canadian building code at related dates: group 1 was buildings built before 1945 and group 8 was buildings built after 2005. Therefore, buildings built before 1945 will get a score of 0.2 while buildings built after 2005 will get an age score of 0. The groups in between will be scaled accordingly (Table 11.4).

iv. ***Living area:*** Living area is classified into five classes according to the Thermal Archetypes project. Group 1 consists of houses that have a living area greater than 230 m² and group 5 consists of houses with a living area less than 83 m². Therefore, houses in group 1 living area will get a weight of 0.15 while houses in group 5 living area will get a weight of 0 (Table 11.5).

For each house, waste heat (Equation 11.7) and Tables 11.3 through 11.5 are used to calculate the corresponding weighted HEAT Scores. This method mitigates the limitations from the other two methods.

TABLE 11.4

Age Groups and Their Weights Used to Define HEAT Scores

Age Group (Year Constructed)	Weight
1 (≤1945)	0.20
2 (>1945–1960)	0.17
3 (>1960–1977)	0.14
4 (>1977–1983)	0.11
5 (>1983–1995)	0.08
6 (>1995–2000)	0.05
7 (>2000–2005)	0.02
8 (>2005)	0

TABLE 11.5
Living Area and Their Weights Used to Define
HEAT Scores

Living Area Group (m²)	Weight
1 (>231)	0.15
2 (>169–231)	0.11
3 (>121–169)	0.07
4 (>83–121)	0.03
5 (≤83)	0

11.3.3.2 Advantages of the Criteria Weights Method

Of all methods developed (thus far), the criteria weights method is considered the most reliable because it has the following advantages:

i. It considers the temperature transfer through roofing materials.
ii. It takes into account the local climatic conditions.
iii. It uses the age and living area of the house.
iv. It allows for a comparison of HEAT Scores across cities.
v. It defines HEAT Scores that are not relative—thus adding, changing, or removing a house will not affect other the scores of other houses—as scores are calculated independently.
vi. A comparison of HEAT Scores across cities is valid since this method will take into account the local weather conditions of the city during the data collection as well as the temperature transfer through the roofing materials common to that city.

Based on these advantages, the criteria weights method is considered the best available method yet developed. Therefore, we evaluated it against the ERS. The following section describes the technique used to evaluate the conceptual validity of this approach.

11.3.4 EVALUATION OF THE CRITERIA WEIGHTS HEAT SCORES METHOD

In Canada, the energy performance of houses is traditionally rated using the ERS. Though service providers increasingly use (handheld) thermal imaging to analyze a house's energy efficiency, to the best of our knowledge, there has been no research carried out using any form of thermal imagery to provide a comparative house-based energy rating system. As a result, HEAT Scores are the first of its kind (at least in Canada), which brings with it the challenge of *validation*. A house's energy efficiency is a complex system, which is affected by many factors including but not limited to lifestyle of the occupants, outdoor weather conditions, type of the house, insulation of the walls, ceiling and foundation, and the type of doors and windows. This is similar to an open natural complex system where uncertainties are inherently present, and where verification and validation of such systems are impossible (Oreskes 1998;

Oreskes et al. 1994). Therefore, HEAT Scores cannot be directly validated. However, the concept or method used to develop HEAT Scores can be evaluated using ERS data. EnerGuide Ratings are calculated with a sophisticated (and complicated) piece of software called *HOT 2000*,* which incorporates many hundreds of parameters, including but not limited to (i) age, (ii) living area, (iii) climatic condition, and (iv) the insulation of ceiling, walls, and foundation. In collaboration with the HEAT project, NRCAN provided a total of 11,940 valid Calgary EnerGuide records. However, because of privacy restrictions, NRCAN did not provide any personal information linking the ratings to specific house addresses; consequently, EnerGuide Ratings cannot be directly related to HEAT Scores. However, a general linear regression is applied to determine the relationship between the EnerGuide Rating of a house and its age, living area, and insulation of ceiling, walls, and foundation. It is *hypothesized* that (the amount of) insulation is directly related to the temperature that is observed on the thermal image. This hypothesis is based on the heat loss equation (Equation 11.10), which suggests that as the insulation decreases, heat loss increases. It must be noted that Allinson (2007) concluded that aerial thermal imagery cannot be used to determine loft insulation thickness. However, the author inferred in their conclusions that there was a small temperature difference recorded on the rooftops between well-insulated and poorly insulated roofs. Therefore, based on this hypothesis, if EnerGuide ratings are inversely related to age and living area, but directly related to the insulation levels of ceiling, walls, and foundations, then the rationale behind the method used to develop HEAT Scores can be accepted as reliable. To test this hypothesis, a linear regression model (Table 11.6) is developed for EnerGuide Ratings.

Results from the regression model (Table 11.6) show that all the independent variables (i.e., age, floor area, etc.) were statistically significant at $p < 0.001$. Even though the R^2 is only 0.31, the F statistic[†] shows that the overall model is significant at the $p < 0.001$ level. Since the model proves to be statistically significant, we next assessed whether it meets the assumptions of linear regression or ordinary least squares. The test results confirm that the linear regression model confirmed to the assumptions of both.

The following can be further deduced from the model (Table 11.6):

  i. As age increases, the EnerGuide Rating decreases.
 ii. As the living area increases, the EnerGuide Rating decreases.
iii. As the insulation of the walls, ceiling, and foundation increases, the EnerGuide Rating also increases.

We suggest that these findings indirectly evaluate the HEAT Scores method. It also needs to be *noted* that while higher EnerGuide Rating numbers are

---

* HOT2000 is a building energy simulation tool that is considered North America's top-of-the-line energy analysis and design software for low-rise residential buildings (source: http://canmetenergy .nrcan.gc.ca/software-tools/hot2000/84).
† Generally, the F statistic is used to test the significance of a model. The p value for this model is less than 0.001, which means that this model is significant at 99.9% confidence level.

TABLE 11.6
Linear Regression Model Defining the Relationship between the EnerGuide House Rating and Age, Living Area, and Insulation of the House

```
Call:
lm(formula = eghrating$eghrating ~ age + floorarea + ceilins +
    mainwallins + fndwallins, subset = eghrating$eghrating >
    0)
Residuals:
    Min      1Q   Median      3Q     Max
-49.143   -3.424   1.035   4.504 102.197

Coefficients:
             Estimate Std. Error t value  Pr(>|t|)
(Intercept) 56.2932579  0.5548025  101.47   <2e-16 ***
age         -0.1115372  0.0051759  -21.55   <2e-16 ***
floorarea   -0.0199596  0.0006694  -29.82   <2e-16 ***
ceilins      1.1135490  0.0479067   23.24   <2e-16 ***
mainwallins  2.5105601  0.1990184   12.62   <2e-16 ***
fndwallins   2.8794770  0.0930716   30.94   <2e-16 ***
---
Signif. codes:  0 '***' 0.001 '**' 0.01 '*' 0.05 '.' 0.1 ' ' 1

Residual standard error: 7.458 on 11934 degrees of freedom
Multiple R-squared: 0.3131, Adjusted R-squared: 0.3129
F-statistic:  1088 on 5 and 11934 DF,  p-value: < 2.2e-16
```

considered as a sign of increasing home energy efficiency, lower HEAT Score values represent more energy-efficient houses (or at least homes wasting less heat). Therefore, as age and living area increase, HEAT Scores will also increase. Insulation values are inversely related to thermal values. Thus, as insulation values decrease, the thermal values recorded in the thermal image will increase. Consequently, HEAT Scores will increase as insulation levels decrease. A limitation of this method is that it requires both structural information (i.e., age and living area) and thermal information (i.e., waste heat and average hotspots)—with structural information seldom accessible from municipalities—especially small ones.

11.4   RESULTS AND DISCUSSION

HEAT Scores can also be calculated for communities and cities. A city HEAT Score is calculated based on the average HEAT Score of all houses in the city, while community HEAT Scores are calculated based on the average HEAT Score for all homes in each community. The following sections describe how HEAT Scores can also be used as stand-alone metrics and visualized as multiscale HEAT maps.

11.4.1   CRITERIA-WEIGHTED MULTISCALE HEAT SCORES
AND USER INTERFACE—CITY LEVEL

HEAT Scores are visualized on a multiscale interface in the HEAT GeoWeb site
(Figure 11.4).

When users log into the HEAT Scores page,* they experience the *City HEAT Map*
(Figure 11.4a) as a smoothly gradated image—with colors varying from red (high
waste heat) to blue (low waste heat). This map is actually an interpolated image of
HEAT Scores using the inverse distance-weighting algorithm. It is created to allow
users to quickly scan over the (pseudo) city and find the hottest locations. That is,
they will be able to quickly see which communities in the city are wasting more heat
and consequently consuming more energy. A city HEAT Score is also provided to
the user, which is a simple average of all the HEAT Scores in the city (Figure 11.4b).
The users will also see a statistical summary of the city including the total number
of homes, total estimated financial cost and GHG emissions, and estimated savings
and GHG reductions for a specific fuel source (Figure 11.4c). They will also see the
distribution of HEAT Scores in the city as a histogram (Figure 11.4d), as well as the
list of all communities in the city (Figure 11.4e).

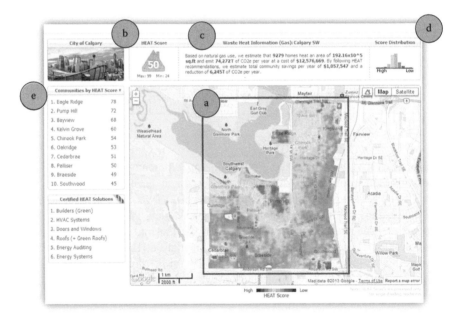

FIGURE 11.4 The HEAT Score web page showing (a) the interpolated city HEAT map with
hot (red) and cold (blue) locations; (b) the selected city/community's HEAT Score; (c) the
associated statistics for estimated total consumption cost, total savings, GHG emissions, and
GHG reductions for space heating—per fuel type, that is, natural gas; (d) the distribution of
HEAT Scores—within the assessed area; and (e) the list of communities in the city ordered
by HEAT Score.

---

* HEAT Scores, http://www.saveheat.co/heat-scores.php

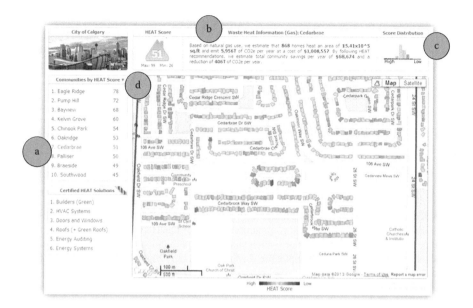

FIGURE 11.5   The HEAT Score web page showing (a) the Cedarbrae community's HEAT Score; (b) community statistics for estimated total cost, savings, GHG emissions, and GHG reductions for space heating—per fuel type, that is, natural gas; (c) the community distribution of HEAT Scores; and (d) the community HEAT map showing houses colored from red to blue representing high–low waste heat per house.

11.4.2   CRITERIA-WEIGHTED MULTISCALE **HEAT** SCORES—COMMUNITY LEVEL

When a user zooms into the next level, they will see a *community HEAT map* (Figure 11.5).

In this map, HEAT Scores are divided into 10 simple equi-interval classes ranging from 0 to 100. These are shown as filled house polygons colored from blue to red: blue represents 0 (i.e., low waste heat), while red represents 100 (i.e., high waste heat). Figure 11.5 shows the houses in the Calgary community of Cedarbrae. Community HEAT maps provide an additional level of detail, allowing a user to quickly scan over individual hot and cold houses within the study area. Similar to the city HEAT map, each community is also provided with a HEAT Score and statistical summary (Figure 11.5a through c).

11.4.3   MULTISCALE **HEAT** SCORES—RESIDENTIAL LEVEL

At the finest scale, when a user clicks on a house, a *pop-up GUI (Graphical User Interface)* is shown with three tabs (Figure 11.6).

The first tab shows the *HEAT Score* of the selected house, along with a comparison of the city and community HEAT Scores. It also provides a relative descriptor and detailed comparative city information related to this ranking. For example, the house in Figure 11.6a has a HEAT Score of 27, which is described as "Moderately

(a)   (b)   (c)

FIGURE 11.6 This figure shows (a) the HEAT Score tab that allows users to compare their house HEAT Score to their city and community scores. (b) The Hotspots tab shows the hottest roof locations on the selected house, which can be linked to Google Street View to identify potential problem areas. For example, the warmest Hotspot (10.3°C) corresponds to the front door and chimney area. (c) The Savings tab shows house specific energy model estimates for savings and GHG reductions—along with environmental equivalents (i.e., *same as planting 44 trees*).

Low Waste Heat." It is also noted that "This house wastes more heat than 290 (3%) other homes in this city." As illustrated in Figure 11.6b, the second tab (*Hotspots*) shows the actual thermal image of the house of interest [colored from red (hot) to cold (blue)] with the 12 hottest locations, shown three at a time (and the ability to scroll through them). Users are also able to see their house with an integrated Google Street view to verify the location of their hotspots. For example, in the Street View image (Figure 11.5b), the corresponding thermal location shows two hotspots and a red-colored area over the front door (and around the chimney), suggesting that these areas are leaking heat and may benefit from further evaluation. The third tab (Figure 11.6c) shows the estimated *Savings* and GHG reductions (210 CAD per year and 1.7 metric tons) if the owner is able to take action that will reduce their heat loss from the average roof (waste heat) temperature (7.8°C), to the minimum roof temperature (6.2°C) shown on the right side of the colored waste heat legend in Figure 11.5b. This information is derived from a house-specific energy model and a specific fuel source (i.e., natural gas)—the detailed description of which is beyond the scope of this paper.

11.4.4 COMPARISON OF THREE DIFFERENT HEAT SCORES

Figure 11.7 shows a different HEAT Score for the same house resulting from the three different methods described in Section 11.3.

This house was constructed in 1910 and has a living area of 330 m². The HEAT Scores developed from the standardized score and logistic regression are very similar, 75 and 76, respectively. However, the HEAT Score developed by assigning weights has rated the house at 89. This is one of the oldest and largest houses in the city (in the absence of information regarding energy efficiency renovations).

(a)   (b)   (c)

FIGURE 11.7 Different HEAT Scores for the same house developed using (a) standard score (z-score), (b) logistic regression, and (c) weights assigned to different factors contributing to the HEAT Score.

Therefore, according to the rationale developed in Section 11.3.3 (regarding age and area), this house will consume a very large amount of energy; thus, it should have a high HEAT Score.

11.4.5 STATISTICAL DISTRIBUTION OF THREE DIFFERENT HEAT SCORES

Figure 11.8 shows the statistical distribution of HEAT Scores developed from the three different methods evaluated.

HEAT Scores developed using the standardized score vary from 1 to 99 and are normally distributed throughout the city (Figure 11.8a). This comes as no surprise as z-scores are normalized scores. The HEAT Scores developed using logistic regression varies from 1 to 76 (Figure 11.8b) and their distribution are skewed to the left—the "cooler" more "energy efficient" portion of the HEAT Score spectrum. The HEAT Scores developed by assigning weights to individual factors vary from

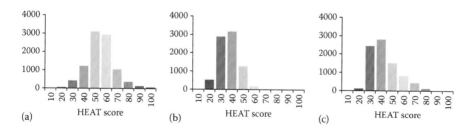

FIGURE 11.8 HEAT Score statistical distributions representing the entire study site derived from (a) standard score or z-score, (b) logistic regression, and (c) weights assigned to different factors.

12 to 89 (Figure 11.8c). This distribution is also skewed to the left but it shows more variation than the HEAT Scores developed using the logistic regression. A well-distributed HEAT Score is preferred as the study site certainly contains a broad range of house ages, living area, and construction types. In addition, it is highly unlikely that any homes will have a perfect score of 1 (as in logistic regression based scores), as all homes waste heat; otherwise, they do not ventilate correctly and can be considered as candidates for the "sick-home" syndrome (Joshi 2008).

From Table 11.7, it can be seen that the community HEAT Scores developed using the *standard score* widely range from 42 to 78, with the community of Eagle Ridge being the "hottest" (i.e., high waste heat). HEAT Scores developed using *logistic regression* range from 31 to 46, with the community of Cedarbrae being the "hottest."

We also note that Pump Hill, which is one of the hottest communities based on *standard scores*, ties for the coldest community according to logistic regression. This is because Pump Hill contains some of the largest houses in the study area and HEAT Scores derived from the standard score were directly related to living area, while HEAT Scores derived from logistic regression are inversely related to living area. HEAT Scores derived from *criteria weighting* range from 31 to 52, resulting in a slightly wider distribution than logistic regression. According to this method, Cedarbrae is the hottest community with a value of 52, which is very close to the value of 51 from the standard score, and ties for the hottest community with a value of 46 from logistic regression.

Based on the HEAT Scores developed from *weights* and from *logistic regression*, Palliser is the coldest community. However, Palliser was not the coldest according to the HEAT Scores developed from standard scores. On further examination, it was revealed that this occurred due to the influence of the hotspots. This is because average hotspot temperatures in this community are closer to the WUFI modeled

---

TABLE 11.7

Community HEAT Scores Developed Using Different Methods: Ranked from Hottest to the Coldest Based on the Standard Scores

Community	No. of Homes	Average HEAT Score Using		
		Standard Score	Logistic Regression	Factors Weights
Eagle ridge	96	78	39	47
Pump hill	405	72	31	41
Bayview	212	68	35	39
Kelvin grove	379	60	36	42
Chinook park	465	54	37	42
Palliser	474	50	31	31
Oakridge	896	53	39	41
Kingsland	753	42	31	32
Haysboro	1868	44	33	37
Braeside	1408	49	38	34
Southwood	1455	45	36	47
Cedarbrae	868	51	46	52

optimal rooftop temperature, which resulted in a low score for hotspots and waste heat, and consequently a low HEAT Score. This could mean that either the WUFI modeled rooftop temperatures for this portion of the study area are in error or that the roofs in this community are not made of asphalt shingles (as assumed)—thus, their related emissivity values and (resulting "true") temperature are incorrectly modeled. By evaluating the rooftops of this community in Google Street view, it is apparent that there are a number of roofs made of cedar; however, there are also many rooftops composed of asphalt shingles. As a result, this issue can only be solved after carefully modeling rooftop temperatures for different kinds of roof systems found in this site and applying an emissivity correction for the corresponding roof materials. Though beyond the scope of this paper, this will require developing a detailed and accurate land cover map (composed of rooftop material classes) and assigning corresponding emissivity values to correct *relative* rooftop temperatures to *true* kinetic temperatures.

Table 11.7 also suggests that there is no clear pattern or trend between the HEAT Scores developed from these methods. This is due to the fact that all these methods are very different from each other. For example, a house that has a very large living area will get a high HEAT Score based on standard score and a low HEAT Score based on logistic regression, while a HEAT Score based on criteria weights assigns only 15% of the weights for living area. Also, age is not considered in HEAT Score based on standard score while the other two methods give age a considerable weight. Therefore, an old house with a large living area is likely to get a high HEAT Score based on criteria weights compared to the other two methods.

11.4.6   Challenges

This project faced a number of challenges. In the following sections, we briefly discuss the effects of (i) geometric correction, (ii) vegetation covering homes, (iii) correct emissivity values, (iv) microclimatic variability, (v) defining rooftop temperature, and (vi) the relationship between hotspots and energy consumption. Though it is beyond the scope of this paper to solve each of these challenges, it is important to recognize the limitations inherent in this research. Furthermore, while the HEAT Score methods described in this paper are intended as improvements over the initial Phase I HEAT proof of concept, their potential for operationalization over larger areas will only be fully realized when solutions to these challenges are implemented.

11.4.6.1   Geometric Correction and HEAT Scores

As described earlier, the thermal data and the cadastral data came from different initial sources acquired at different times and under different conditions. This required them to be geometrically corrected to each other. Even though both data sets were painstakingly geometrically corrected to each other (with 800+ GCPs), there are places where the geometric accuracy is questionable. For example, in Figure 11.9, it can be seen that the overlap between the cadastral data set (gray polygon) and the thermal image (gray tones), while reasonable, is not perfect. These kinds of errors may result in defining hotspots on *the ground* (i.e., the lighter gray "warmer" pixels located within the gray polygon) instead of being located on *the rooftop* (the darker gray "cooler"

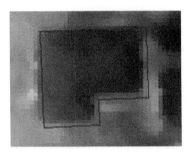

FIGURE 11.9 Errors in geometric correction between the thermal image and cadastral GIS data (images not to scale).

within the gray polygon). This locational error will affect the thermal metrics such as the average hotspot temperatures, which, in turn, will affect the calculation of waste heat and HEAT Scores over the entire image.

11.4.6.2 Vegetation and HEAT Scores

Another challenge is the occurrence of vegetation overlapping many rooftops, thus obscuring the TIR roof signal (Figure 11.10a). In our TIR scene, vegetation appears to be hotter than the rooftops (Figure 11.10b—note the light gray tones of the vegetation located at the bottom of the red polygon). This is due to the high emissivity values of vegetation compared to the rooftops (Brunsell and Gillies 2003). As a result, vegetation covering the roof (and within the GIS polygon) will appear hot, resulting in misidentified hotspots and higher HEAT Scores for these houses.

The solution to this problem is to first remove the trees from the analysis. This could be achieved either by using an NDVI (Normalized Difference Vegetation Index) image created from complementary data to classify the vegetation (such as the 2012 City of Calgary R, G, B, NIR ortho-photo) or by using GEOBIA "feature detection" methodologies to define *house versus tree-objects* instead of using the GIS cadastral data. GEOBIA is the acronym for Geographic Object-Based Image Analysis. "GEOBIA is a sub-discipline of Geographic Information Science (GIScience) devoted to developing automated methods to partition remote sensing imagery into meaningful

(a)   (b)

FIGURE 11.10 A view of trees (bottom right) obscuring part of the (same) rooftop (a) as seen in Google Maps and (b) the thermal image (images not to scale).

image-objects, and assessing their characteristics through spatial, spectral and temporal scales, so as to generate new geographic information in GIS-ready format" (Hay and Castilla 2008). Using GEOBIA methods in combination with vegetation indices and GIS polygons (to automatically define the relative area/object of interest) holds great promise to define house objects and solve the issues with vegetation (as well as with the geometric correction), but it is not a trivial task.

11.4.6.3 Emissivity and HEAT Scores

Based on visual analysis, approximately 80% of all roofs in the study area are estimated to be composed of asphalt shingles. Therefore, a general emissivity value of 0.91 is used for calculating the true temperature of roofs. However, there are some houses with rooftops constructed of metal, cedar shingles, rubber, and concrete. For example, a house that was assigned HEAT Scores of 39 based on asphalt shingles emissivity value was later found to be made of dry concrete. When the emissivity value is adjusted for dry concrete (0.95), the house's HEAT Score changed to 16. We also note that without hyperspectral imagery, it is nontrivial to define a single generic emissivity value for metals, rubber, or concrete roof materials, especially with the variation in building types and the many coatings that exist for these roofing classes, and how they weather over time. In order to solve this problem, a rooftop classification needs to be carried out to define different types/classes of roof materials. Then, each classified roof can be allocated with a more accurate emissivity value. Conversely, hyperspectral imagery could be acquired and used to provide more exact roof material classification—however, there are numerous related costs to these data in terms of increased acquisition and processing costs. Though beyond the scope of this paper, this is also an active area of research by the HEAT research team.

11.4.6.4 Microclimate and HEAT Scores

Another challenge that this project encountered is microclimate variability. Places closer to the valleys and rivers/streams appear colder (especially at night). These places will also have a different amount of humidity compared to other places in the study area. Ji et al. (2012) noted that the effect of an urban river on decreasing local temperatures is high. Therefore, these microclimate conditions need to be mitigated in order to obtain the true kinetic temperature. As previously noted, a novel research solution has recently been proposed (Rahman et al. 2014), which was not available at the time this research was conducted.

11.4.6.5 Defining Optimal Rooftop Temperature for HEAT Scores

Estimates of optimal rooftop temperature for asphalt roof assemblies are modeled using WUFI (see Section 11.3.2). Ideally, these temperatures have to be validated using field measures. Currently, this method produces the best estimates available for this project. In the future, a number of sample field measurements (at least one measurement for each different type of roof material in the city) have to be collected using a TIR camera, ideally while the airborne thermal image is being collected. These measurements need to be observed on new houses where the waste heat is known to be minimal. While taking these measurements, the roof type materials need to be recorded, so that they can be accurately corrected for emissivity.

The WUFI modeled temperatures for different roof assemblies could then be validated with these field measurements.

11.4.6.6 Hotspots and Their Relation to Energy Consumption

The next challenge is that some of the hotspots defined on the rooftops might be due to chimneys or vents whose main purpose is to provide ventilation for hot gases or smoke. In order to minimize their effects, the average of the first six hotspots on the rooftop is used. This smoothes out any high temperatures generated from chimneys/vents. Though it is common sense to appreciate that the hotter the (nighttime) rooftop temperature, the higher the energy consumed by the house, there is no clear relationship between the temperature recorded by the sensor and the energy consumed *inside* the house. In order to overcome this challenge, a TIR camera should observe the rooftop temperatures of two "control" houses over a period of time. The first control house should be an old house, which wastes more heat, and the second one should be a newly constructed house wasting minimal heat. Thermal data recorded over the period of time could then be used in conjunction with the energy consumption data to discover the relationships between them.

11.5 CONCLUSION

On the world scale, Canada is a massive energy consumer (Ménard 2005) and is environmentally ranked second to last among the 25 OECD (Organization for Economic Co-operation and Development) countries reviewed (Gunton et al. 2005). In general, buildings consume 30% of all the energy produced; consequently, they are a major candidate for energy conservation. The majority of this energy is consumed for space and water heating. As a result, space heating (and related heat loss) provides significant opportunities to conserve energy. Behavioral science theories applied to energy analysis suggest that incorporating feedback, social norm, and public commitment into energy efficiency programs is highly effective. Building on behavioral science theories and energy efficiency insights, this paper describes three different methods for implementing HEAT Scores from thermal imagery as a form of feedback to support urban energy efficiency programs. These are (i) the standard score or z-score, (ii) WUFI and logistic regression, and (iii) criteria weighting. The first two methods had important limitations, which led to the development of the third method. In this method, HEAT Scores have been developed by assigning weights to attributes and results have been evaluated against the ERS, which is traditionally used to analyze the performance of houses in Canada. All HEAT Scores have been created within a GeoWeb environment to provide meaningful indirect visual feedback to users in the form of colored maps and simple ranked values (between 0 and 100 representing low and high waste heat, respectively). As a result, individuals will be able to compare their house with all other houses in their community and city. Since the results are easily accessible on a public website, we envision that individuals will follow the social norms of the community/city and encourage healthy competition to reduce their energy consumption. Furthermore, it is expected that people will commit to reduce their energy consumption as nobody wants to be seen wasting energy. City planners may also benefit from using community HEAT Scores to identify high waste heat communities and target their energy retrofit programs. HEAT Scores may also be used to promote web-based "urban energy efficiency" monitoring/competitions between Canadian homeowners, communities, and cities.

11.6 FUTURE WORK

As future work, the weighted criteria HEAT Scores method will be applied to and evaluated over the entire City of Calgary (300,000+ houses). In this paper, the conceptual method of HEAT Scores was evaluated against ERS data. In the future, the HEAT Scores themselves need to be evaluated against ERS data. This may be accomplished either by collecting Volunteered Geographic Information from the users (Abdulkarim et al. 2014) or by obtaining access to ERS data with address information—though currently this is not possible because of privacy rules. In addition to evaluating HEAT Scores, the optimal rooftop temperature defined using WUFI needs to be evaluated using field measurements. These field measurements need to be sampled over a variety of house types including different age groups and roof materials. This study assumed that a majority of the roof materials were asphalt shingles. In the future, a detailed emissivity classification of roof materials needs to be performed in order to generate more accurate estimates of true rooftop kinetic temperature. This will further improve the validity of HEAT Scores, which are based on these temperature values.

ACKNOWLEDGMENTS

The authors would like to acknowledge the Institute for Sustainable Energy, Environment and Economy; Alberta Innovates Technology Futures; the Department of Geography; and the Foothills Facility for Remote Sensing and GIScience at the University of Calgary for providing funding as well as state-of-the-art facilities for conducting this research. The opinions expressed in this work are those of the authors and not necessarily those of the funding agencies.

REFERENCES

Abdulkarim, B., Kamberov, R., and Hay, G. J. 2014. Supporting urban energy efficiency with volunteered roof information and the Google Maps API. *Remote Sens.* 6(10): 9691–9711 (http://www.mdpi.com/2072-4292/6/10/9691).

Aberdeen. 2013. Thermal images of city recorded during flyovers, Aberdeen City Council, England, http://www.aberdeencity.gov.uk/CouncilNews/ci_cns/pr_thermalimages_260213.asp (accessed July 15, 2017).

Allinson, D. 2007. Evaluation of Aerial Thermography to Discriminate Loft Insulation in Residential Housing. PhD Thesis, University of Nottingham, Nottingham, UK. Available online: http://etheses.nottingham.ac.uk/284/, 166 pp.

Ashby, K. V., Nevius, M., Walton, M., and Ceniceros, B. 2010. Behaving ourselves: how behavior change insights are being applied to energy efficiency programs. *American Council for an Energy-Efficient Economy (ACEEE) 2010 Summer Study on Energy Efficiency in Buildings*, pp. 7–16.

Berk, A., Bernstein, L. S., and Robertson, D. C. 1989. MODTRAN: A Moderate Resolution Model for LOWTRA7N, AFGL-TR-89-0122, U.S. Air Force Geophysics Laboratory, Hanscom Air Force Base, Mass.

Blaschke, T., Hay, G. J., Weng, Q., and Resch, B. 2011. Collective sensing: Integrating geo-spatial technologies to understand urban systems—An overview. *Remote Sens.* 3(7): 1743–1776 (http://www.mdpi.com/2072-4292/3/8/1743/).

Briones, J., Greenwood, J., and Vasta, S. 2012. The Market Impact of Accessible Energy Data. MaRS Discovery District. September. Online report, 22 pp. http://www.marsdd .com/wp-content/uploads/2012/10/Accessible-Energy_Report_2012.pdf (accessed July 15, 2017).

Brunsell, N. A. and Gillies, R. R. 2003. Length scale analysis of surface energy fluxes derived from remote sensing. *Journal of Hydrometeorology* 4: 1212–1219.

Calgary Climate Change Accord. 2009. https://www.calgary.ca/UEP/ESM/Documents /ESM-Documents/calgary_climate_change_wecp_cities.pdf?noredirect=1 (accessed July 15, 2017).

Calgary Community Greenhouse Gas (GHG) Reduction Plan. 2011. http://www.calgary .ca/UEP/ESM/Documents/ESM-Documents/Calgary_GHG_Plan_Nov_2011.pdf (accessed July 15, 2017).

Carroll, E., Hatton E., and Brown M. 2009. Residential Energy Use Behavior Change Pilot, Report presented to Joe Plummer, Minnesota Department of Commerce, Office of Energy Security (CMFS project code B21383). Online report, 80 pp.

Cherfas, J. 1991. Skeptics and visionaries examine energy savings. *Science* 251: 54–56.

Cialdini, R. B. 2007. Descriptive social norms as underappreciated sources of social control. *Psychometrika* 72: 263–268.

Colcord, J. 1981. Thermal imagery energy surveys. *Photogrammetric Engineering and Remote Sensing* 47(2): 237–240.

CUI. 2008. Energy Mapping Study. Prepared by the Canadian Urban Institute. December 19, 2008. 92 pp.

Darby, S. 2006. The Effectiveness of Feedback on Energy Consumption. Report published by Environmental Change Institute. University of Oxford, 21 pp.

Denmark. 2011. Thermal Mapping of HEAT-Loss from Buildings, Denmark, http://blomasa .com/news/thermal-mapping-of-heat-loss-from-buildings-denmark.html (accessed July 15, 2017).

Energy Management in Osnabrück. 2016. https://enviropaul.wordpress.com/2016/06/07 /energy-management-in-osnabruck/ (accessed July 15, 2017).

Exeter. 2009. Exeter heat loss survey, Exeter, England, http://www.devonlive.com/exeter -homes-waste-heat/story-11824753-detail/story.html (accessed July 15, 2017).

Goodhew, J., Pahl, S., Auburn, T., and Goodhew, S. 2015. Making heat visible: Promoting energy conservation behaviors through thermal imaging. *Environment and Behavior* 47(10): 1059–1088.

Gunton, T. et al. 2005. The Maple Leaf in the OECD, Comparing Progress toward Sustainability. Vancouver: David Suzuki Foundation.

Haringey. 2007. Home Heat Loss, http://www.haringey.gov.uk/index/housing_and_planning /housing/housingadvice/homeheatloss.htm (accessed July 15, 2017).

Hay, G. J. and Castilla, G. 2008. Geographic Object-Based Image Analysis (GEOBIA): A new name for a new discipline. In *Object-Based Image Analysis–Spatial Concepts for Knowledge-Driven Remote Sensing Applications* (T. Blaschke, S. Lang, and G. J. Hay, editors), Springer-Verlag.

Hay, G. J., Hemachandran, B., and Kyle, C. D. 2010. HEAT (Home Energy Assessment Technologies): Residential waste heat monitoring, Google Maps and airborne thermal imagery. Alberta, Canada. *GIM International* 24(03): 13–15.

Hay, G. J., Kyle, C., Hemachandran, B., Chen, G., Rahman, M. M., Fung, T. S., and Arvai, J. L. 2011. Geospatial technologies to improve urban energy efficiency. *Remote Sens.* 3(7): 1380–1405 (http://www.saveheat.co).

HEAT. 2013. http://www.ucalgary.ca/utoday/issue/2013-11-08/heat-project-takes-grand-prize-mit -conference-global-climate-change?utm_source=UToday&utm_medium=Email&utm _content=textlink&utm_campaign=November-8-2013&utm_term=heat-project-takes -grand-prize-mit-conference-global-climate-change (accessed July 15, 2017).

Hosomura, T. 1994. Geometric correction of NOAA images by triangulation. Proc. SPIE 2357, ISPRS Commission III Symposium: Spatial Information from Digital Photogrammetry and Computer Vision, 387 (August 17, 1994); doi:10.1117/12.182885.

Jensen, J. R. 2007. *Remote Sensing of the Environment: An Earth Resource Perspective 2nd Edition*. Upper Saddle River: Pearson Prentice Hall.

Ji, P., Zhu, C., and Li, S. 2012. Effects of urban river width on the temperature and humidity of nearby green belts in summer. *Yingyong Shengtai Xuebao* 23(3): 679–684. http:// www.ncbi.nlm.nih.gov/pubmed/22720611 (accessed July 15, 2017).

Joshi, S. M. 2008. The sick building syndrome. *Indian J. Occup. Environ. Med.* 12(2): 61–64. doi:10.4103/0019-5278.43262.

Kneizys, F. X., Shettle, E. P., Gallery, W. O. et al. 1983. Atmospheric transmittance/ radiance: Computer code LOWTRAN 6. Air Force Geophysics Laboratory, Environmental Research Paper 846, Hanscom AFB, MA.

Kordjamshidi, M. 2011. House Rating Schemes, Green Energy and Technology, doi:10.1007/978-3-642-15790-5_2, Springer-Verlag Berlin Heidelberg, pp. 9–10.

Mahone, A. and Haley, B. 2011. Overview of Residential Energy Feedback and Behavior based Energy Efficiency, Report Prepared for the Customer Information and Behavior Working Group of the State and Local Energy Efficiency Action Network, Energy and Environmental Economics, Inc., San Francisco, pp. 41–42.

Ménard, M. 2005. Canada, a Big Energy Consumer, a Regional Perspective. Ottawa, Ont.: Statistics Canada. Available from http://www.statcan.gc.ca/pub/11-621-m/11-621 -m2005023-eng.htm (last accessed July 15, 2017).

Nelson, P. E. and Ananian, J. S. D. 2009. Compact asphalt shingle roof systems: Should they be vented? *Journal of ASTM International* 6(4): 6.

Nicole, J. E. 2009. An emissivity modulation method for spatial enhancement of thermal satellite images in urban heat island analysis. *Photogrammetric Engineering & Remote Sensing* 75(5).

Oreskes, N. 1998. Evaluation (not validation) of quantitative models. *Environmental Health Perspectives Supplements* 106(S6): 1453–1460.

Oreskes, N., Shrader-Frechette, K., and Belitz, K. 1994. Verification, validation, and confirmation of numerical models in the Earth Sciences. *Science* 263: 641–646.

Parekh, A. and Kirney, C. 2012. Thermal and Mechanical Systems Descriptors For Simplified Energy Use Evaluation of Canadian Houses, SimBuild Conference Aug 1–3, Madison, Wisconsin, pp. 1–8, http://ibpsa-usa.org/index.php/ibpusa/article/view/441/427 (accessed July 15, 2017).

Paris. 2010. Thermal Mapping, Paris, France, http://thermographie-gpso.webgeoservices .com/viewer/index.php (accessed April 15, 2013).

Peng, C. Y., Lee, K. L., and Ingersoll, G. M. 2002. An introduction to logistic regression analysis and reporting. *The Journal of Educational Research* 96: 1–14.

Portsmouth. 2012. Is your home leaking heat and money? Portsmouth City Council, England, http://www.portsmouththermalmap.bluesky-world.com/ (accessed July 15, 2017).

Rahman, M. M., Hay, G. J., Couloigner, I., and Hemachandaran, B. 2014. Transforming image-objects into multiscale fields: A GEOBIA Approach to Mitigate Urban Microclimatic Variability within H-Res Thermal Infrared Airborne Flight-Lines. *Remote Sens.* 6(12): 9435–9457 (http://www.mdpi.com/2072-4292/6/10/9435).

States of Jersey. 2011. Heat Loss Map of Jersey, States of Jersey, http://www.gov.je/Environment/GenerateEnergy/Energyefficiency/Pages/JerseyHeatLossMap.aspx (accessed July 15, 2017).

Stathopoulou, M., Synnefa, A., Cartalis, C., Santamouris, M., Karlessi, T., and Akbari, H. 2009. A surface heat island study of Athens using high-resolution satellite imagery and measurements of the optical and thermal properties of commonly used building and paving materials. *International Journal of Sustainable Energy* 28: 59–76.

Sugawara, H. and Takamura, T. 2006. Longwave radiation flux from an urban canopy: Evaluation via measurements of directional radiometric temperature. *Remote Sensing of Environment* 104: 226–237.

The 2020 Sustainability Direction. 2010. http://www.calgary.ca/CA/cmo/Documents/2013-0648_ChangesTo2020SusCover_spread_web.pdf (accessed July 15, 2017).

Walpole, R. E., Myers, R. H., Myers, S. L., and Ye, K. 2007. *Probability and Statistics for Engineers & Scientists, Eighth Edition.* Pearson Prentice Hall, pp. 500–503.

Weng, Q. 2001. A remote sensing-GIS evaluation of urban expansion and its impact on surface temperature in the Zhujiang Delta, China. *International Journal of Remote Sensing* 22: 1999–2014.

Worcestershire. 2009. Warmer Worcestershire, Worcestershire Council, England, http://www.warmerworcestershire.com/ (accessed July 15, 2017).

12 Air Quality and Health Monitoring in Urban Areas Using EO and Clinical Data

Andrea Marinoni and Paolo Gamba

CONTENTS

12.1 INTRODUCTION: BACKGROUND AND DRIVING FORCES

In order to retrieve an accurate and reliable quantification of the effects of demographic growth, environmental changes, and development policies, it is required to achieve a precise characterization of human–environment interaction (HEI) [1–3]. In fact, the decision-making process for most welfare policies may take advantage of a better knowledge of these interactions for a proper allocation of the resources on the territory and a more effective environment protection. Similarly, the pharmaceutical sector and social care services may benefit from a more accurate assessment of HEI and achieve a punctual resource planning on the medium and long term, as well as an improved service delivery [2,4].

The main problem is that HEI results from multivariate phenomena. Hence, an accurate HEI understanding may be obtained only by gathering information from several heterogeneous sources. Among them, properly processed Earth observation (EO) data sets (Figure 12.1) play a key role, because of the possibility to extract the

Earth observation framework

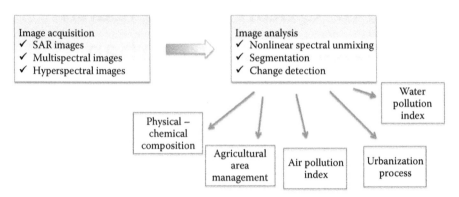

FIGURE 12.1 Outline of Earth Observation (EO) framework. EO processing results can be used to feed a properly designed Big Data architecture for human–environment interaction assessment.

physiochemical composition of the Earth surface from remotely sensed data or to characterize atmosphere constituents and water quality [2,5–7].

As another important example, EO data analysis allows a more accurate assessment of the anthropogenic impact on the environment by continuously monitoring (e.g., on a monthly basis) the urbanization process. This is particularly relevant at the global level because in the last part of the twentieth century, the proportion of the world's population living in urban areas has grown from 14% to more than 50% [8–10]. This dramatic change has affected public hygiene, environmental sanitation, access to healthcare, wealth and employment distribution, and workforce structures. It has also led to clear alterations in disease, dietary, and physical activity patterns. The interaction of remarkable economic and social transformations (associated with greater geographic mobility, lower fertility, longer life expectancy, and population ageing) has led the urbanization processes through history [11–14]. Urbanization has provided several socioeconomic outcomes, such as new jobs and new opportunities for millions of people. Indeed, urban living has been frequently observed to be correlated to, associated with, and affecting other everyday life aspects, such as higher levels of literacy and education, better health, greater access to social services, and enhanced opportunities for cultural and political participation [8,9,12,14].

However, it is also true that urban growth might badly affect sustainable development if it shows up fast and unplanned. In this case, the benefits of city life might not be equitably shared, since key infrastructures and policies might not be properly developed and timely implemented. Urban areas typically record higher inequality levels than rural areas, leading to substandard conditions for hundreds of millions of the world's urban poor people [8,9,15].

Important undesired outcomes of unplanned or inadequately managed urban expansion are unsustainable production and consumption patterns, as well as pollution and environmental degradation. Specifically, rapid urbanization affects the resource base, since it causes the need for energy, water, and sanitation to rise and the demand for public services, education, and healthcare to increase [3,11,12].

Hence, the integral and proper implementation of the three pillars of sustainable development (i.e., economic development, social development, and environmental protection) does require urbanization to be properly assessed and managed [16–18]. Thus, it is fundamental to provide and process accurate, consistent, and timely data on global trends in urbanization and city growth in order to address current and future needs with respect to urban growth. Moreover, these data would be critical to foster inclusive and equitable urban and rural development [6,7,19].

As cities gather the majority of nations' economic activity, commerce, transportation, and government, they are definitely crucial in terms of regional and national development and poverty reduction. Moreover, since urban extents act as links to rural areas, other cities, and across international borders, monitoring (ir)regularities in city sprawl behaviors has implications on welfare characterization [11,12,19,20].

To this aim, investigating regularities and singular patterns in EO data sets enriched with other records (collected from environmental data, clinical and epidemiological records, climatic changes, demographic mapping, population density distribution, availability of commodities and facilities, census, and productivity evaluations) may provide effective characterization of how HEIs are taking place in a given region (Figure 12.2). This approach can be positively considered as an instance of Big Data mining. In fact, it strongly requires inferring information from large-scale high-variability high-variety records [2,4].

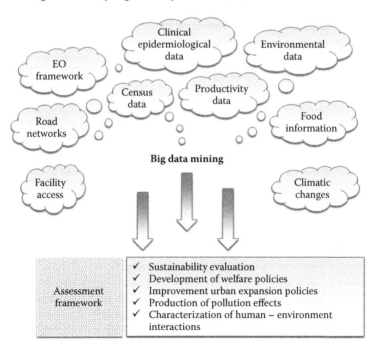

FIGURE 12.2 Outline of the proposed Big Data architecture for human–environment interaction assessment.

As an example, the study of the effects and impacts of exposure factors over the health pattern of people in an urbanized area is a key research issue. This research field is called exposomics (from exposure and –omics) and aims at providing a complete characterization of the external exposure of human being to environmental factors [21]. A complete spatial–temporal description of the whole set of agents that are involved in the analysis is required for the ultimate goal of exposomics, and here, EO/Big Data analysis can be extremely useful [22–25].

In order to effectively implement this analysis, all clinical, medical, and epidemiological data sets must be geo-referenced. Further, exposure investigation requires environmental data to show a very fine resolution over space and time. However, it would be very difficult to match this constraint by using data from ground stations only, since they can outline the environmental situation over geographically limited areas [25,26].

EO is strategic in an exposomics framework. A proper processing of remotely sensed images provides reliable and effective information on wide spatial regions. Hence, EO represents a valid asset in an HEI assessment architecture. In fact, by means of specific techniques for data exploitation, it is possible to retrieve from EOs effective estimations and evaluations on the physical–chemical composition of the given scene, as air quality and water pollution ratios [2,6,7,26,27].

Hence, urban remote sensing plays a key role in accurate HEI assessment over anthropogenic extents, which identifies an important trend and development topic for societal, economic, and welfare challenges. Proper processing of thermal infrared records in remotely sensed data provides information on black particulate concentration. These water/air quality maps can be used to investigate the impact and effects of water/air pollution on several aspects of urban living, such as primary healthcare, active transportation, and quantitative exposure to pollutants. Therefore, sustainable development in urban areas definitely takes advantage of the outcomes produced by remotely sensed data processing and investigation. Further, it identifies a crucial tool to address a fundamental challenge for future urban growth, planning, and evolution [2,3,19,28].

Section 12.2 introduces the framework used to retrieve accurate and reliable estimate of air quality from multispectral EO satellite data. Moreover, it also provides a brief description of an architecture of data-driven discovery used to extract information about HEI from heterogeneous data sets.

12.2 METHODS

In this section, a framework for obtaining precise assessment of air quality over wide areas from multispectral EO satellite data is introduced. Specifically, by properly calibrating and processing the remotely sensed infrared contribution, it is possible to provide a thorough and reliable characterization of the state of atmospheric pollution over the considered area. This outcome is then used to infer information on the role of air quality on special health issues, such as those delivered by patients affected by diabetes-related complications. Further, Section 12.2.2 reports a brief introduction to a data-driven approach for detection of hidden regularities within heterogeneous records. This framework aims at enhancing the investigation and exploration of heterogeneous data clouds with an assumption-free and model-free approach, allowing, in this case, finding a precise description of the regularities that are shared by air quality trends, glycated hemoglobin, and body mass index (BMI) behavior.

12.2.1 Retrieving Air Quality Maps from EO Data

The radiance processes that affect the energy transmission through the atmosphere is at the basis of the correlation between the presence of black particulate and the temperature measured at the level of the ground. In detail, the atmospheric transmission factor decreases with the occurrence of a pollution layer. Because the solar heating is decreased accordingly, this effect can be measured on the thermal infrared contribution of multispectral remotely sensed records. The energy that is radiated upward is dramatically impoverished, because of the absorption provided by the pollution layer itself on the emitted radiance [4,24,25,29].

A plot of temperature records against black particulate concentration measured at geo-referenced ground stations very clearly shows this correlation. Specifically, the mathematical relationship between thermal raw counts and the black particulate concentration is polynomial. By training this model using the selected ground station measures, air quality estimates for every point of a satellite scene can be retrieved [25,26].

It is worth noting that several algorithms using regression on aerosol optical depth (AOD) products have been proposed in technical literature (see, for instance, Ref. [30]). These architectures aim at describing the on-ground air quality distribution by employing machine learning and statistical frameworks from AOD maps over wide areas, where ground stations are used to calibrate the records. However, studies have shown that the relationship between black particulate and AOD is not always suitable for simple regression models. This effect results from the large number of parameters and biases affecting the data. Hence, this approach might lead to inaccurate estimates when scarce prior knowledge of the meteorological and climate conditions of the given area is available [31,32]. The architecture we report in this chapter is able to overcome these issues, since its dependence on geometrical and meteorological parameters is less distinct [24].

12.2.2 Investigating Heterogeneous Data by Information Theory–Based Approach

Once maps of spectral reflectance, thermal maps, and (estimated) black particulate concentrations are available, information extraction from these data becomes of primary importance. To this aim, spatial extraction and recognition may be implemented using pattern mining techniques. Following the technical literature, several models and protocols for data investigation, classification, and clustering may be exploited [4,33–36]. Unfortunately, when applied to complex data sets such as those obtained from high-variability high-variety sources of information, classic architectures show some drawbacks, especially as a result of the trade-off between accuracy and computational complexity, a typical issue in Big Data science [37–41]. Accordingly, a brand new class of algorithms based on assumption-free and model-free data exploration has started to emerge, able to perform pattern recognition while overcoming problems in data investigation attributed to forced linear projections in the samples–labels mapping, to overfitting, or to inadequate regularization and inaccurate dimensionality reduction [2,42].

In order to effectively counteract these issues, we have explored information theory–based pattern recognition that can provide a substantial paradigm shift. Indeed, since information theory–based methods use proper information rate metrics and manage the feature relationships in order to optimize the data inference, they guarantee at the same time that data information content is preserved. Thus, information theory–based pattern recognition is very suited to affinity analysis in complex data sets. To this aim, we propose to build a bipartite graph representation of the interactions among the samples [42].

Let us take into account a data set that consists of P samples and S features. Each record can be represented as a row of a $P \times S$ matrix $\underline{D} = \left\{ \underline{D}_i \right\}_{i=1,\ldots,P}$, $\underline{D}_i = \left[D_{ij} \right]_{j=1,\ldots,S}$, $D_{ij} \in \Gamma \subseteq \Re$. This system can be arranged according to a bipartite graph representation, where the two families of nodes, namely, p-nodes and s-nodes, associated with samples and features, respectively. The edge distribution depends on the elements D_{ij}. Specifically, if D_{ij} is called to be not relevant by data cleansing, the ith p-node is not linked to the jth s-node. Otherwise, they are linked by an edge, whose weight is the value of D_{ij} itself.

Therefore, in order to complete the description of the considered data set, we need to define the degree of each node. The degree of the ith p-node is the number of relevant features that the ith sample shows, that is, $d_{p_i} = \sum_{j=1}^{S} \varphi \left(\left| D_{ij} \right| \right)$, where $\varphi(x) = 1 \leftrightarrow x > 0$. Analogously, the degree of the jth s-node is defined as the number of samples showing relevant records over the jth attribute, that is, $d_{s_j} = \sum_{i=1}^{P} \varphi \left(\left| D_{ij} \right| \right)$. It is also possible to set their average aggregates, that is, the *average p-node degree* and *average s-node degree*, defined as $d_{\bar{p}} = \sum_{i=1}^{P} d_{p_i} / P$ and $d_s = \sum_{j=1}^{S} d_{s_j} / S$, respectively.

Let us now define a measure of the affinity between samples. To this aim, let us consider $\varepsilon_j \in \Re_{\geq 0}$, tolerance factor (i.e., a threshold) on the jth attribute. A set Z of z samples, with $|Z| = z \leq P$, shows affinity over the jth attribute if $\forall \ (k, l) \in Z \subseteq \{1, \ldots, P\}$:

$$\left\| D_{kj} - D_{lj} \right\| \leq \varepsilon_j. \tag{12.1}$$

It is also possible to define an *affinity criterion*, which consists of a set of thresholds $T = \left[\varepsilon_j \right]_{j \in \Sigma}$, $\Sigma \subseteq \{1, \ldots, S\}$. Accordingly, a *local affinity pattern* (LAP) is defined as a set Π of $|\Pi| \leq P$ samples that are affine over a set Σ of attributes [42].

In order to visualize these quantities over a practical example, let us consider the structure in Figure 12.3. It reports a data set $\underline{D}' = \left\{ D'_{ij} \right\}_{(i,j) \in \{1,\ldots,P\} \times \{1,\ldots,S\}}$, with $P = 5$, $S = 10$, $D'_{ij} \in \{0,1,2\}$, where $D'_{ij} = 0$ if the jth attribute of the ith simple has been considered as not significant. In this graph, s-nodes are shown as triangles, while p-nodes

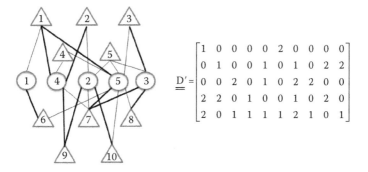

$$D' = \begin{bmatrix} 1 & 0 & 0 & 0 & 0 & 2 & 0 & 0 & 0 & 0 \\ 0 & 1 & 0 & 0 & 1 & 0 & 1 & 0 & 2 & 2 \\ 0 & 0 & 2 & 0 & 1 & 0 & 2 & 2 & 0 & 0 \\ 2 & 2 & 0 & 1 & 0 & 0 & 1 & 0 & 2 & 0 \\ 2 & 0 & 1 & 1 & 1 & 1 & 2 & 1 & 0 & 1 \end{bmatrix}$$

FIGURE 12.3   Bipartite graph representation of data set $\underline{D}'$.

are identified by circles: the edges that connect the nodes show different thickness according to their weight. Note that in this data set, $d_{p_2} = 5$, while $d_{s_5} = 3$.

If we consider the graphical representation of the bipartite graph, we can assume the whole data set as a polymer of two atoms (p- and s-nodes). In this framework, LAPs are "molecules" in this system. Therefore, the approach for global detection of LAPs has been named as "progressive molecule detection algorithm," or PROMODE in short [42].

Once the graph is built, the algorithm for data-driven LAP discovery works progressively over this bipartite graph. In fact, it iterates for each p-node two main functions. Basically, when considering the ith p-node, PROMODE discards all the edges that do not lead to a LAP. This operation can be performed by considering the weights of the edges connecting two p-nodes and an s-node. In fact, it is true that if $\forall t = 1,..., S$, each D_{mt} is not affine to D_{it}, then the mth p-node cannot be involved in any LAP entailing p-node i. In other terms, an s-node cannot be involved in any LAP if it connects two p-nodes that do not show local affinity.

Once this sieving procedure is completed, then PROMODE searches for the LAPs that insist on the ith p-node scanning every possible combination of the surviving p-nodes that fulfill the given affinity criteria. Hence, if an s-node does not show affinity features with respect to the considered combination of p-nodes with the ith p-node, it is discarded from the loop. This process is iteratively performed in order to explore all the LAPs in the data set.

From an information theory standpoint, it is possible to appreciate that the PROMODE workflow is equivalent to the search for the bipartite graph topologies that show the minimum extrinsic message degree. Indeed, the LAP search performed by PROMODE in order to achieve complete affinity mining can be seen as weight balance of the messages that flow within the base bipartite graph. Moreover, in this information theory–based scenario, the nodes that PROMODE discards are those that actually would lead to an increase of the extrinsic message degree within the graph [42,43].

Finally, it is possible to prove that this approach is actually able to outperform classic data mining algorithms, hence boosting the ratio between accuracy and computational complexity [42].

12.3   EXPERIMENTAL RESULTS

In order to demonstrate the power of the previously introduced approach, we applied
the architectures in Section 12.2 to two different test scenarios. The following subsections report the outcomes of the proposed analysis.

12.3.1   ESTIMATING THE EFFECT OF AIR POLLUTION
ON PATIENTS AFFECTED BY DIABETES

The first example is a set of large-scale heterogeneous data consisting of EO multispectral records, environmental data, clinical records, and administrative reports.
These data have been collected over the Province of Pavia, Italy, in the 2009–2014
time frame. The Province of Pavia, a highly urbanized area in the Lombardy region,
is placed in northwestern Italy and covers an area of 2968.64 km². It has 189 municipalities, grouped into 9 districts: Figure 12.4 reports the RGB composite of the considered region obtained from Landsat data in summer.

On this area, we collected more than 100 Landsat images (35 of them with less
than 5% cloud cover). Each Landsat image consists of 2800 × 2800 pixels and has a
30-m spatial resolution. In order to provide reliable estimates of the quantity of black
particulate concentrated over a given area, we considered the thermal infrared contribution of these multispectral records, according to the scheme in Section 12.2.1.
We used the PM10 concentration measured by the ground stations that are located
in Pavia and four towns within its second-order administrative area (i.e., Voghera,
Vigevano, Parona, and Sannazzaro de' Burgondi) to calibrate the polynomial fitting
used to retrieve the air quality maps.

Furthermore, we searched whether it was possible to obtain relevant information
on HEI for the whole Province of Pavia by searching for hidden regularities between

FIGURE 12.4   RGB color composite of a Landsat scene for the province of Pavia acquired
on July 19, 2014.

Landsat-based air pollution maps and clinical data collected on the ground. These clinical data refer to a cohort of patients living in this area who are affected by diabetes and who have been monitored for the whole considered time frame. Among the amount of tests and metrics routinely checked, glycated hemoglobin and BMIs play a key role and were thus used in this research. All the clinical records are time- and geo-referenced at the municipality level. Moreover, statistically relevant measures have been achieved by aggregating glycated hemoglobin and BMI values for each year.

The top part of Figure 12.5 thoroughly reports all the steps of the proposed analysis. At the bottom left corner of this figure, the thermal infrared raw counts of the images collected on four seasons in the 2009–2014 time frame are displayed as a function of the black particulate PM10. As clearly seen, the air pollution affects thermal infrared contribution according to the structure described in Section 12.2.1, since black particulate concentration is indeed inversely proportional to the recorded reflectance signals. Hence, air quality maps can be retrieved by properly processing the multispectral remote sensing images after calibrating the relationship polynomial model between these two quantities.

In more detail, we aim at achieving maps of the air quality on the Province of Pavia according to the calibration graph displayed in Figure 12.5. This figure also reports in a black solid line the polynomial fitting that is obtained when considering the measured black particulate concentration as a function of the raw counts provided by remotely sensed thermal infrared imagery. It is possible to appreciate that this fitting is very precise and reliable, with a confidence coefficient $R^2 = 0.9$.

Moreover, at the bottom center of Figure 12.5, the distribution of the estimated air pollution over the Province of Pavia aggregated over the 2009–2014 time frame is reported. The actual distribution of the air pollution over peculiar areas of the Province is immediately visible. Areas in red, orange, and blue boxes show an air quality situation definitely worse than the rest of the region. These areas are located around particular anthropogenic extents and activities, so it is very interesting to investigate their air quality distribution.

Indeed, the red box encompasses the area around one of the largest oil refinery in the whole northwestern Italy. Similarly, the blue box denotes an area that involves a state highway and a chemical factory. Further, a cheese factory, a concrete factory, and an incinerator are placed in correspondence to the three locations showing bad air quality within the orange box. These examples therefore provide remarkable motivations on grounds of the strong correlation between air pollution and thermal infrared contribution in multispectral images.

Thanks to these data, it is possible to explore the inference analysis over the glycated hemoglobin data set. Air quality estimates retrieved from the thermal infrared contribution processing were pooled at a seasonal scale, so that statistically reliable evaluations can be drawn. Additionally, the outcomes were aggregated on a spatial dimension as well, since the clinic records have been acquired and stored at a district-level scale over the Province of Pavia. Figure 12.5 (bottom right) shows the results of this analysis. Specifically, seasonal air quality estimates for 2011 and 2012 are shown. Hence, it is clear that the estimated air quality trend is able to emphasize the different situations occurring over the region that is taken into account. Further, we point out

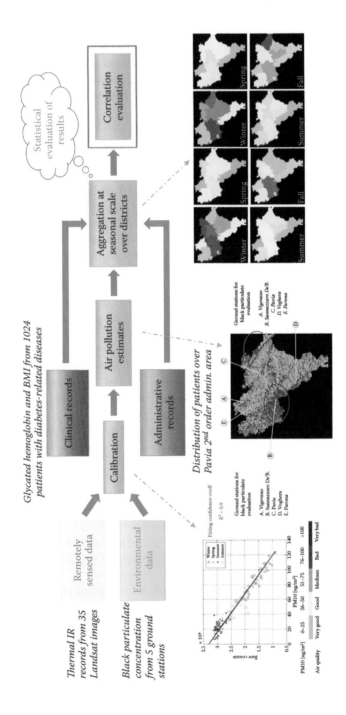

FIGURE 12.5 Basic workflow of the framework used for integration of heterogeneous data sets collecting clinical, administrative, environmental, and remote sensing records to assess the impact of air quality on human health.

that the contribution to air quality provided by anthropogenic extents has an effect on wider geographical areas, as seen in Figure 12.5 (bottom center).

It is possible to note that a remarkable correlation between air pollution computed from remotely sensed data processing and clinical records behavior through the considered time frame can be detected from the experimental results. In fact, it is possible to determine a clear link between the estimate distribution of air pollutants and the pattern of hospitalizations of those patients. Moreover, the onsets of microvascular diseases can be characterized by means of the strong connection between glycated hemoglobin trend and air quality estimate distribution (Figure 12.6). As a matter of fact, these two quantities are coupled with the correlation factor around 91%; therefore, a robust statistical association between air quality estimates and health outcomes can be retrieved by means of the proposed approach. Further, four districts of the Province of Pavia reported a significant drift from the mean values that characterize the clinical records behavior. The fluctuations of the data scores through the clinical data set are studied by computing a p-value significance test from mixed model analysis. This approach shows that the estimated air quality maps deliver a strong support to the behavior of high relevance samples, that is, the records that follow particular paths with respect to the mean distribution of the glycated hemoglobin values.

Hence, the actual impact of air pollution on human health can be understood and quantified by the correlation that has been obtained from the framework for heterogeneous data sets analysis. Further interplays and regularities can be retrieved by exploring and investigating enlarged data sets in an architecture for large-scale heterogeneous records analysis, such as the one in Section 12.2.2. The next section reports the results achieved by running PROMODE algorithm on the Province of Pavia data cloud.

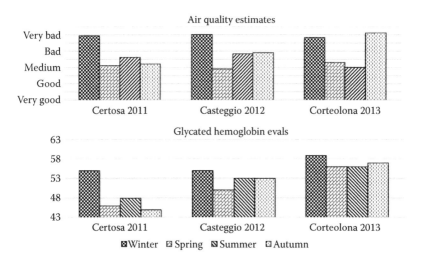

FIGURE 12.6   Air pollution as estimated by remotely sensed data and glycated hemoglobin behaviors over three towns within the second-order administrative area of Pavia, Italy.

12.3.2 AFFINE PATTERN SEARCH IN MULTISPECTRAL REMOTELY SENSED DATA ENLARGED WITH CLINICAL RECORDS

We run PROMODE on the Province of Pavia data set in order to simultaneously characterize the hidden regularities shared among remote sensing, environmental, clinical, and administrative data. Specifically, the data cloud processed by means of PROMODE architecture consists of the inverse of thermal infrared contribution collected by the Landsat sensor, as well as yearly and municipality-aggregated glycated hemoglobin and BMI values of diabetes-suffering patients.

This means that each sample in the data set identifies an area of the Province of Pavia of 900 square meters and is described by 35 spectral features from EOs and 12 clinical and administrative features, summing up to a total of 47 attributes. Taking into account the notation introduced in Section 12.2.2, the average p-node and s-node degree profiles are $(d_{\bar{p}}, d_{\bar{s}}) = (44, 7{,}836{,}822)$.

We expect that the data-driven investigation carried out by extracting affinity patterns will provide relevant information about zones highly affected by air pollution with or without reporting health disease onsets recorded in diabetes-suffering patients by their blood glucose level. It is known that air pollution and black particulate concentration directly affect glycated hemoglobin level and BMI [12,14,26,44]. Thus, critical situations from a HEI perspective can be identified and highlighted by the detection of affinity patterns showing high values for all the features considered in this data set.

To this aim, we take into account the normalized LAP energy in order to employ a reliable metric to accurately classify and segment affinity patterns. In detail, let us consider the αth LAP induced by a set Π_α of samples over a set Σ_α of features through a data set $\underline{D}$. Its average normalized energy E_α is defined as

$$E_\alpha = \frac{1}{|\Pi_\alpha| \cdot |\Sigma_\alpha|} \sum_{(\rho,\sigma) \in \{\Pi_\alpha \times \Sigma_\alpha\}} \left| \frac{D_{\rho\sigma}}{L_\sigma^+} \right|^2, \tag{12.2}$$

where $L_\sigma^+ = \max\limits_{p} \{D_{p\sigma}\}_{p=1,\dots,P}$ and P is the total number of samples.

After running PROMODE to extract every possible pattern, only those characterized by a large normalized LAP energy metric are retained. LAP detection is performed with a tolerance factor that is set according to the distribution of each feature in the data cloud. Specifically, if we consider two samples (namely, a and b) in data set $\underline{D}$, then we call an affinity on the lth feature between them if $\|D_{al} - D_{bl}\| < 0.02 \cdot L_l^-$, where $L_l^- = \min\limits_{p} \{D_{pl}\}_{p=1,\dots,P}$.

When all LAPs have been detected, each LAP energy contribution is computed according to Equation 12.2. Then, the energies are sorted in the interval $\Delta_E = \left[\min(\underline{E}), \; \max(\underline{E}) \right]$, where $\underline{E} = \{E_\alpha\}_{\alpha=1,\dots,\hat{\alpha}}$ and $\hat{\alpha}$ is the total number of detected LAPs. Finally, Δ_E is quantized in δ_E levels that are uniformly drawn. Thus, we assign every sample (pixel) to a power class by applying a majority rule, as we take into account the occurrence of the LAPs that involves it throughout the δ_E slots on Δ_E.

The classification results we achieved over the Province of Pavia data set by set-ting $\delta_E = 20$ are shown in Figure 12.7. Since the proposed method aims at identify-ing the HEI-critical regions (i.e., regions with high values of air pollution, glycated hemoglobin, and BMI) over the considered region, pixels belonging to high energy class are easy to be recognized. Thus, the proposed approach allows the identifi-cation of hidden spatial patterns that emphasize the statistical association between health records and air quality. Indeed, it is possible to appreciate that the LAP high-energy zones show up in correspondence to air pollution agents like oil refineries, factories, highways, and dense urban settlements.

For instance, let us consider the area in the left circle of Figure 12.7. It encompasses the region close to the village of Sannazzaro de' Burgondi, which lies within an area of rice and wheat fields. For this reason, the air quality and the glycated hemoglobin levels are reasonably expected to be very good. However, the LAP energy outcomes report a robust concentration of air pollution as well as values of glycated hemoglobin and BMI that are definitely off the regular charts. Indeed, it is possible to trace back this result to the oil refinery that is located nearby this village. The gas plumes pro-duced by this plant might have induced degradation of the air quality as well as of the glycated hemoglobin scores of the patients living in the region around it.

As another example, let us now consider the other area highlighted in Figure 12.7. It identifies the zone surrounding the town of Landriano. A rural landscape mainly characterizes this area, with a lot of rice fields and small farms. Thus, as in the previ-ous example, good air quality and clinical records are supposed to be found. Instead, robust air pollution estimates and high glycated hemoglobin scores describe the

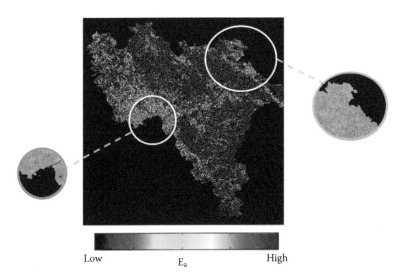

FIGURE 12.7 Classification of the pixels according to the grouping based on the joint like-lihood of air pollution, glycated hemoglobin, and BMIs obtained using the LAP energy. LAP detection is performed by means of adaptive affinity criterion, as in Section 12.3.2. Details of the area around the towns of Sannazzaro de' Burgondi (left) and Landriano (right) are pointed out.

patch. If we focus more on the anthropogenic extents in the area, we can see that the high LAP energy pixels are distributed neatly along the A1 highway and the SS412 freeway, which are typically very crowded and congested. Analogously, HEI-critical pixels are detected around a chemical factory and in the town of Landriano itself.

12.4   CONCLUSIONS AND FINAL REMARKS

Accurate characterization of air quality can be reliably and accurately retrieved by means of the proper processing of remotely sensed data. Reliable maps over wide regions can be retrieved by taking advantage of the physical processes induced by black particulate pollutants to the emitted radiance that is acquired by multispectral sensors. These air quality distributions identify a strong support for precise description of the air pollution in urban systems.

Indeed, relevant information on HEI assessment and Big Data exploration over data sets was collected by using EO, environmental, clinical, and administrative records. In fact, investigating these heterogeneous features over wide regions provides useful insight for a deep understanding of the interplay between air pollution and some specific diseases like diabetes. Thus, this approach represents a valid structure to provide information on the impact of air pollution on human health, as well as on the influence of anthropogenic extents on black particulate concentration.

Experimental investigations have been carried out over real data sets collected on the wide urbanized area of the Province of Pavia, Italy, to prove the actual consistency of the aforesaid interactions. Further, a new data mining paradigm relying on information theory–based data-driven pattern recognition has been used to obtain accurate characterization of the hidden regularities in the considered large-scale heterogeneous data set. The achieved outcomes show that the proposed architecture is able to deliver reliable estimates of the relevant patterns of air quality, glycated hemoglobin, and BMI. Therefore, this scheme identifies an actual framework for HEI assessment, and it can be used to foster and boost precise urban growth, planning, and evolution according to sustainable development.

REFERENCES

1. M. Faber, N. Niemes and G. Stephan, *Entropy, Environment and Resources*, Berlin, D: Springer Verlag, 2nd edition, 1995.
2. A. Marinoni and P. Gamba, Big data for human–environment interaction assessment: Challenges and opportunities, in *Proc. of 2014 ESA Big Data from Space Conference (BiDS)*, Frascati, Italy, 2014.
3. A. Garcia, D. Pindolia, K. Lopiano and A. Tatem, Modeling internal migration flows in sub-saharan Africa using census microdata, *Migration Studies*, vol. 1, pp. 1–22, 2014.
4. R. R. Vatsavai, A. Ganguly, V. Chandola, A. Stefanidis, S. Klasky and S. Shekhar, Spatiotemporal data mining in the era of big spatial data: Algorithms and applications, in *Proc. ACM SIGSPATIAL Int. Workshop Anal. Big Geospatial Data*, Redondo Beach, CA, 2012.
5. Y. Kaufman, A. Wald, L. Remer, B.-C. Gao, R.-R. Li and L. Flynn, The MODIS 2.1-μm channel-correlation with visible reflectance for use in remote sensing of aerosol, *IEEE Trans. Geosci. Remote Sens.*, vol. 35, no. 5, pp. 1286–1298, 1997.

6. A. Marinoni and P. Gamba, A novel approach for efficient p-linear hyperspectral unmixing, *IEEE Journal of Selected Topics on Signal Processing*, vol. 9, no. 6, pp. 1156–1168, 2015.

7. A. Marinoni, J. Plaza, A. Plaza and P. Gamba, Nonlinear hyperspectral unmixing using nonlinearity order estimation and polytope decomposition, *IEEE Journal of Selected Topics in Applied Earth Observations and Remote Sensing*, vol. 8, no. 6, p. 2644–2654, 2015.

8. World Energy Council, World energy issue monitor 2015, WEC-WEI, 2015.

9. UN Department of Economic and Social Affairs, World Urbanization prospects: The 2014 revision, UN Population Division Publications, 2014.

10. UN Department of Economic and Social Affairs, World Economic and Social Survey 2013: Sustainable Development Challenges, UN Population Division Publications, 2013.

11. K. Borowiecki, Geographic clustering and productivity: An instrumental variable approach for classical composers, *Elsevier Journal of Urban Economics*, vol. 73, no. 1, pp. 94–110, 2013.

12. S. Allender, C. Foster, L. Hutchinson and C. Arambepola, Quantification of urbanization in relation to chronic diseases in developing countries: A systematic review, *Journal of Urban Health: Bulletin of the New York Academy of Medicine*, vol. 85, no. 6, pp. 938–951, 2008.

13. G. Meehl and C. Tebaldi, More intense, more frequent, and longer lasting heat waves in the 21st century, *Science*, vol. 305, no. 5686, pp. 994–997, 2004.

14. J. F. Pearson, C. Bachireddy, S. Shyamprasad, A. B. Goldfine and J. S. Brownstein, Association between fine particulate matter and diabetes prevalence in the U.S., *Diabetes Care*, vol. 33, no. 10, pp. 2196–2201, 2010.

15. C. Watson, Trends in world urbanisation, in *International Conference on Insect Pests in the Urban Environment (ICIPUE)*, Cambridge, UK, 1993.

16. R. Lawrance and M. Labus, Early detection of douglas-fir beetle infestation with sub canopy resolution hyperspectral imagery, *Western Journal of Applied Forestry*, vol. 18, pp. 202–206, 2003.

17. D. Pimentel, C. Harvey, P. Resosudarmo, K. Sinclair, D. Kurz, M. McNair, S. Crist, L. Shpritz, L. Fitton, R. Saffouri and R. Blair, Environmental and economic costs of soil erosion and conservation benefits, *Science*, vol. 267, pp. 1117–1123, 1995.

18. P. Pinter, J. Hatfield, J. Schespers, E. Barnes and M. Moran, Remote sensing for crop management, *Photogrammetric Engineering & Remote Sensing*, vol. 69, pp. 647–664, 2003.

19. S. Chakravortty, D. Sinha and A. Bhondekar, Assessment of urbanization of an area with hyperspectral image data, *FICTA*, vol. 2, pp. 315–322, 2014.

20. C. Jacobson, Identification and quantification of the hydrological impacts of imperviousness in urban catchments: A review, *Journal on Environ. Manag.*, vol. 92, no. 6, pp. 1438–1448, 2011.

21. S. Martin, Exposome informatics: Considerations for the design of future biomedical research information systems, *J Am Med Inform Assoc.*, vol. 21, no. 3, pp. 386–390, 2014.

22. C. Yang, Airborne hyperspectral imagery for mapping crop yield variability, *Geography Compass*, vol. 3, no. 5, pp. 1717–1731, 2009.

23. C. Yang and Y. Chang, Assessing disease severity of rice bacterial leaf blight with canopy hyperspectral reflectance, *Precision Agriculture*, vol. 11, pp. 61–81, 2010.

24. L. Wald, L. Basly and J.-M. Baleynaud, Satellite data for the air pollution mapping, in *18th EARSeL Symposium on Operational Remote Sensing for Sustainable Development*, Enschede, The Netherlands, 1998.

25. A. Marinoni, A. Dagliati, R. Bellazzi and P. Gamba, Inferring air quality maps from remotely sensed data to exploit georeferenced clinical onsets: The Pavia 2013 case, in *Proc. IEEE Int. Conf. Geosci. Remote Sens. (IGARSS)*, Milan, Italy, 2015.

26. A. Dagliati, A. Marinoni, C. Cerra, P. Decata, L. Chiovato, P. Gamba and R. Bellazzi, Integration of administrative, clinical, and environmental data to support the management of type 2 diabetes mellitus: From satellites to clinical care, *SAGE Journal of Diabetes Science and Technology,* vol. 10, no. 1, pp. 19–26, 2016.

27. A. Bannari, A. Pacheco, K. Staenz, H. McNairn and K. Omari, Estimating and mapping crop residues cover on agricultural lands using hyperspectral and ikonos data, *Remote Sensing of Environment*, vol. 104, pp. 447–459, 2006.

28. A. Marinoni and P. Gamba, Accurate detection of anthropogenic settlements in hyperspectral images by higher order nonlinear unmixing, *IEEE Journal of Selected Topics in Applied Earth Observations and Remote Sensing*, vol. 9, no. 5, pp. 1792–1801, 2016.

29. L. Wald and J.-M. Baleynaud, Observing air quality over the city of nantes by means of Landsat thermal infrared data, *Int. J. of Remote Sensing,* vol. 20, no. 5, pp. 947–959, 1999.

30. D. J. Lary, T. Lary and B. Sattler, Using machine learning to estimate global PM2.5 for environmental health studies, *Environmental Health Insights*, vol. 9, no. S1, pp. 41–52, 2015.

31. E. J. Hyer, J. S., Reid and J. Zhang, An over-land aerosol optical depth data set for data assimilation by filtering, correction, and aggregation of MODIS collection 5 optical depth retrievals, *Atmos Measure Tech,* vol. 4, no. 3, pp. 379–408, 2011.

32. J. Zhang and J. S. Reid, An analysis of clear sky and contextual biases using an operational over ocean MODIS aerosol product, *Geophys Res Lett*, vol. 36, no. 15, 2009.

33. Y. Chi, X. Wang, P. S. Yu and R. R. Muntz, Moment: Maintaining closed frequent itemsets over a stream sliding window, in *Proc. IEEE Int. Conf. Data Mining*, Brighton, U.K, 2004.

34. A. Ganguly and K. Steinhaeuser, Data mining for climate change and impacts, in *Proc. IEEE Int. Conf. Data Min. Workshops*, Pisa, Italy, 2008.

35. M. Ester, H.-P. Kriegel, J. Sander and X. Xu, A density-based algorithm for discovering clusters in large spatial databases with noise, in *Proc. ACM SIGKDD Int. Conf. Knowl. Discovery Data Min.*, Portland, OR, USA, 1996.

36. J. Yuan, D. Wang and R. Li, Remote sensing image segmentation by combining spectral and texture features, *IEEE Trans. Geosci. Remote Sens.*, vol. 52, no. 1, pp. 16–24, 2013.

37. J. Gama, R. Fernandes and R. Rocha, Decision trees for mining data streams, *Intell. Data Anal.,* vol. 10, pp. 23–45, 2006.

38. G. Hulten, L. Spencer and P. Domingos, Mining time-changing data streams, in *Proc. ACM SIGKDD Int. Conf. Knowl. Discovery Data Min.*, San Francisco, CA, USA, 2001.

39. C. Jiang, F. Coenen and M. Zito, A survey of frequent subgraph mining algorithms, *Knowl. Eng. Rev.*, vol. 28, no. 1, pp. 1–31, 2013.

40. C. Giannella, J. Han, J. Pei, X. Yan and P. S. Yu, Mining frequent patterns in data streams at multiple time granularities, in *Proc. Next Gener. Data Mining (NGDM)*, 2003.

41. G. S. Manku and R. Motwani, Approximate frequency counts over data streams, in *Proc. Int. Conf. Very Large Data Bases*, Hong Kong, China, 2002.

42. A. Marinoni and P. Gamba, An efficient approach for local affinity pattern detection in remotely sensed big data, *IEEE Journal of Selected Topics in Applied Earth Observations and Remote Sensing*, vol. 8, no. 10, pp. 4622–4633, 2015.

43. T. Tian, C. Jones, J. D. Villasenor and R. D. Wesel, Selective avoidance of cycles in irregular LDPC code construction, *IEEE Trans. Commun.*, vol. 52, no. 8, pp. 1242–1247, 2004.

44. S. Rajagopalan and R. D. Brook, Air pollution and type 2 diabetes: Mechanistic insights, *Diabetes*, vol. 61, pp. 3037–3045, 2012.

13 Urban Green Mapping and Valuation

Stefan Lang, Thomas Blaschke,
Gyula Kothencz, and Daniel Hölbling

CONTENTS

13.1 INTRODUCTION

Urban green areas (Bao et al. 2016; Carrus et al. 2015) enhance people's well-being in cities, as areas for recreation or exercise or simply as an antipole to the rush of urban life. Urban green, directly experienced or visually enjoyed, significantly increases the attractiveness of urban areas and the quality of life (QoL) therein (Lee and Maheswaran 2011). This is particularly applicable for small green areas (*pocket parks*) that can be relished for the moment and contribute to restorative effects (Kaplan 1995) in a residential environment. Increasing traffic congestion, ever-present noise, and widespread, ongoing construction diminish free space and necessitate the maintenance of green spaces and the creation of new ones. Ongoing discussions in politics, media, and the public demonstrate the importance of this topic (Kothencz et al. 2017).

Urban green as an integrative concept, beyond the installment of single green elements in a pedestrian zone or a single tree planted in front of a skyscraper, comprises several aspects: (1) urban ecosystems and biodiversity, manifested in the physical green (i.e., vegetative environment); (2) the psychological well-being and QoL, including restorative effects and recovery of attention (Kaplan 1995) induced by green structures and nonmonotonous urban landscapes (Carrus et al. 2015); and (3) green mobility and production and consumption including resource maintenance and efficiency in the sense of the *green* city. Driven by its multifaceted significance, green city policies are implemented in Europe—for example, the European Green

Capital initiative*—and elsewhere. International strategies have also been established for sustainable urban neighborhoods considering global change dynamics (Luederitz et al. 2013), with cities being both the main cause and solution driver for the challenges of sustainable human life (Grimm et al. 2008).

Green areas are increasingly vital for urban societies in sustaining and improving the quality of urban life (QoUL) (Lee and Maheswaran 2011). Therefore, urban planning and management place great demand on both quantitative and qualitative information on urban and peri-urban green areas. A variety of data sources can be used to provide information for decision makers and planners: quantitative information can be derived from public data infrastructures, including local or national statistics (Coutinho-Rodrigues et al. 2011), and from up-to-date Earth observation (EO) data (see Section 13.2). Individual and aggregated qualitative information can be obtained, among other sources, from questionnaire-based surveys of urban citizens (Lo and Jim 2012). Quantitative and qualitative data reveal different aspects of the urban green areas (Leslie et al. 2010). Observations and measurements, and derived spatial indicators (Van de Voorde 2016) represent the actual proportion of physical green infrastructure in cities, the health status of urban vegetation, the location of districts with low green space provision, and so on. On the other hand, such observation-based information is limited in conveying the perceived and used QoUL components of green areas (see Sections 13.3 and 13.4). To assess the significance of the latter, the scientific community and city administrations have started to collect and use citizens' opinions (Nordh et al. 2011). Although both quantitative and qualitative information is available in many cases, they are usually analyzed independently and not linked in a way that enables the actual QoUL contribution of urban green areas to be fully understood (Kothencz and Blaschke 2017; Rhew et al. 2011). This can be attributed to the different nature of these data domains. For example, spatial indicators of urban green spaces and how they are perceived by green space visitors are only partially in agreement (Kothencz and Blaschke 2017). A recent study carried out in Szeged, Hungary (Kothencz et al. 2017), found a partial relationship between quantitative data, frequency of green space visitors' geo-tagged photographs and running trajectories, and the visitors' perception of urban green spaces. Moreover, frequently used quantitative indicators of urban greenness, such as the normalized difference vegetation index (NDVI, see Section 13.2), may not necessarily correlate with citizens' perceptions of urban greenness (Leslie et al. 2010). Consequently, the collective use of both spatial information and human perceptions of urban green areas is vital for understanding the role of urban green areas in the QoUL (Kothencz and Blaschke 2017; Keul et al. 2017). A comprehensive evidence-based green monitoring and valuation strategy is a prerequisite for informed decision making to sustain and increase the QoUL contribution of urban green areas. For this to be successful, both spatial and perceptual information are equally important.

Taking these considerations into account, this chapter discusses several methodological strategies to map both quantitative and qualitative parameters of urban green. It shows how different categories of urban green can be extracted from remote sensing imagery, based on both spectral and structural features, and how qualitative

---

* http://ec.europa.eu/environment/europeangreencapital/index_en.htm

attributes can be added in order to move from structural green classes to units of uniform green valuation. We also address the added value of including height and volumetric information to better represent the visual impression of urban green.

13.2 EO-BASED URBAN GREEN MONITORING

Satellite-based EO is a key enabler for area-wide mapping and monitoring of urban green in multiple nested scales, including the strategic scale(s) of urban planning and management (Nielsen 2015). The degree of greenness can be approximated by indicators derived from vegetation cover (Bao et al. 2016), using thresholding techniques based on the NDVI (Parent et al. 2015) or other related vegetation indices utilizing band ratios of visible and near-infrared spectral bands including the red edge portion of the spectrum (Adamczyk and Osberger 2015). This allows a general overview of the presence of green structures and a quantitative assessment of constancy or change of greenness. A disadvantage is the dependency on the threshold setting for the binary distinction between green and non-green, which also depends on the actual vegetation cover captured by the EO image data. In addition, no distinction can be made between different types of green. More explicitly, urban green can be mapped as biophysical land cover types of specific green objects or composite structures in its formal or constitutive dimension (Bibby and Shepherd 2000; Nielsen 2015). With respect to the functional and purpose-related dimension, a park, a green belt, green infrastructure, and so on represent telic land use features of city planning. The mapping of different categories of urban green can be enhanced with further spatial analysis of, for example, distance, accessibility, neighborhood, and so on, and thus take into account the population distribution in a certain urban context. Van de Voorde (2016) has studied the proximity of public urban green spaces from the perspective of citizens by applying a dasymetric mapping approach in the city of Brussels, Belgium. The European Copernicus Urban Atlas project uses satellite remote sensing to map presence and change of specific urban land use features throughout Europe (Seifert 2009). This remote sensing–based operational information service provides standardized information in regular update intervals for larger urban areas across Europe. Figure 13.1 shows the mapping of urban green features at different scales derived by EO data analysis for the city of Salzburg, Austria. Above, a data set is depicted from the Copernicus Urban Atlas project. *Urban Green Areas* are summarized in the class code 14100. The middle and lower parts show subsets from the Salzburg urban green monitoring (see Section 13.4.1) based on very high resolution (VHR) satellite data since 2005.

The categorization of the urban land cover/use classification has recently benefitted from the option of representing the third dimension of green features and thereby estimating volumetric parameters. This can be done using auxiliary LiDAR data (Hecht et al. 2006) or digital surface model (DSM) data derived from (tri-)stereo imagery (Kothencz et al. submitted). More details are provided in Section 13.4.4. This follows the general trend of urban inventories increasingly moving toward the three-dimensional (3D) representation of the urban space (Awrangjeb et al. 2013; Kawata and Koizumi 2014). Green monitoring and valuation strategies benefit from this. The 3D representation of urban vegetation is still less prominent in urban planning and management than that of 3D building models (Khoshelham et al. 2010; Ma et al. 2015), while urban green monitoring and valuation strategies supplemented with vegetation height can provide added value for decision making. Several studies combine multiple EO data sources to model urban green volumes (Huang et al. 2013; Parent et al. 2015). For many city administrations and research institutes, it is a practical and financial challenge to afford multiple data sets for monitoring activities. Kothencz et al. (submitted) elaborated a methodology that used VHR tri- and bi-stereo Pléiades imagery to delineate the height of urban vegetation for urban green monitoring and valuation in two comparable Central European urban settings, namely, the cities of Salzburg, Austria, and Szeged, Hungary (see Figure 13.2). For a study in Krakow,

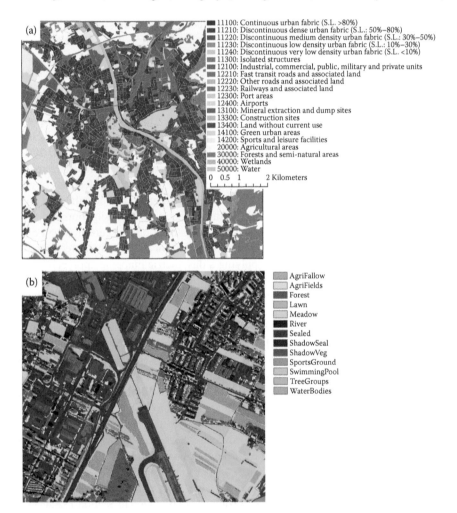

FIGURE 13.1 Urban green mapping using EO data. (a) Detailed view of the European Urban Atlas, city of Salzburg (retrieved from the European Environmental Agency 2017). (b and c) Green structure mapping Salzburg based on VHR data: (b) aggregated view, (c) detailed view.   *(Continued)*

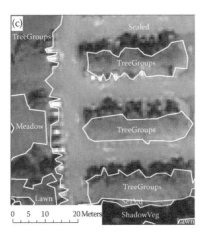

FIGURE 13.1 (CONTINUED) Urban green mapping using EO data. (a) Detailed view of the European Urban Atlas, city of Salzburg (retrieved from the European Environmental Agency 2017). (b and c) Green structure mapping Salzburg based on VHR data: (b) aggregated view, (c) detailed view.

FIGURE 13.2 Distribution of vegetation heights in an urban setting as derived from Pleiades tri-stereo analysis in Salzburg (Kothencz et al. submitted). The top layer represents the surface layer (middle) classified by five vegetation height classes, draped over the elevation model (bottom).

Poland, Tompalsky and Wecyk (2012) classified a VHR optical GeoEye-1 satellite image with object-based image analysis (OBIA) techniques (Blaschke et al. 2014) and derived a DSM from a very dense airborne LiDAR point-cloud with 22 points per square meter. After transforming the 3D green data into voxels, they calculated a *Vegetation Volume to Built-up Volume Index* (VV2BV), which reveals the relationship between building and vegetation volume within a given distance with a focus on high vegetation. A second *Urban Vegetation Index* (UVI) also includes the class of low vegetation (see Section 13.4.4).

13.3 ADDING QUALITATIVE ASPECTS OF URBAN GREEN

As stated above, measuring the amount of green areas per unit based on EO data is straightforward, but these measures may not always correspond to the quality of urban green. Quality can be determined based on biophysical variables such as biomass per square meter, water retention potential, or air cleaning potential. Alternatively, one can take on a citizen perspective and can model the importance of urban green to humans, for example, living in a certain neighborhood. In particular, the latter strategy aims to foster a citizen-centered QoL approach. QoUL with a focus on urban life, and QoL in general, can be assessed subjectively, according to individual perceptions, and objectively, via secondary data sources. Haslauer et al. (2015) investigated how well these two approaches may agree with one another and, in particular, studied the spatial mismatch between perceived satisfaction and objectively measured results for the city of Vienna, Austria, with regard to the public transport and green space availability. Areas of general disagreement, where satisfaction and GIS-derived measurements diverged, were mapped within a GIS, and the characteristics of residents living in these places were assessed. Results show that some variations do exist while the objective and subjective measurements are largely in congruence with one another, thus stressing the spatial heterogeneity in residential QoL perceptions.

Keul et al. (2017) present a mixed methods approach linking objective GIS data with subjective well-being and QoL. It traces negative effects like urban stress (e.g., population density) and positive recovery elements (e.g., green spaces) identified by environmental psychology and the health sciences. The authors present a reliable three-dimensional psychological construct for QoL (environmental/social quality, social roots, and subjective infrastructure). GIS and non-GIS predictors for urban well-being were isolated by multiple linear regressions. The factor *social roots* had no GIS predictors. Significant district differences underline the importance of sociocultural microsystems. Keul et al. (2017) applied this methodology by comparing the city of Salzburg, Austria, with two districts of the city of Vienna, Austria, and one district of the city of Timisoara, Romania. The replication study revealed comparative stability, with the item analytic results supporting the three-dimensional psychological construct of QoL.

The great challenge is to repeatedly and systematically monitor QoL beyond single studies. This requires mapping, analyzing, and monitoring of QoL/QoUL and its determinants and influencing factors in order to incorporate findings and outcomes into future urban planning processes. Depending on particular research aims and background, QoL/QoUL studies typically utilize either subjective (McCrea et al. 2005; Sirgy et al. 2000) or objective indicators (Blomquist et al. 1988; Stover and Leven 1992). While GIS-based methods are particularly useful in evaluating objective indicators in terms of spatial patterns and relationships, methods from the field of environmental psychology are commonly used in the elicitation and evaluation of subjective well-being, for example, based on interview data (Keul and Prinz 2011;

Marans 2003). In the following subsection, the objectively measurable green volume is calculated for the study area of Salzburg, Austria. Vegetation height and volume are derived from airborne LiDAR data, and indices are calculated to quantify the relationship between the volume of buildings and the volume of vegetation. We need to restrict the literature overview here to remote sensing studies related to urban green and QoL. Many more studies on urban morphology have been performed in geography, landscape planning and architecture, environmental psychology, and sociology. Some of them have relied heavily on remote sensing or GIS, or both. We thus follow the conclusion of Pavlovska (2009) that GIS is well suited for the visualization of qualitative research information that cannot be represented by classic quantitative databases.

13.4 QUANTIFYING URBAN GREEN AND ITS PERCEPTION

13.4.1 THE SALZBURG GREEN MONITORING STUDY

The material presented in the remainder of this chapter is taken from a multi-stage, ongoing study aiming to establish an EO-based monitoring scheme of urban green areas for the municipality of Salzburg. The main purpose of this is to establish a monitoring scheme that can serve urban and regional planning purposes (Lang et al. 2007). The core part of the research was done based on the initial study carried out in 2005 and every 5 years since then using VHR optical imagery. In accordance with the evolution of VHR sensors, we have so far used QuickBird (2005), WorldView-2 (2010), and Pléiades (2015) VHR imagery (see Tables 13.1 and 13.2 for details). Here, we discuss results from the analysis based on QuickBird data from 2005 and WorldView-2 data from 2010. The QuickBird data in 2005 were derived for the exact area of the municipality of Salzburg (approximately 65 km²) and the WorldView-2 data in 2010 for an extended functional metropolitan, transboundary (Austria/Germany) region of approximately 146 km². The Pléiades data in 2015 were acquired in tri-stereo mode to enhance the green feature extraction capability through elevation information. This is done based on findings obtained from airborne laser scanning data originating from 2006 and derived surface information, as well as auxiliary data for the 2010 time interval to compensate for cloud coverage.

TABLE 13.1

Characteristics of Satellite Imagery and Preprocessing Steps

| Sensor | AOI Size (km²) | Acquisition Date | Scene Characteristics | | |
			Cloud Cover	GSD	Number of Bands
QuickBird	69	June 21, 2005	<1%	2.4 m	4
WorldView-2	146	September 11, 2010	6.67%	2.0 m	8
Pléiades-A	159	September 2, 2015	<1%	2.0 m	4

AOI, area of interest; GSD, ground sample distance (i.e., spatial resolution).

TABLE 13.2
Preprocessing and Image Enhancement Steps

		Preprocessing				
Sensor	Acquisition Mode	Image Calibration	Ortho/RPC Correction	Pan-sharpening	DSM Generation	Data Imputation
QuickBird	Standard tasking	–	Yes	0.6 m	–	–
WorldView-2	Priority tasking	–	Yes	0.5 m	–	Yes
Pléiades-A	Tri-stereo	Yes	Yes	0.6 m	Yes	–

DSM, digital surface model; RPC, rational polynomial coefficient.

The study area (Figure 13.3) comprises the Austrian city of Salzburg, its outskirts, and surrounding areas, both on Austrian and German territories. It is located at approximately 47°48′ N and 13°02′ E, north of the foothills of the Alps, in the Salzburg basin. One main feature of the area is the Salzach River, which flows through the city of Salzburg and is popular for tourist boat trips and scenic attractions. East of the river, in the midst of the urban area, lies the densely forested Kapuzinerberg,

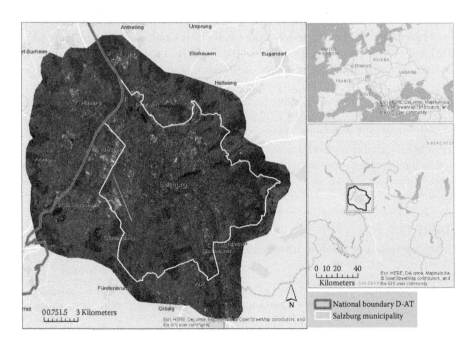

FIGURE 13.3 Location of study area and AOI coverage. The figure shows a false-color representation and demarcates the extent of the WorldView-2 (2010) data matching that of Pléiades from 2015. The QuickBird data from 2005 covered the area of the municipality of Salzburg (strong white line).

and further out the Gaisberg, the highest elevation in the study area with 1287 m. Situated west of the Salzach is the historic center of Salzburg as well as the airport. Further beyond the industrial and commercial areas flows the second major river of the study area, the Saalach, which demarcates the border to Germany. The south-western and the German parts of the area, except for the town of Freilassing and the small community Ainring, are mainly covered by meadows and cropland. The study area is inhabited by a total of approximately 176,000 people (Salzburg: ±150,000, census 2006; Freilassing: ±16,000; Ainring: ±10,000, both census 2009) and is situated on average 424 m above sea level (Anyyeva 2016).

The approach is based on semi-automated OBIA class modeling (Tiede et al. 2010) for preprocessed VHR data. Different representations of urban green have been calculated for the study area (see Figure 13.4). Based on detailed results that represent individual

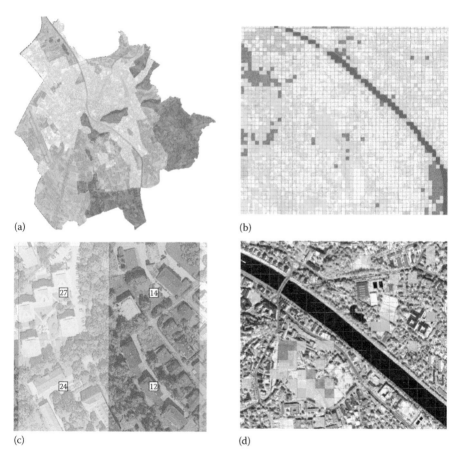

(a)

(b)

(c)

(d)

FIGURE 13.4 Different representations of greenness: (a) Degree of greenness per enumeration unit in 2010 for the municipality of Salzburg (Schöpfer, Lang, and Blaschke 2005); (b) 50 × 50 m grid cell representation (2005); (c) delineation of individual green structures and calculation of green patches, indicated by numeric labels (2005) (Lang et al. 2006); (d) change assessment between 2005 and 2010 per grid cell.

green structures derived by OBIA methods (see Figure 13.4c), the authors were able to generate aggregated information on various reporting units including regular raster grid cells, as well as other administrative (sub-)units of the study area, like enumeration areas. The 50 m × 50 m grid (Figure 13.4b) is congruent to other representations of statistical indicators in Austria. For a general discussion on methodological challenges using different reporting units, please refer to Hagenlocher et al. (2014). See also Figure 13.6.

In the case of the WorldView-2 coverage in 2010, the 6.6% cloud coverage required data to be imputed from other sources. Here, a publicly available orthophoto mosaic facilitated the replacement of cloud-obscured areas (see Figure 13.5). RGB orthophotos that do not include infrared spectral information have been used for visualization and comparison purposes only.

Based on the 2015 Pléiades data, Lang and Csillik (2017) investigated the potential of the superpixel approach (Achanta et al. 2012), which entails merging the first two steps of delineating smallest units and then aggregating them to a regular grid.

Areas affected by cloud shadow (WV-2)   Classification on orthophoto mosaic

Data imputation - merge of classification results

FIGURE 13.5 Data imputation in areas covered by cloud shadow. The circle indicates an artifact left over from the shadow effect. (Modified after Powell, David. 2011. Deriving an urban green index from object-based classification of very high resolution remote sensing imagery with different conditions of illumination. MSc, Department of Geoinformatics—Z_GIS, University of Salzburg.)

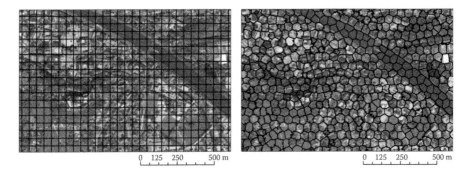

FIGURE 13.6 Results of a superpixel analysis using 50-m grid spacing (right), adhering to the ETRF reference grid (left). (Modified after Lang, Stefan, and Ovidiu Csillik. 2017. ETRF grid-constrained superpixels generation in urban areas using multi-sensor very high resolution imagery. *GI Forum—Journal for Geographic Information Science* 2/2017.)

If the latter is reasonably small, the generated mosaic allows the representation of urban features, including the relevant green structures, in a reasonable way (see Figure 13.6). The approach is limited in its representation of linear structures and very small features though. In this case, the authors tested a grid spacing taken from the European terrestrial reference frame, to constrain the superpixel parametrization. Even if the application of a standardized grid makes the approach transferable to any other area in Europe, the choice of the grid remains arbitrary as any other spacing. As such, it imposes a geometry that in its initial stage is not related to the actual scene content.

13.4.2 MIMICKING HUMANS' PERCEPTIONS WITH EO AND GIS DATA

While most remote sensing studies quantify green based on observable (and classifiable) parameters, an increasing number of studies investigate citizens' perceptions of urban green. Here, we demonstrate how qualitative information has been combined with measured data on urban green. In an initial study for a subset of the city of Salzburg, Schöpfer et al. (2005) developed the ruleset for the object-based classification based on an orthophoto mosaic. Image segmentation was used to provide segments to be classified in the target scale domain of around 1:5000. An NDVI layer supported the initial classification process by providing a threshold and thus allowing a differentiation between vegetated and nonvegetated land cover. The initial study was then transferred to the entire area of the municipality using QuickBird data from June 2005. Surface information was not available for this time slot. The vegetated areas were analyzed to address textured vegetation types like trees, hedgerows, and small forests, and smoother (less textured) types like lawns and gardens. The standard deviation of the sensor's green band was used to represent the internal homogeneity of the generated segments. Less textured vegetation was then split up into low intensity and high intensity to distinguish between meadows and different agricultural areas. As described in Schöpfer et al. (2005), 13 out of 29 initial green structure types were transformed into a cognition network and classified using OBIA rulesets. The network facilitates and serves as a graphical means for communication and interaction with users or stakeholders, who can decide about the actual semantic content of aggregated target classes like *green* or *not-green*. In this case, the class hierarchy contains 11 green structure types plus two classes not considered as green (i.e., sealed surface and shadow on nonvegetation). Depending on the threshold applied to the ranking among the respondents or any other decision rule, the classes are assigned to green/not-green. Cadastral data were used to differentiate between sealed features like houses and streets, which have similar spectral signatures. Schöpfer et al. (2005) additionally tried to ascertain the residents' perspectives under the assumption that green space is perceived differently by the observer depending on their point of view and weighted the results by considering the number of multistory buildings and the distance between buildings.

Lang et al. (2007) strived to mimic citizens' perceptions of urban green structures. They reported the results of the QuickBird satellite image from 2005 on the above-mentioned 50 m × 50 m grid to calculate four different green indices. In parallel, interviews were carried out to detect citizens' preferences of green structures. Twenty-nine different green structure types were validated by 128 participants according to their relative importance on a Likert scale ranging from 1 to 5. Answers were statistically aggregated and not spatially referenced, as the locations of respondents were not relevant for this study. To assist respondents with assigning values to the respective types that were eventually scientifically named, exemplary images (photos) were provided along with a questionnaire. In addition, the interviews contained questions pertaining to the general perception of urban green in the city. The interviews were carried out in June 2005 and coincide with the date of the QuickBird scene. The results of the interviews and the ranks for different types of green structures varied between 2 and 4.5. Based on this information, a weighted green index (GI_w) was calculated and normalized to a scale between 1 and 5, whereby 1 indicates low and 5 indicates high green satisfaction. Statistical analysis of the results revealed that 12.3 km^2 of the city of Salzburg fall into the category above 3.9, indicating high green satisfaction, and 34.9 km^2 fall into the category below 3.0. To represent an averaged green impression per cell, a single value was calculated. The GI_w can be calculated for any spatial reference unit since it is based on the percentages of green structure types and their respective importance. The following metric was used to determine the resulting GI_w of a given unit: first, weights were established by determining area percentages of each structure type occurring in the respective unit, with regard to the entire area of green within the unit. Then, ratings of the structure types were linearly stretched to an interval of [0|1] (GI_{w0}, see Figure 13.7). The ratings occurring in the unit were then multiplied by the assigned weight. Finally, for GI_w, the sums of these products were re-transformed into the original range of the ratings (Lang et al. 2006).

13.4.3 Delineating Green Valuation Units

To represent the telic dimension of urban green, so-called geons (Lang et al. 2014) have been derived. Geons are scale-adapted units that represent areas of policy-relevant phenomena under certain size and homogeneity constraints. Here, they represent units of uniform green valuation. The geons were derived by variance-based

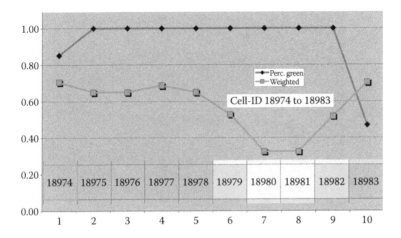

FIGURE 13.7 Calculation of GI_w (Lang et al. 2007) for 10 neighboring grid cells in a line (bottom) and differentiated green profiles percentage of green versus GI_{w0} (Lang et al. 2006). While cells 2 to 9 are all covered by 100% green (percentage of green = 1), the weighted green index differs from 1, owing to the different proportions of green structure types therein.

regionalization techniques based on the GI_w assigned to 50 m × 50 m grid cells (Lang submitted). They were quality assessed through local auto-correlation measures and field validated by 360° photo-documentation at the intersection of the units. The average size of the generated units was 13.15 hectares, with a range between 2 and 50 hectares and a left-skewed frequency distribution. Figure 13.8 shows the generated geons. Dots indicate the places where ground validation was performed. The figure also contains a detailed view of geon #4, which represents the old town of Salzburg.

13.4.4 ADDING HEIGHT AND VOLUMETRIC INFORMATION

This subsection describes an attempt to calculate objective and subjective dimensions of the QoL indicator of urban green while integrating heights and volumes of the vegetation from airborne LiDAR data. Indices are calculated to express the relationship between the volume of buildings or groups of buildings, respectively, and the volume of their surrounding vegetation. Results indicate that by adding height and volume information to the QoL indicator of urban green, the perceptions of citizens can be better represented as compared to standard measures based on two-dimensional vegetation coverage.

Based on the QuickBird 2005 data, Heugenhauser (2014) applied the approach of Tompalsky and Wecyk (2012) to the classification results and combined it with the fraction of surrounding vegetation approach by Möller and Blaschke (2006) while calculating several distance fringes. Then, a DSM was derived from an airborne LiDAR point-cloud with approximately 2 points per square meter.

For this add-on study, the QuickBird image from June 2005 was reduced to the size of the LiDAR data set, which covers about 62% of the total 65.64 km² of the city of Salzburg. The data were classified into 13 classes, of which 9 are used within this

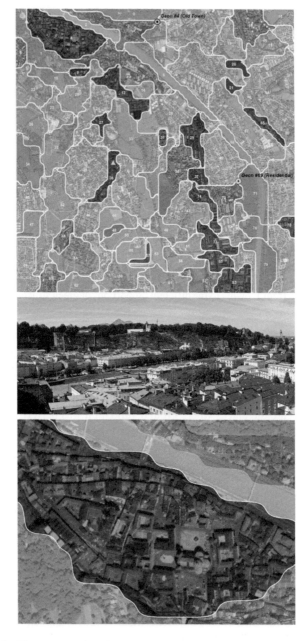

FIGURE 13.8 Top: geon delineation in the inner part of the city of Salzburg. (Modified from Lang, Stefan. (in press). Urban green valuation integrating biophysical and qualitative aspects. *European Journal of Remote Sensing*.) ID = 4 marks the intersection of four geon types, representing semi-dense building style, and a woody hill to the north. ID = 69 is at the intersection of two types of residential areas and the riparian forest to the east. Center: view of old town from geon ID = 4, photograph: A. Binn. Bottom: geon representation of old town (same area as captured by photograph).

study, using an OBIA approach. An accuracy assessment yields an overall accuracy of 89% and a κ of 0.85. A DSM and a digital terrain model were provided by the government of the province of Salzburg. These were originally created from another LiDAR mission from 2006. Additionally, a cadastral data set in vector format was provided for the whole city. In order to calculate the VV2BV and UVI, the volumes of buildings from the cadaster and the corresponding classification objects were calculated. This was achieved by multiplying the calculated area by the mean DSM values of the image classification objects. For all houses and for groups of houses in cases where buildings directly border each other or are closer than the estimated mapping accuracy, a buffer analysis was applied with 20 m, 50 m, and 100 m radius, respectively. In all three cases, more than 13,000 overlapping buffers were generated. Based on this, the surface and volume of building units were calculated.

The index values were assigned to every single building unit or groups of buildings where structures touch each other or are smaller than the minimum mapping unit. The indices were calculated as a percentage and as absolute values. The three buffer distances applied to the indices were chosen for the following reasons. It was assumed that vegetation values within the 20-m buffer distance would probably be an indicator for private green, which would partially belong to the respective property. The 50-m buffer reflects close-by greenery, which can still be perceived as green surroundings by the buildings' inhabitants. The 100-m buffer approximates the distance at which green structures are in walking distance. The latter two distance buffers have a higher relevance to planning purposes. Chen et al. (2014) revealed that 50-m buffer zones adequately describe areas of green structures in urban settings. The outcomes of the percentage calculations were divided into five equal-interval classes (0%–20%, 20%–40%, 40%–60% 60%–80%, and 80%–100%; see Figure 13.9 for details). The overall study area yields an average VV2BV percentage of 26.7, which indicates a ratio of 0.27 m^3 of vegetation volume for each cubic meter of building.

Recently, the 2005 study change to is complemented by DSM information derived from Pléiades tri-stereo data from 2015. The advantage of DSM generation from VHR satellite data is the simultaneous acquisition of both spectral and height information (Kulessa and Lang 2016).

13.5   CONCLUSION

This chapter focused on the challenge of including quality information in urban green studies, whereby the quality of urban green was mainly ascertained through citizens' perceptions. This is certainly only one possible strategy and requires a mixed methods approach. Despite all methodological challenges, the authors believe that these qualitative aspects (rather than biophysical vegetation parameters alone) better allow citizen involvement in spatial planning. It is challenging, as this integration adds more uncertainty to an otherwise objective remote sensing approach. In the "Salzburg Green" study in 2005, a weighted index based on citizens' perceptions attaches the collective public opinion to spatial entities. In other words, it reveals how a certain spatial unit is publicly valuated in terms of its greenness. This illustrates the notion that although satellite image data can serve as one means for the monitoring of urban green, it benefits from being complemented by subjectively perceived attributes

FIGURE 13.9 Results of volumetric analysis. Yellow classes indicate a more balanced relationship between vegetated and sealed surfaces (40% and 60%). Light green color values represent a slight majority of green (60%–80%); bright green buildings have a maximum of 20% of sealed classes within buffers. Red buildings indicate less than 20% vegetation volume and vegetated surface around the buffered 50 m, and more than 80% of the surrounding area is covered by either the built-up and/or sealed classes. Orange buildings represent the class between 20% and 40%.

of the residential environment. Indicators of this type may be called collectively weighted indicators (Lang et al. 2007). Such indicators may be considered more citizen based and a step toward in enhancing the acceptance of spatial decision making. The particular combination of quantifiable information and transparent and repeatable spatial analysis with interview-based subjective information—yet still derived in a scientifically sound and repeatable way—may lead to a fully repeatable monitoring scheme for analyzing urban green. Such a combination goes far beyond what most scholars in remote sensing do and is part of GIScience. Blaschke et al. (2011) have argued that GIScience is a relatively new interdisciplinary field of research based on the understanding that basic and applied research must be reflected within society, and Blaschke and Merschdorf (2014) stated that GIS has been well established in many different economic sectors, like natural resource management, real estate, and insurance. New fields for GIScience research have arisen, for example, in the health care sector, concerning epidemiology, hospital management, and patient care logistics. Likewise, the authors of this chapter hope to have contributed to a broader use of EO data and remote sensing techniques for quality of (urban) life applicatiions. Urban green and its beneficial effect may be subjectively conceived by the citizens. In this respect, EO and spatial analysis coupled with qualitative survey-based assessments increase the capacity of addressing specific aspects of urban green beyond well-established, functional land use classifications.

ACKNOWLEDGMENTS

The authors like to thank all collaborators and colleagues, including Master and PhD students, as well as interns, who contributed to the Salzburg monitoring study over the last 12 years. Dirk Tiede has worked as a co-supervisor on LiDAR pre-processing, and Petra Füreder and Florian Albrecht have supported VHR data acquisition and analysis. This ongoing study integrating various academic contributions has received initial funding from the municipality of Salzburg through the "Green" project in 2005, while ongoing financial support is provided by the Department of Geoinformatics and its Doctoral College GIScience (DK W1237-N23), funded by the Austrian Science Fund. The authors are grateful to the reviewers and proofreaders for spending their time and effort to improve and streamline the manuscript.

REFERENCES

Achanta, Radhakrishna, Appu Shaji, Kevin Smith, Aurelien Lucchi, Pascal Fua, and Sabine Süsstrunk. 2012. SLIC superpixels compared to state-of-the-art superpixel methods. *IEEE Transactions on Pattern Analysis and Machine Intelligence* 34:2274–2282.

Adamczyk, Joanna, and Antonia Osberger. 2015. Red-edge vegetation indices for detecting and assessing disturbances in Norway spruce dominated mountain forests. *International Journal of Applied Earth Observation and Geoinformation* (37 Special Issue, Earth observation for habitat mapping and biodiversity monitoring, ed. by S. Lang et al.):90–99.

Anyyeva, Aynabat. 2016. Extraction and 3D characterization of green features in urban areas from Pléiades image. MSc, Department of Geoinformatics - Z_GIS, University of Salzburg.

Awrangjeb, Mohammad, Chunsun Zhang, and Clive S. Fraser. 2013. Automatic extraction of building roofs using LIDAR data and multispectral imagery. *ISPRS Journal of Photogrammetry and Remote Sensing* 83:1–18. doi: http://dx.doi.org/10.1016/j.isprsjprs.2013.05.006.

Bao, Tongliga, Xueming Li, Jing Zhang, Yingjia Zhang, and Shenzhen Tian. 2016. Assessing the distribution of urban green spaces and its anisotropic cooling distance on urban heat island pattern in Baotou, China. *ISPRS International Journal of Geo-Informatiion* 5 (2). doi: doi: 10.3390/ijgi5020012.

Bibby, Peter, and John Shepherd. 2000. GIS, land use, and representation. *Environment and Planning B: Planning and Design* 27 (4):583–598.

Blaschke, Thomas, Geoff J. Hay, Maggi Kelly, Stefan Lang, Peter Hofmann, Elisabeth Addink, Raul Queiroz Feitosa, Freek Van der Meer, Harald Van der Werff, Frieke Van Coillie, and Dirk Tiede. 2014. Geographic Object-based Image Analysis: A new paradigm in Remote Sensing and Geographic Information Science. *International Journal of Photogrammetry and Remote Sensing* 87 (1):180–191 doi: 10.1016/j.isprsjprs.2013.09.014.

Blaschke, Thomas, and Helena Merschdorf. 2014. Geographic Information Science as a multidisciplinary and multi-paradigmatic field. *Cartography and Geographic Information Science* 41 (3):196–213.

Blaschke, Thomas, Josef Strobl, and Karl Donert. 2011. Geographic Information Science: Building a doctoral programme integrating interdisciplinary concepts and methods. *Procedia—Social and Behavioral Sciences* 21:139–146.

Blomquist, G. C., M. C. Berger, and J. P. Hoehn. 1988. New estimates of quality of life in urban areas. *The American Economic Review*:89–107.

Carrus, Giuseppe, Massimilio Scopelliti, Raffaele Lafortezza, Giuseppe Colangelo, Francesco Ferrini, Fabio Salbitano, Mariagrazia Agrimi, Luigi Portoghesi, Paolo Semenzato, and Giovanni Sanesi. 2015. Go greener, feel better? The positive effects of biodiversity on the well-being of individuals visiting urban and peri-urban green areas. *Landscape and Urban Planning* (134):221–228.

Chen, Ailian X., Angela Yao, Ranho Sun, Liding Chen. 2014. Effect of urban green patterns on surface urban cool islands and its seasonal variations. *Urban Forestry & Urban Greening* 13:646–654.

Coutinho-Rodrigues, João, Ana Simão, and Carlos Henggeler Antunes. 2011. A GIS-based multicriteria spatial decision support system for planning urban infrastructures. *Decision Support Systems* 51 (3):720–726. doi: http://dx.doi.org/10.1016/j.dss.2011.02.010.

Grimm, Nancy B., Stanley H. Faeth, Nancy E. Golubiewski, Charles L. Redman, Jiangwo Wu, Xuemei Bai, and John M. Briggs. 2008. Global change and the ecology of cities. *Science* 319 (5864):756–760.

Hagenlocher, Michael, Stefan Kienberger, Stefan Lang, and Thomas Blaschke. 2014. Implications of spatial scales and reporting units for the spatial modelling of vulnerability to vector-borne diseases. In *GI_Forum 2014. Geospatial Innovation for Society*, edited by Robert Vogler, Adrijana Car, Josef Strobl and Gerald Griesebner, 197–206. Berlin: Wichmann Verlag.

Haslauer, Eva, Eric Delmelle, Alexander G. Keul, Thomas Blaschke, and Thomas Prinz. 2015. Comparing subjective and objective quality of life criteria: A case study of green space and public transport in Vienna, Austria. *Social Indicators Research* 124:911–924.

Hecht, Robert, Gotthard Meinel, and Manfred Buchroithner. 2006. Estimation of urban green volume based on last pulse Lidar data at leaf-off aerial flight times. First Workshop of the EARSeL Special Interest Group on Urban Remote Sensing, "Challenges and Solutions" on CD-ROM, Berlin.

Heugenhauser, Andreas. 2014. Quantifying urban green with LiDAR and high resolution optical remote sensing. A 2.5D per-building approach for the city of Salzburg. Master Thesis, Department of Geoinformatics, Salzburg.

Huang, Yan, Bailang Yu, Jianhua Zhou, Chunlin Hu, Wenqi Tan, Zhiming Hu, and Jianping Wu. 2013. Toward automatic estimation of urban green volume using airborne LiDAR data and high resolution Remote Sensing images. *Frontiers of Earth Science* 7 (1):43–54. doi: 10.1007/s11707-012-0339-6.

Kaplan, Stephen. 1995. The restorative benefit of nature: Toward an integrative framework. *Journal of Environmental Psychology* 15:169–182.

Kawata, Yoshiyuki, and Kohei Koizumi. 2014. Automatic reconstruction of 3D urban landscape by computing connected regions and assigning them an average altitude from LiDAR point cloud image. Proceedings Volume 9245, Earth Resources and Environmental Remote Sensing/GIS Applications V. doi: 10.1117/12.2066622.

Keul, Alexander G., Thomas Blaschke, and B. Brunner. 2017. Urban quality of life: Towards a general objective and subjective core descriptor set. *Momentum Quarterly* 6 (2):123–137.

Keul, Alexander, and Thomas Prinz. 2011. The Salzburg quality of urban life study with GIS support. In *Investigating Quality of Urban Life—Theory, Methods, and Empirical Research*, edited by R. W. Marans and R. J. Stimson, 273–293. Dordrecht: Springer.

Khoshelham, Kourosh, Carla Nardinocchi, Emanuele Frontoni, Adriano Mancini, and Primo Zingaretti. 2010. Performance evaluation of automated approaches to building detection in multi-source aerial data. *ISPRS Journal of Photogrammetry and Remote Sensing* 65 (1):123–133. doi: http://dx.doi.org/10.1016/j.isprsjprs.2009.09.005.

Kothencz, Gyula, and Thomas Blaschke. 2017. Urban parks: Visitors' perceptions versus spatial indicators. *Land Use Policy* 64:233–244. doi: http://dx.doi.org/10.1016/j.landusepol.2017.02.012.

Kothencz, Gyula, Ronald Kolcsár, Pablo Cabrera-Barona, and Péter Szilassi. 2017. Urban green space perception and its contribution to well-being. *International Journal of Environmental Research and Public Healt* 14 (7).

Kothencz, Gyula, Kerstin Kulessa, Aynabat Anyyeva, and Stefan Lang. submitted. Urban vegetation extraction from VHR (tri-)stereo imagery—A comparative study in two central European cities. *European Journal of Remote Sensing*.

Kulessa, Kerstin, and Stefan Lang. 2016. 3D object-based feature extraction from 3-stereo DSM in urban context. GEOBIA 2016: Solutions and Synergies, Twente.

Lang, Stefan. (in press). Urban green valuation integrating biophysical and qualitative aspects. *European Journal of Remote Sensing*.

Lang, Stefan, and Ovidiu Csillik. 2017. ETRF grid-constrained superpixels generation in urban areas using multi-sensor very high resolution imagery. *GI Forum—Journal for Geographic Information Science* 2/2017.

Lang, Stefan, Thomas Jekel, Daniel Hölbling, Elisabeth Schöpfer, Thomas Prinz, Elisabeth Kloyber, and Thomas Blaschke. 2006. Where the grass is greener—Mapping of urban green structures according to relative importance in the eyes of the citizens. First Workshop of the EARSeL Special Interest Group on Urban Remote Sensing, "Challenges and Solutions" on CD-ROM.

Lang, Stefan, Stefan Kienberger, Dirk Tiede, Michael Hagenlocher, and Lena Pernkopf. 2014. Geons—Domain-specific regionalization of space. *Cartography and Geographic Information Science* 41 (3):214–226. doi: 10.1080/15230406.2014.902755.

Lang, Stefan, Elisabeth Schöpfer, Daniel Hölbling, Thomas Blaschke, Matthias Moeller, Thomas Jekel, and Elisabeth Kloyber. 2007. Quantifying and qualifying urban green by integrating remote sensing, GIS, and social science method. In *Use of Landscape Sciences for the Assessment of Environmental Security*, edited by Irene Petrosillo, Felix Müller, K. B. Jones, Giovanni Zurlini, Kinga Krauze, Sergey Victorov, Bai-Lian Li and William G. Kepner, 93–105. Springer.

Lee, Andrew. C. K., and R. Maheswaran. 2011. The health benefits of urban green spaces: A review of the evidence. *Journal of Public Health* 33 (2):212–222. doi: 10.1093/pubmed/fdq068.

Leslie, Eva, Takemi Sugiyama, Daniel Ierodiaconou, and Peter Kremer. 2010. Perceived and objectively measured greenness of neighbourhoods: Are they measuring the same thing? *Landscape and Urban Planning* 95 (1–2):28–33. doi: http://dx.doi.org/10.1016/j.landurbplan.2009.11.002.

Lo, Alex Y. H., and C. Y. Jim. 2012. Citizen attitude and expectation towards greenspace provision in compact urban milieu. *Land Use Policy* 29 (3):577–586. doi: http://dx.doi.org/10.1016/j.landusepol.2011.09.011.

Luederitz, Christopher, Daniel J. Lang, and Henrik Von Wehrden. 2013. A systematic review of guiding principles for sustainable urban neighborhood development. *Landscape and Urban Planning* 113:40–52.

Ma, Ligang, Di Wu, Jingsong Deng, Ke Wang, Jun Li, Qing Gu, and Yunzhuo Dai. 2015. Discrimination of residential and industrial buildings using LiDAR data and an effective spatial-neighbor algorithm in a typical urban industrial park. *European Journal of Remote Sensing* 48 (1):1–15. doi: 10.5721/EuJRS20154801.

Marans, Robert W. 2003. Understanding environmental quality through quality of life studies: The 2001 DAS and its use of subjective and objective indicators. *Landscape and Urban Planning* 65 (1):73–83.

McCrea, Rod, Robert Stimson, and John Western. 2005. A moderated model of satisfaction with urban living using data for Brisbane–South East Queensland, Australia. *Social Indicators Research* 72 (2):121–152.

Möller, Matthias, and Thomas Blaschke. 2006. A new index for the differentiation of vegetation fractions in urban neighborhoods based on satellite imagery. ASPRS 2006 Conference, Reno, Nevada.

Nielsen, Michael M. 2015. Remote sensing for urban planning and management: The use of window-independent context segmentation to extract urban features in Stockholm. *Computers, Environment and Urban System* 52:1–9.

Nordh, Helena, Chaham Alalouch, and Terry Hartig. 2011. Assessing restorative components of small urban parks using conjoint methodology. *Urban Forestry & Urban Greening* 10 (2):95–103. doi: http://dx.doi.org/10.1016/j.ufug.2010.12.003.

Parent, Jason R., John C. Volin, and Daniel L. Civco. 2015. A fully-automated approach to land cover mapping with airborne LiDAR and high resolution multispectral imagery in a forested suburban landscape. *ISPRS Journal of Photogrammetry and Remote Sensing* 104:18–29. doi: http://dx.doi.org/10.1016/j.isprsjprs.2015.02.012.

Pavlovska, Marianna. 2009. Non-quantitative GIS. In *Qualitative GIS: A Mixed Methods Approach to Integrating Qualitative Research and Geographic Information Systems*, edited by S. Elwood and M. Cope, 13–37. London: Sage.

Powell, David. 2011. Deriving an urban green index from object-based classification of very high resolution remote sensing imagery with different conditions of illumination. MSc, Department of Geoinformatics—Z_GIS, University of Salzburg.

Rhew, Isaac C., Ann Vander Stoep, Anne Kearney, Nicholas L. Smith, and Matthew D. Dunbar. 2011. Validation of the normalized difference vegetation index as a measure of neighborhood greenness. *Annals of Epidemiology* 21 (12):946–952. doi: http://dx.doi.org/10.1016/j.annepidem.2011.09.001.

Schöpfer, Elisabeth, Stefan Lang, and Thomas Blaschke. 2005. A "Green Index" incorporating remote sensing and citizen's perception of green space. *International Archives of Photogramm., Remote Sensing and Spatial Information Sciences*.

Seifert, Frank M. 2009. Improving urban monitoring toward a European Urban Atlas. In *Global Mapping of Human Settlement—Experiences, Datasets, and Prospects*, edited by Paolo Gamba and Martin Herold, 231–248. Boca Raton: CRC Press.

Sirgy, M. Joseph, Don R. Rahtz, Muris Cicic, and Robert Underwood. 2000. A method for assessing residents' satisfaction with community-based services: A quality-of-life perspective. *Social Indicators Research* 49 (3):279–316.

Stover, Mark Edward, and Charles L. Leven. 1992. Methodological issues in the determination of the quality of life in urban area. *Urban Studies* 29 (5):737–754.

Tiede, Dirk, Stefan Lang, Florian Albrecht, and Daniel Hölbling. 2010. Object-based class modeling for cadastre constrained delineation of geo-objects. *Photogrammetric Engineering & Remote Sensing* 76 (2):193–202. doi: 10.14358/PERS.76.2.193.

Tompalsky, Piotr, and Piotr Wecyk. 2012. LIDAR and VHRS data for assessing living quality in cities. An approach based on 3D spatial indices. *ISPRS International Journal of Photogrammetry, Remote Sensing and Spatial Information Sciences* XXXIX-B6:173–176.

Van de Voorde, Tim. 2016. Spatially explicit urban green indicators for characterizing vegetation cover and public green space proximity: A case study on Brussels, Belgium. *International Journal of Digital Earth*. doi: doi: 10.1080/17538947.2016.1252434.

Index

Printed and bound by CPI Group (UK) Ltd, Croydon, CR0 4YY

24/10/2024

01778304-0010